| Otters

Otters
ecology, behaviour and conservation

HANS KRUUK

Department of Zoology, Aberdeen, Scotland

OXFORD
UNIVERSITY PRESS

Great Clarendon Street, Oxford OX2 6DP

Oxford University Press is a department of the University of Oxford.
It furthers the University's objective of excellence in research, scholarship,
and education by publishing worldwide in

Oxford New York

Auckland Cape Town Dar es Salaam Hong Kong Karachi
Kuala Lumpur Madrid Melbourne Mexico City Nairobi
New Delhi Shanghai Taipei Toronto

With offices in

Argentina Austria Brazil Chile Czech Republic France Greece
Guatemala Hungary Italy Japan Poland Portugal Singapore
South Korea Switzerland Thailand Turkey Ukraine Vietnam

Oxford is a registered trade mark of Oxford University Press
in the UK and in certain other countries

Published in the United States
by Oxford University Press Inc., New York

© Oxford University Press 2006

The moral rights of the author have been asserted
Database right Oxford University Press (maker)

First published 2006

All rights reserved. No part of this publication may be reproduced,
stored in a retrieval system, or transmitted, in any form or by any means,
without the prior permission in writing of Oxford University Press,
or as expressly permitted by law, or under terms agreed with the appropriate
reprographics rights organization. Enquiries concerning reproduction
outside the scope of the above should be sent to the Rights Department,
Oxford University Press, at the address above

You must not circulate this book in any other binding or cover
and you must impose the same condition on any acquirer

British Library Cataloguing in Publication Data

Data available

Library of Congress Cataloging in Publication Data
Kruuk, H. (Hans)
  Otters: ecology, behaviour, and conservation / Hans Kruuk.
    p. cm.
  ISBN-13: 978–0–19–856586–4 (alk. paper)
  ISBN-10: 0–19–856586–0 (alk. paper)
  ISBN-13: 978–0–19–856587–1 (alk. paper)
  ISBN-10: 0–19–856587–9 (alk. paper)
 1. Otters. I. Title.
  QL737.C25K785 2006
  599.769—dc22   2006010273

Typeset by Newgen Imaging Systems (P) Ltd., Chennai, India
Printed in Great Britain
on acid-free paper by
Antony Rowe, Chippenham

ISBN 0–19–856586–0 978–0–19–856586–4
ISBN 0–19–856587–9 (Pbk.) 978–0–19–856587–1 (Pbk.)

10 9 8 7 6 5 4 3 2 1

To Saskia Jane

Preface

My main, honest excuse for writing this book is that I very much like otters and the places where they live, and I like writing about them. This book is for naturalists, scientists and conservationists, in fact for anyone who takes an interest in these animals and their vicissitudes, in Europe, America or anywhere in the world. I started it as a revision and update of my earlier book *Wild otters, predation and populations*, which deals predominantly with the Eurasian otter, mostly in Shetland, and which is now out of print. However, since I wrote that book I have been involved in studies of several other species of otter, and consequently the material has expanded so much beyond the bounds of that first one, widening its scope, doubling the text and numbers of references, that it deserves a different title. The present book deals with all otters worldwide, with research and observations of many scientists and naturalists as well as my own; it includes recent research in genetics as well as ecology and behaviour, and the latest, worrying threats to otter survival. Yet it also covers all the main topics of *Wild otters*.

Naturalists and scientists have a common interest, but there is a great deal that separates them. Especially when charismatic animals such as otters are involved, I am aware that naturalists want nice observations, descriptions, generalizations and principles, whereas scientists need sharply defined questions, research methods and quantification. I have tried to satisfy the needs of both, but have skimped somewhat on quantification, with the excuse that it can be traced easily by following up the many references. This is a book about animals, not just figures, and statistical statements are therefore confined to the legends of graphs, where most of the statistical treatment is based on Siegel and Siegel (1988).

One of the problems encountered when using references to other studies and writing about otters is the existence of different categories of literature on the subject. Because of the appeal of these animals to the general public, there is a large body of popular writing, mostly from naturalists, in books, magazines and unrefereed journals. Scientists call this, somewhat condescendingly, the 'grey literature'. It may contain excellent and useful observations, but often also a large pinch of salt is needed, and it does not reach the science reference programmes on the internet. To extract the useful, from the bad and the merely entertaining, is difficult. At the other extreme are the properly refereed scientific papers in respectable journals, with heavy statistics and complicated models—some of them need translation into a language that is understandable to everybody. I hope that I have been succesful in the use of all such literature, although I often found it quite difficult to navigate between Scylla and Charibdis.

Most of the different otters have much in common, but especially for conservation management purposes, as well for the understanding of otters' evolution, individual species interests are relevant. To facilitate interests in particular species under the various general headings in the text, they are referred to in **bold** type at the beginning of sections referring to them. Finally, and as one of the important aims of this book, I hope that the use of more than 650 references points the way and opens doors into the very extensive literature on otters.

My own research on the animals reported here was part of a joint effort of teams, of frequently changing composition. Many friends, colleagues and students have helped, either by their observations, by their assistance in the field or laboratory, with their advice or their comments and criticisms, by personal communications about their work, with the writing or illustration of this book, or in some other way. It is impossible to mention everybody, and many people involved are named in our joint papers. However, I would like to express special gratitude to some of them for the many services rendered: Merav Ben-David, Gail Blundell, Luigi Boitani, Sim Broekhuizen, David Carss, Jim Conroy, Peter Dale, Nicole Duplaix, Leon Durbin, Jim Estes, Ulco Glimmerveen, Kees Goudswaard, Kris Ingram, Hélène Jacques, Addie de Jongh, Budsabong Kanchanasaka, Manoel Muanis, Andrew Moorhouse, Jan Nel, Bart Nolet, Erik-Jan Ouwerkerk, Jan Reed-Smith, Michael Somers, Helen Waldemarin, Sawai Wanghongsa and Dennis Wansink.

Preface

I am extremely grateful for the generous contribution of photographs to this book, by many excellent photographers, first of all Nicole Duplaix, and including Bryant Austin, André Bärtschi, Miguel Bellosta, Angie Berchielli, Sharon Blaziek, Richard Brucich, Alice Courage, Ainul Hussain, Budsabong Kanchanasaka, Reza Lubis, Gonzalo Medina, Andreas Norin, Chris Reynolds, Thomas Serfass, Christof Schenck, Michael Somers, John Stahl, Elke Staib, Jane Vargas and Chris Wood. I can only apologize for the use of their beautiful colour images in black and white, which does little justice to the photographers. All photographs without attribution are my own.

Oxford University Press granted permission for the use of figures 4.5, 4.6, 6.4, 6.5, 6.11, 8.11, 10.1, 11.5, 11.6 and 12.13. Blackwell Publishing gave permission to use figures 12.4, 12.5, 12.8, 12.11 and 12.12. Elsevier gave permission to use figure 5.12.

As always, my family *et al.*, with their many contributions large and small, have made it all possible: a massive thanks to Jane, Loeske, Johnny, Alice and Patrick.

Hans Kruuk
Aboyne, Aberdeenshire

Contents

1	**Otter ecology and its background**	**1**
2	**Pen pictures. Thirteen otters of the world: some natural history**	**6**
	Europe	6
	Common or Eurasian otter, Lutra lutra	6
	North America	8
	River otter, or North American otter, Lontra canadensis	8
	Sea otter, Enhydra lutra	10
	Latin America	13
	Giant otter, Pteronura brasiliensis	13
	Neotropical otter, Lontra longicaudis	15
	Southern river otter, or huillin, Lontra provocax	17
	Marine otter, or sea cat, Lontra felina	17
	Asia	19
	Hairy-nosed otter, Lutra sumatrana	19
	Smooth-coated or smooth otter, Lutrogale perspicilata	20
	Small-clawed otter, Aonyx cinereus	21
	Africa	23
	Cape clawless otter, Aonyx capensis	23
	Congo clawless otter, or swamp otter, Aonyx congicus	25
	Spotted-necked otter, Lutra maculicollis	26
3	**Evolutionary relationships, questions and methods of otter ecology**	**29**
	Phylogeny and evolution	29
	Questions of behaviour and ecology of otters	31
	Research methods and field techniques	33
	Study areas	36
4	**Habitats**	**39**
	Introduction	39
	Methods	39
	Fresh or salt water?	40
	The aquatic part of the habitat	41
	Terrestrial habitat along sea coasts	43
	Terrestrial habitats along fresh water: banks and bogs	46
	Freshwater habitat in Latin America, Asia and Africa	49
	Otter habitat: some general comments	53
5	**Groups and loners: social organization**	**55**
	Introduction	55
	Shetland: organization of Eurasian otters in a marine habitat	56

Organization of Eurasian otters in freshwater areas	61
Spatial organization of Eurasian otters: some generalizations	62
The social system of North American river otters	62
Social organization of the sea otter	64
Social systems of Latin American otters	66
Social organization of otters in Asia	68
Otter social systems in Africa	69
Social systems of otters: general comments	71
Otter dens, or 'holts'	73

6 Scent marking and interactions: social behaviour — 78

Introduction	78
Sprainting behaviour and other scent communication	79
Aggressive behaviour	86
Sexual behaviour	88
Family life and parental behaviour	90
Some generalizations on social behaviour of otters	98

7 Diet — 99

Introduction	99
Methods of analysis	100
Diet of Eurasian otters in Shetland	102
Eurasian otter diet elsewhere	107
Quantities and calories consumed by Eurasian otters	111
The New World: diet of the river otter	112
Diet of the sea otter	112
The food of giant otters	114
Diets of other Latin American otters	115
Diets of otters of Asia	117
African otter diets	117
Diets: some generalizations and comments	119

8 Resources: about fish and other prey — 120

Introduction	120
Fish numbers and behaviour in Shetland	120
Observations of fish behaviour in Shetland	123
Trapping and counting fish	124
Selection from prey populations in Shetland	128
Eurasian otters and prey populations in freshwater habitats	130
Effects of Eurasian otters on their prey	132
Prey availability for American otters	134
Ecosystems, and effects of predation by sea otters	134
Prey availability in Asia and Africa	136
Conclusions, and global changes in otters' resources	137

9 Otters fishing: hunting behaviour and strategies — **140**
Introduction — 140
Fishing behaviour of Eurasian otters — 140
Fishing and foraging by North American otters — 143
Fishing and foraging by other otter species — 145
When to feed: time and tide — 148
Foraging success — 150
Patch-fishing, prey sites and prey replacement — 152
Depth of dives — 153
Energetics of foraging — 155
The development of foraging behaviour — 160
Otter foraging: some conclusions — 161

10 Thermo-insulation: a limiting factor — **162**
Introduction — 162
Body temperature — 162
Thermo-insulation — 165
Washing in fresh water along sea coasts — 170

11 Populations, recruitment and competition — **173**
Introduction — 173
Populations: numbers, changes and genetic diversity — 173
Population size — 177
Gene flow in otter populations — 180
Otter numbers rejuvenated: recruitment — 181
Some conclusions on recruitment — 189
Otters in ecosystems, and competition with other species — 189

12 Survival and mortality — **193**
Age structure, life expectancy and rates of mortality — 193
Causes of mortality: introduction — 196
Proximate causes of death, and body condition — 196
Effects of pollution — 200
Food availability, starvation and population limitation — 204
Predation on otters — 206
Some conclusions — 209

13 Syntheses: challenges to otter survival — **211**
What is special about otters? — 211
Habitats — 212
Foraging — 215
Social life — 218
Populations — 220
Life at the edge of a precipice — 222
Further research — 223

14 Otters, people and conservation — **224**
- A changing world — 224
- Otter vulnerabilities — 225
- Otter nuisance — 226
- Exploitation of otters — 228
- To catch an otter — 230
- Reintroductions of otters, and conservation genetics — 232
- Conservation needs of ecosystems — 233
- Conservation of otter species — 237

References — **239**
Index — **261**

CHAPTER 1

Otter ecology and its background

The following chapters are about science—about the ecology and other disciplines involved in the understanding of some of the most intriguing mammals in our environment. They are also about the nature of the animals themselves, and about the rewards of studying them in their magnificent context.

Otters, of one species or another, have been part of my own environment for many years, and I have spent a significant part of my life watching them. Yet, almost every single occasion with one of the animals is still engraved on my mind, with the otter in its habitat, and with what happened before and after. Otters have an intrinsic beauty, and they are exciting. At the same time, these animals have been a scientific inspiration to me on numerous occasions, as they have to many other researchers. Just in watching them, questions arise about their survival, and an ecologist cannot fail to be struck by the possibilities of study that suggest themselves in the magnificent surroundings in which these animals live in. This inspiration is to be found in Alaska, in Scotland or England, in Germany, Brazil or Africa.

In Scotland, I found Shetland to be a particularly rich vein of experience. The seas around Shetland are a source of a large variety of wildlife, but when I was there in the 1980s and 1990s I tried to concentrate on otters, which in Shetland is the Eurasian otter (Fig. 1.1). I focused on that animal, though fully aware that it represented only a small aspect of the ecosystem. In the Shetland setting, I was able to reach an understanding of otters that might be difficult in many other places. I saw the animals with problems that were by no means unique to Shetland, but somehow more tractable. And there was the

Figure 1.1 Eurasian otter (female) in Scotland.

wonderful environment: the rich background for some fascinating science. Field experiences there moulded my studies of otters elsewhere in Scotland, as they did in many other parts of the world. Perhaps I can illustrate this with a few examples.

A gannet flies in front of me, a huge white bird with long, pointed wings. It follows the waves at some height above, while I shelter from the cold wind against a rock on the shore. The bird slows, circles, then forms into a pointed projectile, rocketing down. The waters part with a dramatic splash, the gannet disappears then bobs to the surface, giving no clues as to whether it has captured a fish or not. After a few seconds it lifts off again, with heavy, deep wing-beats.

I am so absorbed by the bird that several minutes pass before I notice an otter, just in front of me in the seaweed. It is messing about in the bladder wrack and, when I see a flash of white, I realize that it has a fish. Like a cat with a mouse, it is playing with the unfortunate prey, a good-sized flatfish. The otter grabs the fish, shakes, lets go again, waits a second, darts after it, grabs it again. The otter rolls on to its back, holding the fish in its mouth and paws, then the fish wriggles off again, and the otter pounces. So it goes on for eight minutes, until the predator has had enough of the game, lands the fish and eats—every scrap of it.

What I saw may have been play, but it could also have functioned as exercise, as practice for the big game. Catching fish is difficult, even for a gannet, and for an otter. How exactly does an otter do it, and how many fish are there? Is there anything that makes fishing particularly difficult for an otter?

Elsewhere, and close to my home in Scotland, there are two small lochs, lush in summer, often iced over in winter. Here, I see the same species that I watch in Shetland, the Eurasian otter, climbing on to the ice in winter, diving beneath it, and emerging with an eel in its mouth, the fish curling around the otter's head. Or the otter may come up empty-mouthed. It makes one shiver to think what cold these animals have to endure to get their prey, a perception that is substantiated by interesting research. Even during the excitement of the observation, one wonders what this seemingly effortless pastime takes out of an otter.

Along the shores of the Prince William Sound, in Alaska, I walk between huge trees and rocks on a steep slope. Well worn trails connect the waterline with what looks like an otter city, a collection of entrances to tunnels and small caves, dripping with rainwater and with piles of fish-smelling droppings. While I take in the details, a bald eagle flies past with a fish in its claws, majestic and beautiful, landing on a dead branch high up. Minutes later, two North American river otters (Fig. 1.2) come past, just below me. They are the reason for my being there, I can see how they paddle with their hindlegs, their heads on the surface and tails streaming behind them. There is almost nothing to distinguish them from the Eurasian otter, although this similarity is deceptive. The animals are there only briefly, then they continue along the coast, quite ignoring the large den next to me, and swim out of sight. It flashes through my mind that we know these river otters to be more social than the Eurasians, and it would be interesting to find out why.

Then comes a strange addition to this wonderful setting. About a hundred metres out on the water, I suddenly notice another animal, a sea otter, a very different species, floating on its back and paddling, carrying a cub on its belly—something I had never

Figure 1.2 North American river otter. © Andreas Norin.

seen before. The sea otter mother's shaggy face is raised and keeps looking around, quietly. It is a peaceful scene, but laden with questions. For a start, here are two species of otter, in the same landscape. How do they exist together, how do they compare in their behaviour and ecology? And at the same time in these primeval surroundings, one cannot help but remember that this is Prince William Sound, the setting of one of the best known environmental disasters of the past few decades, the *Exxon Valdez* oil spill. The area abounds with problems of conservation, and my simple observation reminded me of the large body of excellent research that was inspired by the sea otter, as well as its dramatic, present-day fight for survival.

The different otters that I have seen in Europe and North America present a few faces of the same problems worldwide. I also marvelled at Cape clawless and spotted-necked otters on an unspoilt island in Lake Victoria. I saw our Eurasian otter together with smooth and small-clawed otters in a beautifully forested large river in Thailand, and one day I found tracks of all three species, and of tiger (the latter in the footprints of an elephant on a sandbank). Then there were the giant otters in the Brazilian Pantanal, sharing rivers with neotropical otters, both having their own method of dealing with the problem of piranhas. Otters have established themselves throughout the world, and their problems of survival and conservation are as diverse as their habitats.

Nevertheless, the various species have much in common. One aspect that all otters share is the opportunity they provide to watch and study them, to address important ecological questions against the rich background of the edges of their aquatic world.

There is a public fascination with all carnivores that I have analysed in some detail in a previous book, *Hunter and Hunted* (Kruuk 2002). It extends especially to those carnivores that live both on land and in water. Some scientists have argued that our attraction to things aquatic may come from our evolutionary past, and perhaps we imagine ourselves in the animals' watery position. Whatever the origin of their appeal, its effect is that most of us are immensely charmed by an otter, or a seal. Indeed, an otter, of whatever species, is even more attractive than a seal, being sleek and fast, apparently playful, and frequently in and out of its watery element—as we would be ourselves if given half a chance.

It is hardly surprising, then, that some of our classical natural history books are about otters. Henry Williamson's *Tarka the Otter*, first published in 1927, has been and still is highly successful, causing many an eye to shed some tears. Gavin Maxwell's *Ring of Bright Water* (1960) is equally well known, and it is quite appropriate that in the scientific nomenclature an otter subspecies is named after him (the Iraqi subspecies of the smooth otter, *Lutrogale perspicillata maxwelli*).

Ironically, however popular as these animals may be, they have also suffered the consequences of the advances of our civilization more than most other mammals. It is only through detailed research that we now realize the many dangers that these animals are exposed to. The reasons why otters are so vulnerable will emerge in the following chapters. As an example, literally hundreds of thousands of otters, of different species, have been killed for their fur, and many still are to this day. Thousands of North American river otters, or Eurasian otters in Russia, are legally trapped every year, as are many of the others on other continents, legal or otherwise. Because of this, large areas were devoid of otters, though some are now recovering after action by conservationists. The beautiful quality of the otters' fur, which is their undoing, is, of course, an adaptation to their watery habitat.

After serious pollution of waters in Europe and North America in the mid-1900s, with compounds such as dieldrin and DDT, otters disappeared totally from entire countries and states. They suffered more than most other wildlife, and it was a major victory for conservation that these highly damaging organochlorines could be legislated away. Now, in many places, Eurasian and river otters are back again. Their numbers may not be quite the same as they were, however. The following chapters will show that fish populations can be crucial, and in many places fish numbers are seriously depleted. Humankind is to blame, directly or indirectly.

Another major drama between our own species and one of the otters is still being played out today, and I will often refer to it. Sea otters (Fig. 1.3) are tremendously appealing animals, a totally fascinating branch of the otter evolutionary tree with a

Figure 1.3 Sea otter with clam. © Bryant Austin.

highly complicated and well researched biology. Their history is marked by disaster and recovery. Once they were massively abundant along the northern Pacific coasts, but the coat of these shaggy animals is even more attractive and protective than that of other otters, and when they were 'discovered' it seemed the end was neigh. They were subject to an unrelenting onslaught; hundreds of thousands were killed (as were many of the Indians who got caught up in the trade), and within a few decades the species had all but gone.

Once sea otters became protected—after almost none was left—the species started to recover. Some seventy or eighty years later, towards the end of the twentieth century, tens of thousands of sea otters were back again, and conservation congratulated itself: 'Sea otters are now the least threatened of all otter species' (Estes 1990). However, the claim of victory was made too soon.

In the early 1990s, biologists began to notice sharp declines in numbers, and there were incidents of predation by killer whales. The twenty-first century began with sea otter numbers only a small fraction of what they had been twenty years earlier, and declining fast. The cause was an upset of the entire ecosystem, brought about by whaling and overfishing. At the time of writing, the sea otter probably is the most threatened of all otter species.

Conservation issues may prompt research into the ecology of otters, but even without such emergencies these intriguing animals suggest a host of questions. Their aquatic lifestyle has important consequences, not the least of which is that it is energetically unusually expensive. We will see that this reverberates throughout the behaviour and ecology of otters—they may be feeding on often large, nutritious prey, but this costs a great deal of energy and needs many special adaptations. Energetically, they are living on a knife-edge.

Another point of interest is the similarity of all the otter species, in appearance, ecology, and behaviour. Of course there are differences and, especially when several species occur sympatrically, one is confronted with various ways of exploiting the environment, or differences in response to challenges. But overall, the otters of the world are remarkably alike, and often even a good naturalist has to fall back on tiny distinguishing characteristics to recognize a species, such as the shape of the nose-pad.

From a pragmatic viewpoint, the similarity between species is useful, because it enables extrapolation from one to another—perhaps not of exact research results, but at least of hypotheses. Having found the answer to a question in one species, the similarities suggest looking for this also in another. For instance, the Eurasian otters' need for fresh water along sea coasts, just for ablution to rid the fur of salt, is mirrored in the ecology and behaviour of the North American river otter and the Cape clawless otter in Africa. The diving and foraging behaviour of a neotropical otter appears similar to that of the river otter, and explains some of its habitat preferences. This potential for extrapolation means that, when confronted with conservation problems in poorly studied species, one can make careful, but fruitful, use of research experience on better known species elsewhere. Partly, this is what this book is about.

Before discussing aspects of ecology, behaviour and physiology in some detail, in the next chapter I have outlined the natural history of 13 species of otter. The purpose is to provide an overview of the entire subfamily, to get the flavour of the ecology, diet, behaviour and threats to the existence of the different species, and the extent of our knowledge and ignorance about each of them. I have provided this in general terms, with few references except for some of the basic, general publications on each of the species concerned. Detailed references to literature will follow later.

The main concern of this book is directed at answering the ecological question that is essential to the survival of every species: what limits numbers? Why are there so few along the banks of a clean river somewhere in Europe, or, for that matter, why are there not even more otters in Shetland than the ones I see around me? What causes a decline in the numbers of sea otters, and can trapping of river otters be sustained? How can three species of otter inhabit one single river in Thailand? These kinds of question can be asked about otters anywhere, but they are relevant especially there where the problems bite, where a species is slipping away.

What is frustrating, but at the same time challenging, is that to answer my questions for otters we need to know much more about the basic biology of the animals than we do. There is an urgent requirement to get to know more than just lists of prey species. This was evident especially in the earlier years of my study, when it was necessary to collect extensive baseline information on one species, in the few areas where the animals were safe and plentiful. Now, with knowledge gained on Eurasian otters in Shetland and elsewhere, with the detailed research on sea otters in North America, the giant otters in South America, the Cape clawless otter in South Africa, and many more studies, we are in a better position to proceed with questions about otters anywhere in the world.

The urgent conservation problem is one reason for my interest in otters. There are others too; I make no excuse for being somewhat infatuated with the animals, as is almost everyone who has watched them. Apart from that, though, there is a great scientific interest in the way they cope with the problems of their environment, especially when juxtaposed with other species of carnivore.

To approach the general problem of what limits otters, of any species, I have tried to describe the way in which otters are organized, their social system and their numbers, and the methods we used to assess these. To explain what we found, to answer the central question, I then looked at the resources used by the animals, at the peculiar need of some sea-foraging otters for fresh water, how they use their holts (or not), and examined in detail their relationship with prey species. Energetics have cast an interesting light on the otters' trade-offs during foraging. This information is combined to gain insight in the otters' population characteristics, their mortality and reproduction, and the role of pollutants. Recent developments in genetics, using otter DNA, have provided new knowledge about populations that previously could not have been imagined.

Some of our insights are gained by direct comparison between sea-living and freshwater otters, focusing on relations with prey but also including observations on habitat and pollution problems. Many of these observations have implications for the way in which we should manage the environment, in order to secure otter populations where there still are any, or wherever the animals were once common and are now gone. Frequently, research on one species appears to be useful in understanding another.

Perhaps most importantly, I think this book shows that we are learning to see otters as just one aspect of much larger ecosystems. Their existence is closely dependent on what happens to fish and other prey populations, and therefore to the insects in the streams and to the kelp beds in the sea. The nature of these interactions needs to be exposed, and we are beginning to make some headway with that.

As one of the results, otters show themselves as curiously vulnerable, with a life history so insecure that in some ways it is surprising, not that many have gone from their haunts, but that they are still present at all, along banks and coasts. One can but hope that what sustained them until now will continue, if need be with some well considered assistance through conservation management.

CHAPTER 2

Pen pictures. Thirteen otters of the world: some natural history

Europe

One species.

Common or Eurasian otter, *Lutra lutra* (Fig. 2.1)

Shetland 1988, a balmy April morning. The sea was like a sheet of blue silk, with just the distant murmur of moving water as a reminder that there were tides here. Miles away across the huge expanse lay the island of Yell, with its gently curving outline of low yellow hills. I could hear the crooning of a raft of eider ducks, and a gannet hurried past—a scene close to perfection.

An otter swam out, a small dark head at the sharp end of a V in the water. A dive, the tail flipping vertically, then an ever increasing circle with a few bubbles gently spluttering up in the centre. I knew the otter would come up again in about the same place, perhaps in fifteen or twenty seconds, perhaps earlier if it had caught a fish. There was something totally predictable about the animal: it belonged intrinsically, beautifully shaped as it was for this expanse of water, and one of many otters in Shetland.

Yet, at about the same time that I was sitting there along the Yell Sound, the last Eurasian otter disappeared from Holland, the land where I grew up with its sea, lakes, dykes and rivers and, as I knew it, a paradise for otters if ever there was one. At that time, in 1988, only a handful of otters remained in England, although they were still fairly common in countries along the periphery of Europe. But in hundreds of miles throughout the continent, where the animals were abundant a few decades earlier, none was left.

This was several years ago. Now, in the early twenty-first century, the animals are back in many of their former haunts, including Holland, in a triumph over the ravages of human-made pollution (see Chapter 14). Outside Europe, the species occurs in North Africa and throughout Asia (except on many islands in the south-east), but since the 1980s has been extinct in Japan (Fig. 2.2).

Figure 2.1 Eurasian otter (male). Note broad head.

Figure 2.2 Geographical range of the Eurasian otter. The species has the widest distribution of all otters, but is now extinct in Japan.

The observation I have described above is perhaps the easiest and most common way to see this animal, as a head and a bit of tail above the surface of a smooth sea. Where the common or Eurasian otter lives in fresh water, it is more difficult to spot and is often overlooked (especially in fast-flowing waters). However, it occurs in many rivers and lakes, in reed-beds, even in the smallest of streams, as well as along suitable rocky shores of the sea.

The overall colour is dark brown, somewhat lighter underneath, in some areas with light patches on the throat. Compared with other otters, the Eurasian is one of the smaller, with a total length (from nose to tail tip) of about 1.0 m for the female, 1.2 m for a large male, and weight averaging 7 kg for a female (called 'bitch'), 10 kg for a male ('dog'), with regional variation (e.g. in Shetland, females average only 5.1 kg, males 7.3 kg, in Norway 5.9 and 8.5 kg respectively). The sleek, whiskered animal with its long tail ('rudder') is proverbially agile, with almost snake-like abilities to twist and turn. It has almost no fat.

The feet are webbed, but not conspicuously so (unlike some of the other otters); as well as swimming, Eurasian otters make overland trips of many kilometres (looking rather clumsy, bobbing along). Their tracks in mud or snow are highly characteristic, footprints in groups of four, usually with all five nailed toes showing. In snow otters often toboggan, chin and belly on the snow, efficiently propelling themselves with their hindlegs, and producing beautiful snow slides.

Where the Eurasian otter lives in the sea it is active mostly in daytime, in fresh water mostly at night, depending on habitat: it is daytime active in areas where it feeds mostly on nocturnal prey, and vice versa. The Eurasian otter is *the* classical otter, fish its main food, but also taking frogs, crayfish, crabs, birds and mammals.

It fishes by swimming along the surface, then diving with a tail-flip, and searching for prey along the bottom, coming up mostly with bottom-dwelling or bottom-resting fishes. Small fishes are eaten on the surface, and larger ones taken ashore. But the animals are quite versatile in their foraging, and they may catch frogs when available, or rabbits after following them inside the warrens, or water birds on the surface from underneath—or my own farmyard ducks when they are asleep—with many other variations.

The Eurasian otter is solitary, yet there is a complicated otter society. When one sees more than one of these otters together, it will usually be a mother with cubs, often adult-sized ones. Males and females have little to do with one another, apart from mating, and males are kept away from cubs. Several (probably closely related) females may live together in a 'group territory', defended against other females, but within such a group territory the occupants keep well out of each others' way. The territory is divided up into 'core areas' for each of the females, where they rear their young and where they are avoided by the other group territory owners. Male otters have huge ranges that may overlap with several of the female group-territories, the largest recorded male territory being 80 km of stream.

Although Eurasian otters use the territory on their own (or with cubs), they do communicate with one another, and faeces ('spraints') are important for signalling. Spraints are conspicuous with a rather pleasant fishy smell, left on prominent small rocks, logs or patches of grass which may be permanent 'spraint sites' (also for urination), with accumulations over weeks or longer serving as message boards. Each spraint is usually very small for the size of the animal; in particular, the larger males leave tiny ones, dozens of them every day. People know that there are otters in an area because they see the spraints, rarely the animals themselves. These spraints appear to have little to do with territorial defence, but more with passing information between any otters that use the same area (meaning something like 'This is my fishing patch. You'd better move on because I am familiar with the place and at an advantage.').

Eurasian otter mothers look after their cubs and feed them for about a year, far longer than comparable mammals such as badgers, foxes or cats. They have one to four cubs, which are born blind with a very thin coat of hair. The natal den ('holt') is often a long way from water, and difficult to recognize for what it is. The mother stays with the cubs for most of the time, often going out for only one quick fishing trip per day, but when the cubs are about two months old she takes them down to the water, carrying them one by one—a fabulous sight. From then on the family stays together in holts close to the water (these dens are often, mistakenly, referred to as 'natal holts'), and they start going out fishing together. Once the young ones become independent at about one year of age, it still takes them another half year to become as proficient at fishing as their parents.

After that, their life expectancy is rather short, on average around four years. Although we often find otters killed on roads, most of them die a non-violent death, with food shortage probably a major, ultimate cause. Even in the most favourable habitats, the numbers of Eurasian otters are small: they live in very low density.

Cubs may be born at any time of the year, but in some areas (e.g. Shetland, or in northern Europe) births are rather seasonal, probably related to the seasonality of food resources. In years when food is abundant, there are more otter cubs. The gestation period is about 63 days, and there is no delayed implantation: the fetus starts developing immediately after fertilization (unlike, for instance, the North American river otter or the Eurasian badger).

The long period of dependence of the cubs may be necessary to develop fishing skills, and because of the need to catch a prey quickly, to keep the huge energy drain in cold water to a minimum. The Eurasian otter's coat is a good insulator, yet the animal cools down quickly when submerged. So, when otters catch a fish this may be a big prize, but it comes at a large cost of energy and has to be caught quickly and efficiently.

When we watch them, playing and diving and shaking themselves, their life looks easy, with lots of energy to spare. This is appearance only. In fact, Eurasian otters have a high mortality rate, low life expectancy and a low reproductive rate. They are vulnerable to human-made disasters in the environment, such as traffic on roads along rivers, and pollution, hence their almost complete disappearance over large areas in Europe during the twentieth century.

Water pollution, with organochlorines such as dieldrin, DDT, PCBs and heavy metals such as mercury, is now less of a problem in Europe than it was. In most countries otters have returned to the rivers, or they have been successfully reintroduced, they are fully protected by law, and they are even found right inside towns or industrial complexes. Their recovery has been successful even to the extent of becoming a problem to fish farms in several areas. The consequences need serious thought from managers and conservationists, but for most people it is a tremendous satisfaction to have this fascinating animal in heir environment again.

General references for Eurasian otters

Chanin 1985; Kruuk 1995; Mason and Macdonald 1986; Woodroffe 2001.

North America

Two species.

River otter, or North American otter, *Lontra canadensis* (Fig. 2.3)

It may be called a river otter, but there was no river anywhere near the place where I was watching this animal fishing. Leaning over a fallen hemlock along the shore of the Prince William Sound in Alaska, in the middle of the day, I had the otter in close-up, floating on the surface of a mirror-like sea. A tail-flip, a dive, and I could see the animal swimming along the bottom, some two metres deep. Twenty seconds later it appeared with a small fish and chewed its prize afloat, belly-down, nose pointing skywards. I could hear the crunching noises. It was fishing exactly like a Eurasian otter does. In fact, the North American and Eurasian otters resemble one another in numerous ways—in appearance, behaviour and ecology—and one has to keep remembering that they don't even belong to the same genus. However, there are differences, some quite important.

The total length of both river otter males and females is about 1.25 m, but males are heavier (9 kg, females 8 kg). This also shows that river otters are quite slender compared with the Eurasians, which are shorter but about the same weight, and in river otters there is less difference between the sexes. Colour and general appearance are similar to that of the Eurasian otter: dark brown with little variation between the upperside and underside. However, a river otter's face looks somewhat different, a bit rounder, just as bewhiskered but with a larger nose

Figure 2.3 North American river otter. Note large rhinarium (nose-pad). © Nicole Duplaix.

(rhinarium), and its neck appears longer. When the river otter looks at you, it holds its nose higher up than the Eurasian does, giving an impression of suspicion. The feet are fully webbed, especially the hindfeet, and, as an interesting difference with the Eurasian, the males have a small, circular scent gland in the middle of the pads of the inner toes on the hindfeet.

Like the Eurasian otter, the river otter usually propels itself in water by paddling with the hindlegs moving alternately, and/or with vertical body and tail undulation. It walks, often gallops just like the Eurasian, and also slides in snow. On the North American continent it gets the opportunity for sliding more often than its Eurasian counterpart, sometimes over hundreds of meters, especially downhill.

River otters occur throughout most parts of North America, as far north as there are trees, just like the beaver (with which they are often closely associated). They are found from the Atlantic to the Pacific coast, and from the deep south of Florida and Texas to northern Alaska (Fig. 2.4). They may be in the wildest parts of forests or in agricultural areas, even in towns and harbours, and in a wide range of habitats, from rivers, small streams, lakes and bogs to rocky sea coasts. They may sleep above ground, or live in dens (large complexes of tunnels with many exits), or in beaver dams; or they may construct covered 'nests' of reeds. The animals are highly flexible in the way they use their environment. They disappeared from several states in the mid twentieth century, owing to pollution and intensive trapping for fur, but have now returned to many of their old haunts, often aided by large-scale reintroduction programmes and trapping regulated by legislation.

Fish is the normal fare for river otters, especially the slower, fairly small and bottom-living fishes, but they also take frogs and crayfish, as well as small mammals, birds and reptiles occasionally. Sometimes, several otters fish together, clearly cooperating.

River otters are often alone, but can be much more gregarious than the Eurasian otter. People have seen

10 Otters: ecology, behaviour and conservation

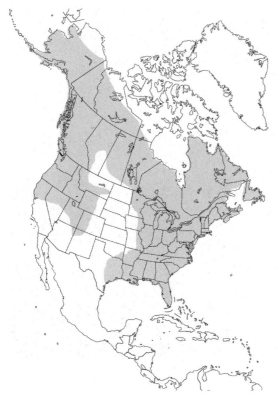

Figure 2.4 Geographical range of the North American river otter, coastal as well as inland.

as many as eighteen of them together, but more often just three or four. Groups usually are either family parties, a mother with her offspring (which may be fully grown), or two (probably related) mothers together, each with her cubs, or they may be large aggregations of unrelated adult males. The larger groups are seen most often in the sea, and this may be related to the presence of predators or to what the animals are feeding on.

Males and females range more or less independently of each other, with large overlaps between individual home ranges, especially between males and females. Male home ranges in fresh water may be as long as 250 km of river, but in the sea they are much smaller. Communication between the animals, through scent-marking, is much in evidence; the scent glands under the feet must play a role when the otters are scratching together heaps of moss and soil. Most conspicuous are the collections of scats (which in Europe would be called 'sprainrs'),

smelling strongly of fish and 'musk', quite different from the Eurasian otter. These marking sites are especially impressive along sea coasts where otters live in high density, and where the scats are closely associated with large dens (with fresh water inside them, used for washing and drinking). What messages are conveyed with these scats is as yet unclear; they may be similar to those of the Eurasian otter.

River otters have up to five cubs at a time, but more usually one to three. Interestingly, actual gestation is 60–63 days, just like the Eurasian otter, but after copulation river otters may delay implantation by as much as 8 months before gestation starts (Eurasian otters do not do this). We have little understanding of what advantages this might bring, and why one species should do it and another not. The cubs stay with their mother for a long time, 10 months or more, although they are weaned at 3–4 four months.

From analysis of the age at death of otters (counting dentine rings in teeth) we know that the mortality rate is high (25–50% per year in many areas), and trapping with leg-hold traps, for fur, takes a large toll. River otters also have their predators, especially wolves, but in addition there are bobcats, coyotes, alligators and even killer whales. Oil pollution in the sea, and chemical pollution in lakes and rivers, has killed many, and with all such causes life expectancy is low. Conservationists in America are most worried about the effects of trapping, quite apart from the cruelty involved; many populations have been wiped out in the past. Yet, river otters are back!

General references for river otters
Melquist and Dronkert 1987; Larivière 1998.

Sea otter, *Enhydra lutra* (Fig. 2.5)

In the 1980s, along many of the magically beautiful coasts of north-east Asia, Alaska and British Columbia, and along the much busier shores as far south as California, sea otters were a common occurrence once more. At any time of day, right inshore or often far out at sea, one might come across a mother otter afloat on her back, carrying a cub on her chest, sometimes leaving it briefly on the surface while she dived deep, deep down, perhaps some 40 metres to

Figure 2.5 Sea otter in California. Note thick, dense fur. © Jane Vargas.

the bottom. Now, early in the twenty-first century, the animals face serious problems again.

Sea otters are very different from all the other otters, and spectacular in many ways. Recognized as such by many scientists, they are also studied more extensively than any of the others, and there is a large amount of information about them.

To mention a few of the unique points of the sea otters that excite researchers: they are more aquatic even than seals, not necessarily coming ashore, usually even giving birth in water. They drink sea water, dive to great depths, use tools to access their food, mothers carry their pup on their belly whilst floating on the surface, they may aggregate in huge 'rafts' of several hundred animals, and their shaggy, extremely dense, fur is quite different from that of any other otter. The abundant twentieth-century populations have a substantial effect on their marine environment and on the human use of resources therein. Recently, however, a dramatic decline set in.

Sea otters are larger than other otters, average adult males and females weighing 29 and 20 kg respectively (the largest male recorded was 42 kg), with mean total lengths of 129 and 120 cm. They are the largest existing members of the mustelid family. Their tail is short, less than one-third of body length, the hind feet are proper flippers, whereas the front feet have sharp, partly retractable, claws not found as such in any other otter. The teeth are blunt, with large flattened molars (for crunching molluscs). The extremely thick fur gives the animals a rather grizzled appearance.

Active in daytime and spending almost all of their life out at sea, mostly floating belly-up on the surface, the animals are very vulnerable to human exploitation—and exploited they have been. The original geographical range is along coasts of the northern Pacific Ocean, from Mexico to Japan; it is somewhat smaller now (Fig. 2.6). After they were 'discovered' and from about 1780 onwards, they were 'harvested' until, in about 1820, they were more or less gone. At the beginning of the twentieth century only about 1000 to 2000 sea otters remained in their entirely vast range of coast, in small pockets here and there, with little hope for their survival.

Figure 2.6 Geographical range of the sea otter: along coasts from north of Japan to Baja, California.

South of Alaska there was only one remnant population, in central California.

Sea otters belong to a species that has influenced history more than any of the other otters. Their fur was reputed to be the best of all animals, and over a fairly short time in the eighteenth century a vast trade built up along the American north-west shores, motivated by large profits from the one species. Hundreds of thousands of otters were slaughtered by Indians, and sold to British and American trading ships. These then progressed to Canton to sell the furs at vast gain in China. The Indian tribes, in the meantime, were betrayed and harassed, persecuted and literally decimated by smallpox, syphilis and tuberculosis brought in by the sailors. Sea otters were driven almost to extinction in the course of just a few decades.

After an effective alarm was raised, legal protection took hold in 1911, and the sea otter population expanded again immediately, in many places at a rate of around 20% per year. In the 1990s it was estimated that sea otters were back again in numbers of approximately 50,000. At the end of the century numbers were falling again, with good evidence that this was caused by a large increase in predation on the otters by killer whales.

Sea otters do not go into fresh water, nor do they go much into oceanic depths further than about 30 metres (although dives of about 100 metres have been recorded). They are often in and around kelp beds, where they sleep and forage. Rocky coasts are their favourites, especially where there are large underwater reefs, and they may haul out on land on rocky points, staying close to the shore. However, they do not wash in freshwater pools, as other otters do where they live in the sea, nor do they have to rely on fresh water for drinking: sea water will do. Perhaps to make up for the fact that sea otters live without access to fresh water for washing, they spend an inordinate amount of time grooming whilst afloat in the sea, even blowing bubbles into their fur to retain insulation.

Feeding behaviour is unusual, and has profound implications for the environment. Their diet consists mostly of sea urchins, crabs and large molluscs such as abalones and clams, and fish, but it varies hugely between regions, with availability of the different foods strongly affected by the sea otter itself. For instance, sea otters have decimated populations of sea urchins in places, which in turn had the effect of inducing the growth of large kelp beds (sea urchins can remove these completely), followed by further changes in prey populations (see Chapter 8). Interestingly, some shellfish appear to be protected against sea otter predation by toxins from algal blooms, and there is evidence that sea otters avoid such areas.

Clams and abalones are carried several at a time in the otter's armpit, often from great depths to the surface, opened and consumed whilst the predator is floating belly-up. To open shells, the animals often use a tool, a small rock, which is also carried up from the ocean floor in the armpit. Even a long way away one can hear the shells being bashed by the otters.

Dive times of sea otters can be up to 4 minutes (average 74 seconds). Because of the large amount of heat loss from the body in this environment, their rate of heat production is two to three times higher than that of land mammals of similar size, and food consumption is correspondingly high at 20–30% of bodyweight per day (some 5–6 kg per day for an average otter).

Sea otters are usually alone, but they may be very gregarious, especially males; amazingly, the largest group ('raft') reported contained about 2000 animals. Sometimes females form 'nursery groups' of mothers and pups, of more than 100 otters. These groups have a very fluid composition with otters coming and going, but clearly the animals are highly social at times, even to the extent of actively defending one another against predators, or against people trying to capture them. The animals are quite vocal, with many different and loud calls. With their kind of lifestyle, scent communication is out of order, and, although I found their faeces at haul-out sites, they are probably not left as scent marks, as 'spraints'—they are mere eliminations.

Males and females live in more or less separate areas and groupings, but some solitary males defend territories, in areas with rafts of females, and they attempt to herd females to keep them in their area. Females take a long time to become sexually mature, some 4–5 years, then have a gestation period that varies in length from 4 to 12 months (because of delayed implantation). The animals copulate in the water, and occasionally the mating process is so vigorous that the female drowns, or gets badly injured as the male bites her face.

Once the single pup has been born (often in water, rarely on land), the length of time that the mother looks after it is also highly variable, from 4 to 9 months in California (where they feed mostly on shellfish and crabs) to well over a year in Alaska (significantly, in some areas feeding mostly on fish there). The mother is in constant attendance of the pup, carrying it on her belly, vigorously defending it against all comers, but she has to abandon it when diving for food. That is the time when some of the pups are taken by bald eagles. Orphaned pups are sometimes adopted by other otters, even by males.

At times, sea otter populations showed significant mortality rates from oil pollution, drowning in nets, and starvation. In California, disease has caused long-term declines. However, all of these changes are small compared with the more recent disaster that has overtaken their populations in Alaska, with killer whales causing a reduction in places of more than 90%. It is difficult to be optimistic about the fate of these wonderful animals.

General references for sea otters

Estes 2002; Gibson 1992; Kenyon 1969; Riedman and Estes 1990; VanBlaricom and Estes 1988.

Latin America

Four species.

Giant otter, *Pteronura brasiliensis* (Fig. 2.7)

Midday, on the wide, slow-flowing Rio Negro in the Pantanal of Brazil. It is hot and remote, surrounded by walls of forest alive with birds and insects. Slowly paddling upstream I meet that distinct, ottery smell of a fish-and-urine mix, very strong, which guides me to a large open patch on the bank under the trees. I know this is the work of the giant otter, a patch that stinks rather than smells, stronger than I have met in any other species of otter. The place is called a 'campsite', some 10 metres across, a scent-marking site that effectively spreads its message almost across the entire width of the river.

After a while, five heads come swimming towards me, a lone observer in a canoe. Once they see me, they call loudly in an impressive display of 'periscoping', showing their strikingly patched throats and their unusual heads. The giant otters are the noisiest and most conspicuous of all otter species, and, partly because of their lack of reticence, their existence has been under severe threat for some time. We now

Figure 2.7 Giant otters in the Amazon. Note flat tail and very webbed forefeet. © André Bärtschi.

know a considerable amount about their ecology and behaviour, owing to several excellent research efforts.

Giant otters occur in many countries of South America, but their geographical range has been much reduced, thanks to the popularity of their fur. Whilst in the north of their range they go as far as the Caribbean (without actually using the sea), their main distribution is in the Brazilian Amazon, and the regions immediately around that, but they are now extinct in Argentina (Fig. 2.8). The animals live in densely forested areas in slowly moving rivers, especially in oxbow lakes along these rivers and in the small creeks, as well as in the forests themselves, when they are flooded during the rains.

The appearance of giant otters is somewhat at odds with that of the other species. With a total length of up to 2 metres, they are the longest (about twice as long as our Eurasian otter), though not the heaviest (weighing up to 32 kg, which is less than a sea otter). Their heads are curiously rounded, ears sticking out sideways, with large and somewhat bulging eyes, and for an otter quite small whiskers. The head is all the more striking because of the unusually long neck, with conspicuous, large and irregular white–yellow throat patches, different in each individual. In addition, the other end is also different, with a long, broad and flat tail that enables giant otters to swim with a very impressive speed, faster than I have seen other otters move. With that tail they may not be as manoeuvrable as the others, but this could be a worthwhile concession to speed in piranha-infested rivers.

Giant otters are gregarious, and are typically seen in a group of up to 20 animals (most commonly fewer than ten). There are itinerant single individuals, but basically they live in groups, forage in groups, and defend themselves in groups as well as having a group territory. In another variant on social existence in otters, giant otter groups consist of a resident pair of animals with their offspring of several generations (so not just a female with offspring).

Like many group-living mammals, and unlike most other otters, the system of communication is highly differentiated, including many loud calls (even when underwater) and striking postures. Scent communication is hugely important: the giant otters' 'campsites' along the water's edge, from which all vegetation is removed, spread olfactory messages over a large area of water. On the campsites, all group members spend much time sprainting, urinating, depositing anal gland secretions, scraping, trampling, kneading mud, rolling, and tearing down vegetation and rubbing it over the body. Not surprisingly, however, almost nothing is known about the many messages contained in these behaviours. However, any other group of giant otters, cruising down the river at speed, cannot fail to notice them.

Giant otters are active only in daytime. At night they sleep in holes along the bank, which are also used for breeding. There are usually two cubs in a litter (very occasionally up to five), rarely more than one female breeds, and several members of the group help with provisioning and protection. Aggressive protection is required not just against predators, such as cayman or anaconda, but also against roaming male giant otters from outside the group, which were shown to be cannibals. Cubs are fully grown by 10 months of age.

Food consists almost only of fish, especially the slow-moving species, but also piranhas and others. There are many species of fish in the giant otter habitat that are active at night, and that lie still in daytime—easy prey for an otter (hence otters' diurnal existence). Often, giant otters take quite large fish,

Figure 2.8 Geographical range of the giant otter, an inland species. It is now extinct in Argentina.

and eat them on a tree branch, or on a bank. Their consumption of fish is estimated as about 3 kg per otter per day, equivalent to 10% of their bodyweight.

The much smaller neotropical otter often lives in the same waters as the giant otter, but there appears to be little overlap of diet between them: the species and sizes of fish and other food that they take are quite different. The two otters are not likely to affect each other much, and I noticed them largely to ignore one another.

Because of the partly impenetrable habitat and huge distances involved, numbers are difficult to assess in a species such as the giant otter, despite it being relatively conspicuous and individually distinct. Densities are always quite low, and even in ideal habitats in the best protected places they rarely occur at more than one animal per 5 km of river.

Conservationists estimate that there are only a few thousand of the animals left in the entire vastness of South America, threatened by the inexorable destruction of the rainforests, by mercury pollution from the many gold mining operations, by hunting, and by disturbance from tourists in motor boats.

Giant otters are hugely spectacular, but efforts to maintain them in the wild are an uphill struggle.

General references for giant otters

Carter and Rosas 1997; Duplaix 1980; Groenendijk *et al*. 2005; Schenck 1997; Staib 2002.

Neotropical otter, *Lontra longicaudis* (Fig. 2.9)

From my canoe in the Rio Negro, Pantanal, Brazil, I can smell the catfish that is being crunched by a neotropical otter only a few metres away. In the middle of the morning, on the edge of the water and half submerged, the otter shows little fear, like most of its conspecifics, and the fact that it is common and strictly diurnal here is very useful to me. I had seen my animal catch its fish characteristically close to the bank; almost all their fish are caught between branches and vegetation next to the shore. Otters that venture further out have a serious menace to contend with: piranhas. The animal that I am watching has part of its tail missing, a very common injury here.

Figure 2.9 Neotropical otter, eating a catfish in Pantanal, Brazil.

Although this species has a wide distribution throughout most of Latin America, it has been little studied, and we are rather ignorant of its biology. What has been published shows it to be similar in ecological and behaviour characteristics to the North American river otter, but clearly it is badly in need of more detailed research. In the field, to all intents and purposes, it looks like a Eurasian otter.

Few animals have been measured. The neotropical otter appears to be similar in size to the river otter (total length up to 1.2 m, weight less than 12 kg), but with a greater difference between males and females: males are up to 25% heavier. Compared with the river otter there are small differences in the skull and teeth, and it has the usual dark brown otter colour, slightly lighter underneath, and without white throat patches. Its nose is hairier than that of the river otter, and its feet are smaller (in fact, for an otter, quite strikingly small).

The very large geographical range covers many different climates (Fig. 2.10), and habitats include fast-flowing rivers as well as lakes, marshes, rocky sea coasts, and mangroves and wetlands in between; it is a species with a wide ecological tolerance. Like other otter species, the neotropical otter may live in or close to human settlement and towns. In the centre of its range, as in the Pantanal where I watched it, it overlaps extensively with the giant otter, occurring in the same creeks and often within sight, but it is not sympatric with any of the other American otters.

Just like the river otter and the Eurasian otter, it has its holt (den) in natural cavities or it digs tunnels for itself, or makes a large covered couch in dense vegetation. Some holts may be a long way from water. The few research projects that have been done have been restricted largely to analysis of sprains (faeces) to analyse diet. These showed another likeness with the two species mentioned above: it is predominantly a fish eater, but also takes crustaceans (crabs and shrimps), amphibians and mammals. Typically, it prefers the slow-moving, bottom-living fish species, and in places it is not averse to taking prey from fish farms.

Its sprains are used also for surveys of distribution; sprainting is seasonal, and associated with physical features of the habitat such as conspicuous rocks and logs, and I found sprains under bridges. Quite likely, the biological function of sprainting is similar to that

Figure 2.10 Geographical range of the neotropical otter: inland and coastal in Central and South America.

in other species such as the Eurasian otter, such as signalling the utilization of resources to other otters. The neotropical otter is solitary, but several individuals may inhabit and forage in the same area. There is no pronounced breeding season, but, like the river otter, this species has the option of delayed implantation, which enables individuals to breed at favourable times of year. Family groups consist of females and up to five, sometimes adult-sized, cubs, and the male appears to have little or no contact.

There are no data on populations of this species; along the Rio Negro in the Pantanal I estimated that there was about one animal per 2–3 km of river bank. In some regions this species has been heavily exploited for its fur, with thousands of pelts being traded, for example, in Peru and Brazil. Predators include piranhas, anacondas, cayman and various large carnivores, with piranhas the important ones, at least in some rivers. Interestingly, in some areas, if a neotropical otter is disturbed in the water it will often head for dry land, unlike most other otters, suggesting that the main dangers are aquatic. In the

Pantanal the many individuals with scars, probably healed injuries caused by piranha, support this.

General references for neotropical otters
Gora *et al.* 2003; Larivière 1999*b*; Pardini 1998.

Southern river otter, or huillin, *Lontra provocax* (Fig. 2.11)

The huillin is a species hanging on to the smallest geographical range of all otters, in the very south of South America. It occurs only in a short, narrow strip of Chile and Argentina (Fig. 2.12). Not surprisingly, it has had little scientific attention and, like the previous species, what we do know of its appearance, ecology and behaviour is similar to that of the better known North American and European *Lontra* and *Lutra*. The animals are about 1 metre in total length or slightly more, males only a little larger than females, dark brown above, lighter underneath with grayish throat. The rhinarium (nose) is bare, not hairy.

This otter is mostly a freshwater species, but it also occurs along the rocky ocean shores, in saltwater, in the same areas as the marine otter *Lontra felina* (see below). Its freshwater habitat includes lakes and rivers, especially where there is suitable bank vegetation such as forest. Preferred sea coasts tend to be sheltered from wave action, especially where there is forest or other vegetation—perhaps this relates to the on-shore and underground presence of freshwater for washing and drinking, as in Eurasian or northern river otters. Den use is similar to that of

Figure 2.11 Southern river otter, or huillin (male). © Gonzalo Medina.

Figure 2.12 Geographical range of the southern river otter: a very small inland and coastal range on the southern tip of South America.

these other species. It is mostly solitary, which probably implies that males do not associate with females and cubs, and its activities are largely nocturnal.

The food consists mostly of crabs, although fish and bivalve shellfish are also taken, and in some areas fish (small, <10 cm) predominate in its diet. Its feeding on molluscs is unusual amongst otters (except by sea otters), and, as a nasty side-effect of this, locals use this habit by catching the animals on shellfish-baited hooks. The animals are badly persecuted for their fur, although they are now officially protected.

General references for southern river otters
Ebensperger and Botto-Mahan 1997; Larivière 1999*a*; Medina 1996, 1998; Medina-Vogel *et al.* 2003.

Marine otter, or sea cat, *Lontra felina* (Fig. 2.13)

This is the only otter species, other than the sea otter, that has adapted completely to a saltwater habitat. It

Figure 2.13 Marine otter, or sea cat. © Gonzalo Medina.

Figure 2.14 Geographical range of the marine otter: in patches along the Pacific coast only, from Peru to the southern tip.

has some interesting characteristics that set it apart from the other *Lontra* and *Lutra*. The marine otter occurs in many patches of coast along the Pacific Ocean, from Peru along all of Chile, and at the southern point of Argentina in Tierra del Fuego (Fig. 2.14). It has been known to use freshwater rivers, but essentially it is a seawater animal, favouring exposed coasts with large rocks.

When other *Lontra*, *Lutra* and *Aonyx* species occur in the sea, they are dependent on sources of fresh water for drinking and for washing the salt out of their fur. This does not appear to be the case for the marine otter, and, although there is no certainty on this point, they occur along coasts with virtually no rainfall and where freshwater is absent. Their fur is different from that of other otters, perhaps releasing them from the need to wash salt from it (to maintain thermoinsulation); it is rough and coarse.

Otherwise, the marine otter has the usual otter appearance: dark brown on top and somewhat lighter underneath. It has rather large molar teeth, and the nose is bare. Interestingly, some of the webbing of the feet is hairy underneath (perhaps to provide extra insulation, or as protection when landing on the extremely rough coasts). The tail is flattened, unlike that of most other otters.

Marine otters are small as otters go, averaging only about 90 cm in total length and weighing between 3 and 6 kg, with males and females about the same. They are amongst the smallest sea-living mammals.

Being so small in a cold environment necessitates a high metabolism, which the animals appear to maintain by eating large quantities of food, especially crabs, as well as smaller numbers of fish and the occasional mollusc. Thus, the diet shows much overlap with that of the southern river otter, which occurs along some of the same coasts (but in more sheltered sites). The food has been assessed by analysing spraints, which are large and conspicuous, and left near the holts and in rocky shelters.

Marine otters may occur in high densities, of up to ten individuals per kilometre of coast. However, they forage on their own, finding their crabs and fish in long, quite deep, dives, often or perhaps only in daytime. They live in holts underneath large boulders or in caves, fairly often with more than one adult as well as two or three cubs. More than one adult may take food to the cubs, and it has been suggested that this means monogamous pairs. However, I think it is more likely that this involves a female with grown-up offspring from previous litters; further research will tell. At present we have no information about territoriality, numbers of animals within one home range or other aspects of social organization.

General references for marine otters

Ebensperger and Botto-Mahan 1997; Larivière 1998; Ostfeld *et al.* 1989.

Asia

Four species, of which the Eurasian otter occurs throughout, and has been described above under *Europe*.

Hairy-nosed otter, *Lutra sumatrana* (Fig. 2.15)

This is one of the least known otters, very nocturnal and living in difficult areas in a rather small part of Asia. It is remarkably similar to the Eurasian otter, its hair-covered rhinarium (Fig. 2.16) being one of the main differences, but they are equal in size. The overall coloration is also like that of the Eurasian otter, except that the throat as well as the upper lips are usually white, with a sharp delineation between the dark upper side and the light underneath, especially on the throat and neck.

Evidence of its presence, such as tracks and spraints, is similar to that of the Eurasian otter, and in consequence our knowledge of its exact geographical range is scanty. As far as is known at present, it occurs on the Malaysian Peninsula and bordering southern parts of Thailand, as well as on the Indonesian islands of Sumatra and Borneo, probably also on Java, and in Cambodia and Vietnam (especially in the Mekong delta) (Fig. 2.17). It shares many of these areas with the short-clawed

Figure 2.16 Hairy nose of the hairy-nosed otter. © Reza Lubis.

Figure 2.15 Hairy-nosed otter sprainting on tree in Melaleuca forest, south Thailand. © Budsabong Kanchanasaka.

Figure 2.17 Geographical range of the hairy-nosed otter, possibly inland only. Distribution on the Indonesian islands of Sumatra, Java and Borneo is uncertain.

and the smooth otter, and at least in Sumatra also the Eurasian otter.

The hairy-nosed otter lives in swamps and swamp forests, sometimes close to and possibly actually in the sea, as well as in rivers and mountain streams. Its food consists largely of small fish, with also some snakes and frogs, and a few small mammals, crabs and insects—similar to that of the Eurasian otter, but different from the diet of other otter species in the general region.

At the time of writing nothing is known of the social behaviour and organization of the hairy-nosed otter; it seems likely that there will be many similarities with the behaviour of the Eurasian otter. Like all the otter species in south-east Asia, it is subjected to quite intense hunting pressure with traps (for skins, meat and medicines), partly also to prevent its raids on local fish ponds, and in some places it has been trained for use in fisheries. It is a species in dire need of further research.

General references for hairy-nosed otters

Kanchanasaka 2001; Nguyen 2001; Sivasothi and Nor 1994.

Smooth-coated or smooth otter, *Lutrogale perspicillata* (Fig. 2.18)

The usual English name for this species being rather clumsy, I prefer to call it simply the smooth otter. I met it for the first time in Thailand, where, on the ample sandbanks of the wide, shallow Huai Kha Khaeng river close to the Burmese border, I had found the tracks of tigers, peacocks and elephants, as well as those of three different species of otter: Eurasian, small-clawed and smooth, clearly distinct. One late afternoon, when I was quietly walking along the bank, I suddenly noticed two otters swimming upstream, large and quite conspicuous. They were unmistakably the smooth otter, large and rather boisterous. They also put on a display of cooperative hunting such as I had never seen in any other otter species.

In the shallow current across the sands, fish were darting across, fast and quite uncatchable. The otters swam parallel, rapidly under and at the surface, one animal slightly behind, zigzagging across the wide river. Along the bank opposite was a narrow zone of reeds, and, as a joint strategy, the otters with their boisterous approach scared the fish in front of them into the reeds. Once the otters themselves arrived there, they dived into the vegetation, several times catching a fish and eating it on the spot, then off again in another zigzag, arriving at the reeds a bit further upstream for another meal. Two kingfishers accompanied the otters, picking off escaping small fish in the shallows.

The smooth otter is known only sparingly; remarkably, it shows similarities with the giant otter, on a different continent, and the two may be closely related. It occurs throughout most of the Indian sub-

Figure 2.18 Smooth or smooth-coated otter in a river in Thailand. © Budsabong Kanchanasaka.

continent and south-east Asia, as far down as Sumatra, and with a curiously separate population in the Iraqi marshes (Fig. 2.19). It is an animal of larger water bodies, especially rivers and dams, favouring rocky shores or banks with dense vegetation, also occurring along sea shores and in mangrove areas. Where I found it together with other otter species, the smooth otter was the one most often in the lower, slow-flowing parts of the river, and in dams.

Smooth otters are robust animals, quite a bit larger than the Eurasian otter, though not the size of the giant otter, with a total length of up to 130 cm. They have unusually large feet, a broad, somewhat flattened tail, and a large nose. The guard hairs in the fur are quite short and smooth, and the colour is that of the usual otter, a dark brown that is rather variable between populations, light underneath, often with a clear demarcation. Also they have a well developed set of whiskers.

Smooth otters share their geographical range with several others, but when present in an area they are the most conspicuous. Partly this is because of their social habits: they are rarely alone, but usually go around in groups of up to ten animals (occasionally more; on average about five). The composition of such groups varies, in variations on the theme of a female with her cubs, sometimes from several years. There may be a male with them, but not always, and sometimes the group consists just of sub-adults and cubs of the year. The litter size is two to four, but little is known of reproduction in the wild.

I noticed that, when slightly alarmed, smooth otters strikingly rear up on their haunches and look around. The Eurasian otter is never seen doing this, except in captivity.

Rather than seeing the animals themselves, in any one area the give-away for the presence of smooth otters is the scats along the banks, as for most otter species. In this case the 'spraint' sites are difficult to overlook, as often they are large and the spraints themselves remarkably smelly (of rotting fish), even over a considerable distance. The larger spraint sites have quantities of faeces, spread out and flattened by otters rolling, rubbing and scraping, somewhat reminiscent of the giant otters' campsites.

On such spraint sites the nature of the diet of smooth otters is evident: there are fish scales everywhere. The food is dominated by fishes, and quite large ones at that, up to 45 cm long (mean size >15 cm), but they also take crabs, frogs and the odd snake. For preference, smooth otters take the rather slow-moving fishes, or those living in large, dense shoals. Where they occur in the same areas as Eurasian and small-clawed otters, the smooth otter is the most piscivorous, taking the largest prey, often as socially organized foragers, more cooperative than any of the others.

Little is known of their populations, numbers, mortality and reproduction. By all accounts, however, the number of smooth otters in any one area is small and their density low, but, as with the giant otters, the animals are conspicuous. It is likely, therefore, that they are even more vulnerable to human interference than the others. They are poached for their fur, and skins are traded in large numbers. In many areas people eat them, and the animals make themselves unpopular by tearing nets and taking the fish. They face a world of human dangers.

General references for smooth otters
Hussain and Choudhury 1995, 1996, 1997, 1998; Kruuk *et al.* 1994a; Sivasothi and Nor 1994.

Small-clawed otter, *Aonyx cinereus* (Fig. 2.20)

This is a midget of an otter, weighing less than 3.5 kg and hardly more than 80 cm long. However, it is the one that many people see, because it is kept in more

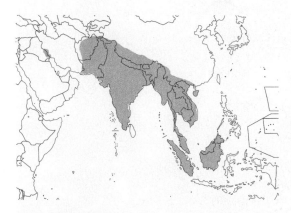

Figure 2.19 Geographical range of the smooth otter: inland and coastal Pakistan, India and south-east Asia, with a small population in the Iraqi marshes.

zoo collections than any of the others, and is often featured on postcards. The reason for this, apart from its size, is that it makes a better exhibit in a zoo, being a more docile animal, more gregarious and playful than most, and, unlike the others, breeds well in zoos.

The small-clawed otter is dark brown above, with a whitish throat and sides of mouth, a good set of whiskers, a bare nose and a somewhat flattened tail. Its feet are small, with only rudimentary nails on the long fingers. It occurs in a fairly large area of south-east Asia, in southern India, into southern China and countries south from there, including Malaysia, a large part of Indonesia (Sumatra, Java and Borneo) and even into the Philippines, Palawan (Fig. 2.21). However, is has been very little studied in the wild, and we know little of its ecology and social organization.

This is an animal of small streams and rivers, of marshes and rice paddies (even high in the mountains), and it often occurs along sea coasts and in mangroves. It rarely goes alone, but lives in often

Figure 2.21 Geographical range of the small-clawed otter: inland and coastal south-east Asia, with a small population in southern India.

Figure 2.20 Small-clawed otter, a very gregarious species. © Nicole Duplaix.

large groups of up to fifteen otters—the average group size is about five. Just like the smooth otter, it often rears up on its haunches to look around, probably to detect possible predators. It is not, however, a cooperative hunter: small-clawed otters are probably the least piscivorous of all, concentrating on the capture of crabs (*Potamon spp.*) and other invertebrates, each individual otter working on its own. These animals do take some frogs and the occasional fish, but where I studied them in Thailand, where they occur together with Eurasian and smooth otters, the small-clawed were the ones least interested in fish. They wandered further from the river than any of the other otters, foraging for crabs amongst river debris.

Searching for crabs in heaps of debris or under stones is done largely with the forelegs; frequently the animals poke their hands into likely places, and their long, sensitive fingers with very little nail on them must be particularly suitable for this. Their unmistakable small, round footprints in mud show these long fingers very clearly, without any claw marks. Another characteristic field sign are the scats, because of the many crabs in the diet; these are conspicuous not only because they are full of white bits of carapax, but also because of their size (some 8 cm long and 3 cm in diameter), much larger than those of other otters in the same habitat, and also larger than the rather similar scats of the crab-eating mongoose. They are often left some distance from the water, not necessarily on a high point but in a hollow or saddle between boulders.

Small-clawed otters have fairly large litters, up to seven cubs at a time, and they may reach sexual maturity within a year. It is likely that a large group consists basically of a female with one or more litters, but much more research is needed.

This is a species with a very wide habitat tolerance. It has adapted to the activities of people better than most other otters, for instance by its use of rice paddies; it often forages in eutrophic waters (when there may be many crabs) and it is considered a nuisance around shrimp and fish farms. It suffers the usual persecution for its fur and meat, and some of its parts are thought to have medical properties. However, because of its habitat tolerance and quite large geographical range, it appears to be less vulnerable than most other otters.

General references for small-clawed otters
Foster-Turley *et al.* 1990; Kruuk *et al.* 1994a; Larivière 2003; Sivasothi and Nor 1994.

Africa

Four species, of which the Eurasian otter occurs only along the fringe of the African continent north of the Sahara; it has been described under *Europe*.

Cape clawless otter, *Aonyx capensis* (Fig. 2.22)

Just below a large dam in South Africa, the Olifants River runs as a beautiful clear, fast flowing stream through arid country, with bushy vegetation and trees along its banks. When, on a sunny early morning, I creep up to one of its rather stagnant channels, there are two clawless otters foraging in the submerged, grassy vegetation. I manage to get quite close without disturbing them; both are large animals as otters go, and their huge whiskers glisten in the sun when they surface. Obviously they cannot see much in the densely matted grass underwater, but they are both poking around in it with their forefeet, moving fast along the surface, and frequently plunging under. Every 20 seconds or so, one emerges with a

Figure 2.22 Cape clawless otter—white underneath and a patch between eye and nose. © Michael Somers.

small crab and audibly crunches it with its mouth pointing skywards, then goes down again.

This is the usual way in which the Cape clawless otter acquires its calories, feeding on the very common freshwater crabs, from dense vegetation or from under stones. The animals often catch their prey in shallow water, but they may also use deeper sites, for instance when foraging in 3-metres deep places in the sea, with average dive lengths of 32 seconds.

The Cape clawless otter is common in suitable habitat in large areas anywhere in Africa south of the Sahara, except in the Congo basin (Fig. 2.23). The animal ranges widely in all kinds of wetlands (rivers, lakes, marshes, tiny streams—wherever there are crabs). It also occurs frequently along suitable sea coasts, where it needs access to fresh water for washing and drinking like the Eurasian otter, but where it feeds on marine prey. Remarkably, these otters make long trips overland to isolated dams, and are quite capable of moving far and fast when out of the water.

The Cape clawless otter grows large, in some inland areas almost as big as the South American giant otter (up to 1.8 m in length), weighing up to 18 kg, but smaller along sea coasts, and with only a small difference between male and female. It is brown on top with a rather round head, a clear white throat, and especially conspicuous large, abundant whiskers. The forefeet are without webs between the long fingers, with rounded fingertips and no nails, obviously important during foraging. Once caught, crabs are crunched with large, broad molars in the otter's massive skull.

Like its much smaller counterpart in Asia (the small-clawed otter), the Cape clawless' mainstay food in freshwater is crabs of various species, *Potamon* or *Potamonautes*, which are abundant in African freshwaters, especially in nitrogen-rich rivers and lakes. Along the coast numerous crab species are taken, such as *Plagusia* and *Cyclographus*, and eaten in their entirety. The diet also includes fish (in some places it is the main food), frogs and even some molluscs; the fish are mostly slow bottom dwellers such as eels and catfish. All of these are usually detected with the forefeet, the otter poking around in vegetation and under stones. The whiskers are probably used especially to detect escaping prey.

Once caught, a prey is held either in the mouth or in the forepaws, and, when eating, the long-fingered forepaws are used almost like the hands of primates; for instance, an otter floating in the water and eating a crab guides the food with its hands. Larger prey are taken ashore. There is no evidence of cooperative prey-catching by the Cape clawless, although it is by no means a totally solitary species. These animals often forage on their own, but one sees groups of up to eight, family parties of mother and (even adult-sized) cubs, or gangs of males only. As yet we know very little about these social arrangements, and what their function is.

The home ranges of individual otters overlap, with males covering stretches of river of more than 50 km (but usually less) and including the areas of several females, and female ranges probably including offspring but excluding non-related females. Along sea coasts, clawless otters have group territories with distinct borders between them, of four to six animals in each, whilst within these group territories individuals may be solitary, or form temporary coalitions. Each group territory contains several active holts. There is evidence that the size of the range is related to the distances between suitable crab sites, for instance reed beds in freshwater areas. The social system appears rather similar to that of the Eurasian otter, but the Cape clawless is somewhat more gregarious, including the small gangs of males as for the North American river otter.

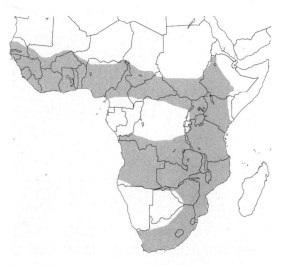

Figure 2.23 Geographical range of the Cape clawless otter: coasts and inland Africa south of the Sahara, not in the Congo basin, and in desert areas.

Each individual within a home range uses a number of 'couches' (sleeping sites), usually above ground in thick vegetation or between rocks, but the otters also dig dens themselves (burrows up to 3 m long, with grass-lined nests) or they live in small cavities between rocks. Breeding occurs at all times of year, with 63 days' gestation (no delayed implantation), one to three cubs per litter, and young becoming independent of the mother after about 1 year. Otter densities along sea coasts have been estimated at about seven animals per 10 km of coast in one study, one otter per 2 km of coast in another, and elsewhere three groups inhabited a 5-km section of river. Usually, however, densities are much lower.

The tracks of this species are unmistakable, with large footprints showing the long, clawless fingers. Similarly characteristic are the faeces, which are large (often >3 cm in diameter) and full of bits of crab, deposited on sites that are usually some distance from the water's edge, and often close to a den or couch. Clawless otters often spraint (defaecate) immediately after foraging. As for the Eurasian otter, spraint sites are distributed throughout the range, and probably serve as communication between animals using the same area. In addition, scent marking is also one of the functions of the otters' urinating, rolling, scraping and rubbing. Another means of communication is vocalization: Cape clawless otters have a range of whistles, huffs, growls and screams that have not yet been studied in detail.

In many inland waters of Africa the clawless otter is in competition, at least to some extent, with the spotted-necked otter and the water mongoose, which both also eat crabs at times. However, there are substantial differences in their diets and habitat preferences. Predators of the otters include crocodiles and seals in the water, and various large carnivores and humans on land.

As is to be expected, these animals are persecuted for their fur, for various parts that are used as medicine, and because of their occasional slaughter of domestic ducks and hens, or their depredations in fish farms, which can be substantial. Along many coasts their favourite habitat is being built over, and disappearance of riverine vegetation inland also has detrimental effects. However, the otters' reliance on crabs, which often associate with somewhat eutrophic waters, also means that this species derives certain benefits from human land use, and their wide distribution throughout Africa suggests a tolerance for environmental conditions that may stand them in good stead.

General references for Cape clawless otters
Arden-Clarke 1986; Rowe-Rowe 1977; Skinner and Smithers 1990; Somers and Nel 2003, 2004.

Congo clawless otter, or swamp otter, *Aonyx congicus* (Fig. 2.24)

A single large otter is foraging in one of the damp clearings in the Congolese rainforests. It walks a few steps, then pushes one of its front feet deep into the soft mud. It is feeling around, its gaze averted and, after a quarter of a minute, it pulls out a most un-otter-like prey: a very large earthworm. It holds the worm firmly in its fingered paw, then eats, chomping quickly, before moving a little bit forward and trying the same thing again—an amazing but normal foraging bout of the Congo clawless, eating worms at a rate of three per minute.

Very little is known about the Congo clawless otter. It lives in difficult places in the Congo basin of

Figure 2.24 Congo clawless otter—a greyish sheen on its brown coat, and dark patches in front of the eyes. © Miguel Bellosta.

Africa (Zaire and adjoining areas), looks very similar to the Cape clawless, and the geographical ranges of these two species are more or less exclusive of each other (Fig. 2.25). Also called the swamp otter, it lives in rainforest, in open clearings or muddy banks, or in dense swamps, seen occasionally but studied rarely.

Its teeth are better adapted for dealing with earthworms; the molars are narrower than those of the Cape clawless otter, and they have more pronounced, sharp cusps. The entire skull is massive, but not as broad as that of the Cape species; however, the animal itself is of similar large size and shape, up to 150 cm long and weighing up to 25 kg. Its colour is slightly different, the hair on top of the head, neck and shoulders showing a grey 'frosting', and with more white on the face than the Cape clawless, white edges to the ears, and in particular a strikingly large dark patch in front of each eye.

The animal is active at night and during the early morning, foraging singly or in family parties. Even its food has not yet been studied properly, but is said to include crabs, fish, frogs, insects and various other prey, as well as the many large earthworms.

Congo clawless otters are widely hunted for 'bush meat' and for their fur throughout their range, and their rainforest habitat is diminishing daily. Yet they still appear to be quite common—a fascinating challenge for a research project.

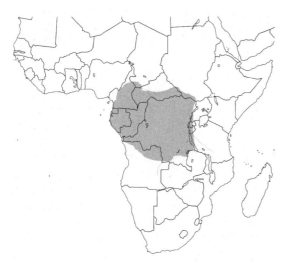

Figure 2.25 Geographical range of the Congo clawless otter: Congo basin only, not along coasts.

General references for Congo clawless otters

Jacques *et al.* (in press).

Spotted-necked otter, *Lutra maculicollis* (Fig. 2.26)

Rubondo is a forest-covered island in Lake Victoria, little affected by people. One early morning, quietly from my canoe, I paddled along with a group of four spotted-necked otters fishing along the shore, a group I had been watching frequently for the past few weeks. Dense masses of branches extended into the murky water, and the otters were in and out of the tangles, coming up for quick gulps of air between long, agitated dives. They were working separately, each one frequently catching small fishes, cichlids only few centimetres long, chomping quickly and continuing again with foraging. After almost half an hour of this, the otters swam out into more open water, now bunched closely together, and crossed the large bay to the other side, where they climbed out on to the rocks, well away from the crocodiles along the shore.

The spotted-necked otter looks quite different from the two clawless ones, which also occur in its geographical range. It is much more like the Eurasian otter, or the North American river otter. However, also compared with those species there are interesting differences. It is a smallish animal, less than 1 metre long in total, weighing only 3–6 kg, with males somewhat larger than females. When it periscopes and looks at you from the water, the head seems square rather than round, with short whiskers—almost like a very small hyena (in Swahili it is called 'fisi maji', meaning 'water hyena'). It is a deep chocolate colour above, looking black when wet, but the underside of many of these animals have strikingly large, irregular, white patches, often right across the belly and the inside of the legs. This feature is especially striking in populations of Lake Malawi and further north, but much less so in South Africa.

The spotted-necked is an otter especially of open lakes, dams and rivers, occurring in a large part of Africa south of the Sahara, down to the Zambezi River, and beyond that in only a few small areas in eastern South Africa (Fig. 2.27). In southern Africa it is found mostly in small rivers, elsewhere in the large East African lakes; to my knowledge, these otters do

Figure 2.26 Spotted-necked otter—in southern areas less spotted than further north.
© Gallo Images (John Harris).

not venture along sea shores anywhere in their range. They have a quite specific habitat: the animals are largely dependent on rocky shores and places with dense shrubbery and forest, preferably where branches and rocks go right into the water. With their large, webbed and well clawed feet, they are clumsy on land and rarely go far from the shore (unlike the clawless).

Between the rocks and branches of their favourite haunts, spotted-neckeds hunt for fish, in some areas almost exclusively so. In South Africa they often eat crabs as well as fish, some frogs and even insects, but the main prey in Lakes Victoria and Malawi are cichlid fishes, mostly less than 10 cm long, and also large ones, for instance in Lake Victoria the introduced *Tilapia*. They appear to specialize in catching fish not out in open water, but between sticks and rocks, in daytime. I noticed that the water may be very muddy (Lake Victoria), or clear as glass (Lake Malawi). During swimming the spotted-necked otters appear even more agile than, for instance, the somewhat larger Eurasian species, with fast twists and turns during their 15–20-second dives, and very brief breathing times on the surface. There is a nice description of a small fish escaping from a spotted-necked by jumping out of the water, dropping down again near the otter's tail, but the otter caught it before the fish hit the water.

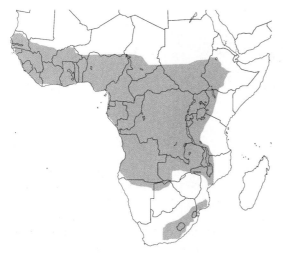

Figure 2.27 Geographical range of the spotted-necked otter: inland only, in Africa south of the Sahara, but not in deserts.

When spotted-necked otters are foraging, each individual hunts for itself, even though it may be one of a pack. Such packs may include as many as twenty animals, but on average no more than about five, and sometimes the otters are entirely on their own. Interestingly, the larger packs are probably all male, and the smaller ones are family parties of a female with cubs. Just like some of the social mongooses, they do not behave as a coordinated group during foraging, but they bunch up especially when travelling, and crossing stretches of open water, and when resting on land. Often they share their habitat with large crocodiles and eagles, so perhaps the group organization has an anti-predator function.

There is no evidence that spotted-necks have more than three cubs per litter, so the fairly large packs must be more than just one family. Watching them, I had the impression that they were noisier than the Eurasian otters, with their different calls being quite similar in whistles, chatters and screams. Perhaps the animals were interacting more often than the Eurasians, just because there were more of them together.

In South Africa the spotted-necked otters have home ranges of about 15 km of stream, on average, with those of males larger than those of females, and several females sharing a group range that is exclusive of other female group ranges. Males may share a range with several other males, and overlap with several females. However, within these ranges each animal often roams about on its own. They usually sleep in couches in dense vegetation, or in cavities between large rocks, but they may also dig dens themselves.

Population densities of this species can be quite high, especially along suitable rocky and well vegetated coasts of the large African lakes. In Lake Victoria numbers were estimated as around one otter per kilometre, in Lake Muhazi of Rwanda about two per kilometre, but in the very clear waters of Lake Malawi numbers were far smaller. Along rivers in South Africa researchers have estimated numbers of spotted-necked otters variously as about one per 10 km or one per 2 km.

Along the coasts of the East African lakes there are many artisanal fishermen, who often set their nylon gill nets along or over the dense underwater masses of branches where the otters hunt. On the one hand, animals frequently get caught and drown, which must be a major mortality factor. On the other hand, otters remove an estimated 10% of the fish from the nets, eat parts of the fish and damage the nets, so fishermen will kill otters wherever possible. Moreover, wherever there are fish farms within their range, spotted-necked otters make themselves unpopular, with dire consequences for themselves. The animals are also hunted for their fur, meat and body parts for medicinal purposes.

Perhaps the spotted-necked otters' main problem is the disappearance of their quite specific habitat along the shores; certainly, along some large East African lakes much of that has now gone. Yet it is encouraging that, wherever the habitat still exists, there are still many of the animals about—even to the extent that in some areas, such as Lake Malawi, they are an important attraction for visitors.

General references for spotted-necked otters

Kruuk and Goudswaard 1990; Perrin *et al.* 2000; Procter 1963.

CHAPTER 3

Evolutionary relationships, questions and methods of otter ecology

Phylogeny and evolution

All otters belong to the largest family of the Carnivora, the Mustelidae (martens). It is divided into subfamilies, the otters being the Lutrinae. Others are the Mustelinae (weasels, martens and minks), Melinae (badgers), Mellivorinae (honey- badger), Taxidiinae (American badger) and Memphitinae (skunks). Of all these, the Mustelinae are the closest relatives of the otters, and are their ancestral branch (Koepfli and Wayne 1998). The elongated body shape of the Mustelinae provided an excellent start from which to evolve an aquatic existence.

The otter lineage of 13 species goes back a relatively long time, compared with that of many other animals. The earliest genus that was recognized as otter (*Mionictis*) occurred right at the beginning of carnivore evolution, in the early Miocene (Willemsen 1992). The directions in which the various present-day species have evolved from there are important as background to the differences in their behaviour and ecology, as will become evident later.

When describing the thirteen different species as in Chapter 2, the natural inclination was to emphasize the differences between them. The variation is striking, and significant. However, what may be at least as important, if not more so, is their similarity: they are all, without doubt and hesitation, clearly otter. Seeing a giant or a smooth otter for the first time after being acquainted with the river otter, one immediately knows that here is another one, another typical otter. Such similarity between species may be due either to relatedness or to convergence in evolution when different species adapt to a habitat in similar ways. In this case, there is good evidence that all otters are closely related.

There is a considerable literature on otter phylogeny (*Oxford Dictionary*: 'racial evolution'), and over the years interpretations have changed markedly with new data. Previous knowledge was based largely on evaluation of morphological differences, and on fossil evidence. There was large variation in interpretation between authors (e.g. Van Zyll de Jong 1987; Willemsen 1992). More recently, however, some of the uncertainty has been taken away by the advent of molecular biology, which has provided a much clearer picture of the relationships between species, and of the course and timing of their evolution.

Analysis of mitochondrial DNA suggested strongly that the divergence of several otter species, from common ancestry, took place some 11–14 million years ago in the middle Miocene (Koepfli and Wayne 1998). The conclusion is based on fairly accurate knowledge of the speed of changes within one particular gene, cytochrome b. The molecular comparisons put the date of the first divergences of otter species (i.e. the beginning of their evolution) somewhat earlier than fossil and other evidence had suggested (Bininda-Emonds *et al.* 1999; Willemsen 1992), which perhaps is not surprising, as the chance of finding fossils of the earliest animals is small.

The two main branches on the otter tree are *Lutra* (Eurasian) and *Lontra* (American), and however similar they may look in the field, Koepfli and Wayne (1998) argued that they are genetically sufficiently dissimilar to warrant these different generic names. These researchers are supported in this by most other recent authors, summarised by Bininda-Emonds *et al.* (1999). Briefly, as far as we now know, the earliest *Lutra* species occurred in Europe during the Pliocene (Willemsen 1992), and *Lontra* appears to be descended from the fossil species *Lutra licenti*, which occurred in the early Pleistocene in China and migrated from there into North America during the Pleistocene (Van Zyll de Jong 1972; Willemsen 1992). *Lontra* then colonized the Americas, as recently as about 1.7 million years ago, resulting in the river otter *Lontra canadensis*, the neotropical otter *L. longicaudis*, the huillin or southern river otter *L. provocax*, and the sea-cat or marine otter *L. felina*. There are also many fossil species, now extinct.

It appears likely that the geographical origin of the *Lutra/Lontra* species was in southern Asia. It is from that general region that the different species groups are branching out, and where we now find the greatest species diversity.

The main Eurasian/African branch *Lutra* produced the present-day common otter *Lutra lutra*, as well as the hairy-nosed otter *L. sumatrana*, and in Africa the spotted-necked otter *L. maculicollis*, although the position of this last species is still problematic (Koepfli and Wayne 1998; Van Zyll de Jong 1987; Willemsen 1992). The data are as yet insufficient to give a date for the divergence of these three species. Present-day species of both *Lutra* in the Old World and *Lontra* in the New World were preceded in previous ages by several species now extinct.

Then there are the three clawless and small-clawed otters: in Africa the Cape clawless *Aonyx capensis* and the very similar Congo clawless *Aonyx congicus*, in south-east Asia the small-clawed otter *Aonyx* (or *Amblonyx*) *cinereus*. The work of Koepfli and Wayne (1998) strongly suggested that these three are descended from the *Lutra* branch, some 5–8 million years ago. These authors also found that the three between them are so closely related that they should all be put into one genus, *Aonyx*, which is the name I shall also use for the small-clawed otter. It is likely that their common ancestor evolved the finger adaptation and dental structure needed to exploit the rich resource of crustaceans in the Asian and African rivers.

This leaves the phylogeny of three species. All three of them are odd ones out: the sea otter *Enhydra lutris*, the giant otter *Pteronura brasiliensis* and the smooth-coated, here called smooth otter, *Lutrogale perspicillata*. All three have flat tails; they are large and gregarious. The sea otter is the most aberrant of all. It is the species that looks least like any of the other otters, and it has a very different lifestyle. It is not surprising, then, that its lineage probably diverged at a very early date from the *Lutra/Lontra* species (Koepfli and Wayne 1998). Earlier, there were also several other sea otter-related species (*Enhydritherium*, *Enhydriodon*; Riedman and Estes 1990), which are now fossil. The actual species *Enhydra lutra* evolved quite recently, in the early Pleistocene; another close relative was *Enhydra macrodonta*, also now extinct (Kilmer 1972).

In addition, the giant otter split from the others at a relatively early stage (Koepfli and Wayne 1998). It is related only distantly to the American *Lontra*, and the evidence of Willemsen (1992) suggests that its story began not somewhere around Brazil, but with the now extinct otter genus *Satherium*, emigrating from Asia into North America and spreading south. All members of that genus became extinct, but it did give rise to *Pteronura*, in South America.

Many open questions remain over the evolution of the various species. At present the most enigmatic is the smooth otter *Lutrogale perspicillata* in south-east Asia. I found no published records based on its DNA, but there have been suggestions that its closest living relative is the giant otter, because of similarities in the morphology of their skulls (Willemsen 1992) and in their behaviour (Duplaix 1980). Further research will tell.

In general, the pattern of evolution of the Lutrinae, as suggested by DNA analysis as well as fossils, appears to be a very rapid divergence into several different branches at an early time in the history of the Carnivores, in the Asian Miocene. This rapid diversification was then followed by long-persisting genera, which between them covered all terrestrial parts of the globe, except Australia and Antarctica. There are no marsupial otters, which, given the aquatic lifestyle of these animals, is perhaps not surprising.

We are left with a scenario of two main otter lineages, the three Eurasian *Lutra* and the four American *Lontra*, the three *Aonyx* as a more recent offshoot of *Lutra*, and the three 'odd ones out' (*Enhydra*, *Pteronura* and *Lutragale*). In addition, the position of *Lutra maculicollis* is still uncertain (Fig. 3.1).

Within each species there often are characteristic genetic differences between animals from populations in different parts of their range. Such geographical variability can sometimes be quite obvious, for instance when we compare the otters in Shetland with those from mainland Britain. The Shetland animals almost always have clear, individually shaped, white patches on their throat (Fig. 3.2), a trait that occurs only rarely in otters from the

Evolutionary relationships, questions and methods of otter ecology 31

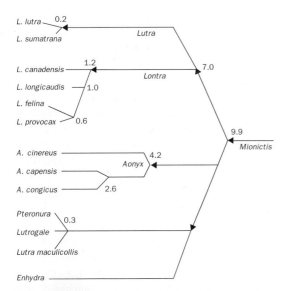

Figure 3.1 Family tree of otters. Species and genera of the subfamily Lutrinae, with approximate ages of separation in millions of years. Based on data from Koepfli and Wayne (1998) and Bininda-Emonds *et al.* (1999).

Figure 3.2 The individual yellowish-white throat patch of a Eurasian otter in Shetland.

British mainland, although Irish otters also have it. Shetland otters are also smaller, but that may be a phenomenon controlled by differences in their environment (e.g. food), and not necessarily genetically fixed. Intraspecific variation has been studied rigorously with molecular methods, and the results have important implications for conservation management, such as decisions on transplantations (Blundell *et al.* 2002). It will be discussed in more detail in Chapters 11 and 14.

Questions of behaviour and ecology of otters

A conservationist raises questions about otter numbers in a place or region, and about mechanisms that limit numbers. How many are there and where, in which habitats? If numbers are not what they are expected to be, what are the reasons, or, perhaps, why is there a decline? Fortunately, it is not difficult for researchers of ecology and behaviour to phrase their interests in such a way that their studies have a direct relevance to this. There is no need for conflict between the demands of conservation and science, and a strong incentive to put our questions in such a way that they address both.

In my own research, mostly on Eurasian otters, my chief objectives were to describe the densities and organization of otters, and the behavioural mechanisms through which they are maintained, as well as factors affecting mortality and reproduction. I wanted to discover the effects of food on numbers and organization, by studying diet, foraging behaviour and strategies, the costs and benefits of foraging, and by assessing populations and behaviour of prey. Finally, I was interested in the effects of habitat on numbers and dispersion.

Equally relevant to our understanding of otter populations are recent molecular studies on the genetics of the animals (Dallas *et al.* 2002). Similarly, advances in our knowledge of animal diseases have shown their potential and actual importance in otter mortality (Kimber and Kollias 2000), and new insights in the top-down effects of predators on populations have begun to dominate studies in some of the ecosystems in which otters are involved (Jackson *et al.* 2001). This will be evident in the following chapters.

Taking a wider, more academic, view, there are broader biological principles underlying many of the questions that are asked here. It may be the relationship between territoriality and resource utilization, the regression between animal numbers and availability of their food, the genetic variation and viability of populations, the biological function of scent marking, the role of parental behaviour in the development of hunting skills, the energetics of hunting, the roles of disease and pollution as a cause of mortality, and many more. It is important that the contributions made by various otter studies to our knowledge of these principles escape from the clutches of 'species-ism', and become a further endowment of our knowledge of biology in general.

One of the central problems to be addressed is the relationship between resources and animal numbers. This has been a focus in ecology for many years, put forward at an early stage by David Lack in his studies on birds (1954), and since then developed in great detail. The spatial organization, that is, the pattern of dispersion of sex and offspring (in this case of otters), in relation to their environment is part of the mechanism, and certainly helps our understanding of it. Especially in carnivores, in predatory species exploiting prey with complicated anti-predator systems and organizations, the spatial aspects are crucial for the understanding of the impacts of resources.

Carnivores are highly variable in the way in which they distribute themselves in their habitat, with large differences between, but also strong variation within, species. The resource dispersion hypothesis (Macdonald 1983) attempts to explain such variation, with the idea that the sizes of home ranges and the numbers of members of the same species are determined not just by overall density of food, but by the heterogeneity, the patchiness of the prey. Accordingly, range size would be determined by the distances between individual patches, numbers of animals inhabiting a range would relate to the productivity of patches. One of the aims of my own study of Eurasian otters was to test some of the predictions of this hypothesis in a solitary carnivore. As it turned out, however, this had limited predictive value, and we still are a long way from understanding how the environment affects social differences (Kruuk 2002).

Given the otters' semi-aquatic way of life and their special adaptations (Estes 1989), there are several interesting consequences that feature in the relationship between the animal and its resources. These include the fact that the home range or territory of otters has an almost linear shape along the water's edge (Erlinge 1967; Melquist and Hornocker 1983). In addition, they need to dive into a distinctly hostile environment to get food (where cold threatens and oxygen is short; Kramer 1988), so foraging is difficult and energetically extremely demanding. Furthermore, otters have to adapt to the extreme mobility of their prey species, which replenish a niche soon after an otter has emptied it. There are several more peculiar features in otters' population dynamics. As I will demonstrate, all of these phenomena have important bearings on resource utilization.

The question of what limits numbers of otters in any one area can be approached from two different angles. From one direction, we study the effects of factors such as food by estimating availability and quantifying foraging effort, and we determine habitat utilization and possible restrictive mechanisms therein. From another angle, we study otter populations, patterns and causes of mortality (including disease, human-made hazards, food shortage and predation), changes in reproductive success, and deduce possible mechanisms in population dynamics from age structures. I have attempted to use both approaches, and to bring these together.

Is one justified in extrapolating from one species of otter to another? Extrapolation remains an eternal, practical question in biology and medicine. In the case of otters, I have answered it with a qualified 'yes', but am aware of pitfalls. Knowledge of causal effects, of ecological and behavioural mechanisms in one species, opens our eyes to possibilities in others, but it is always possible that different species do things in other ways. On the one hand, I have little doubt that, for example, the consequences of thermoregulation in the Eurasian otter are in principle similar to those in the river otter, or even the giant otter. This does need confirmation, of course, but the urgency of such confirmation is not as great as that for many other problems that need to be studied.

On the other hand, many of the ecological and physiological problems, for instance of the sea otter, are unique, as are the threats to its existence. The species deserves an intensive and dedicated research effort—which it has had, with admirable results.

Research methods and field techniques

How does one study otters in the field? Obviously, this depends on the questions asked, but in my own research an ability to watch the animals, to follow them closely and record their behaviour was a common denominator. In addition, it is clear from the many publications on otters that the main research tool, used by many researchers on all species of otters (except sea otter), is provided by the spraints, scats or faeces, which are used for various purposes. Spraints are often easy to find, their presence or absence can be recorded, and their contents identified. They can be a great help, but there also are serious problems and limitations.

If we are interested in behaviour, then the best option is at least to start by watching otters in a place of diurnal activity; for Eurasian and river otters that includes coastal areas, or lakes. Some other otter species are always active by day (sea, giant and marine otters), but most are nocturnal in some areas and diurnal in others (river, Eurasian, neotropical, smooth, spotted-necked, Cape clawless otter).

In Shetland the Eurasian otters were shy, but with some care in that quite open country they could be approached quite closely. The animals are extremely sensitive to smell, so I had to be careful not to get upwind of them, especially when the weather was calm, when they could smell a person 200 metres away or more. I learned to use what little cover there was, not to show myself against the skyline, and to remain motionless when within sight of an otter. I could sit or lie in full view of the animals, as long as I remained absolutely still, wearing rather dark clothing and with a bank, rock or whatever behind me. I also learned to make good use of the times when the otters were underwater, when I could make a move and approach closely, to hide in cover nearer the otter.

The same observation technique works well for river otters along the North American coasts, for Cape clawless in South Africa, smooth otters in Thailand or spotted-necked in Lake Victoria. Neotropical and giant otters in many places are far less shy, and one can see a great deal of their behaviour from a canoe.

Often one first sees an otter, of any species, when it is swimming along a smooth surface. When the water is turbulent, otters are much more difficult to detect. Unfortunately, and despite their reputation, all otters (except the sea otter) spend only a small proportion of their time in water, often less than a quarter, and when they do swim they are much longer under than up. In other words, even when otters are there, the chance of seeing them in any given period is small.

In the end, much of my own observation technique depended simply on intuition, on a feel for the animal, where it would emerge, what it would tolerate and where it would go next. There are no guidebooks for this, nor is it easy to copy from other fieldworkers, because otter watching is often best done alone, without a companion to consider. Like many other types of fieldwork with animals, it is something of an art, of which the technical details have to be learned by practice.

My most essential equipment is a pair of binoculars (ideally 7×50), and I often used a telescope, with zoom magnification $\times 15\text{–}40$, mounted on a tripod. The usual distance for behavioural observations of otters in Shetland or along coasts elsewhere was somewhere between 50 and 150 metres, and with binoculars and telescope I could see many details of the animals and what they were doing or eating. A small dictaphone recorded all observations, mumbled softly from underneath the binoculars. Timing was done with an ordinary watch or a small electrical clock; stopwatches usually were too complicated to manipulate at the same time as keeping up with all the otters' antics. Other species and other areas often required modification of observation methods. In rivers and streams on mainland Scotland, the animals were much more nocturnal, and collecting behaviour observations could be done only with radio-tracking.

Individual recognition is a great help in many behaviour studies. To most people, all otters look alike. However, in all species mature males are often visibly larger than females, with broader heads (see, for example, Fig. 2.1), and at least in the Eurasian otters they often swim more conspicuously, with the tail trailing along the surface. However, a youngish male (see Fig. 5.1) is difficult to recognize from a female, and in some of the other species the sex difference is less obvious. One rarely sees the genitals.

There may also be clear differences between individuals, most strikingly in giant otters with their bright, large and variable throat patches which they display to all the world when they 'periscope' from the surface. Spotted-necked otters have similar markings, but in the field they are less conspicuous. In Shetland I was particularly fortunate because the Eurasian otters also had bright, irregular white or light yellow patches on their throats (see Fig. 3.2), as they do in Ireland. In addition, elsewhere just a few Eurasian otters have these patches, but rarely, and usually much smaller. In itself, the different marking of the Shetland and Irish otters is an interesting phenomenon. It suggests a degree of isolation between populations, confirmed by genetic studies (see Chapter 11). It enabled me to identify otters when I could see them reasonably near, and with their throats exposed. Eurasian otters often show their throats when foraging at sea, both immediately after they pop up from a dive and when floating on the surface with a small prey, which they eat with their nose pointing skywards (see Fig. 9.1).

There sometimes are other individual markings, apart from throat patches: white on the upper lips, a missing tail tip, a scar on the nose. About half of the neotropical otters that I watched in the Pantanal seemed to have pieces of tail missing, injuries reputedly inflicted by piranhas. Or, in any of the species, one repeatedly sees in a given area a female with three cubs, for instance, or a group of the same size every time, which implies identity (with some reservations).

Unfortunately, even in Shetland on many occasions I could not identify an animal—when it was too far away, or just swimming along, or when I caught only a brief glimpse of it. Similarly, in freshwater studies in Scotland it was almost never possible to recognize an otter from its markings. As it was all-important to recognize individuals, I decided to use coloured ear tags as an additional aid in Shetland. This meant that I had to disturb the animals in order to attach them, but the tags themselves did not bother the otters in the slightest. Ear tags made them immediately recognizable for as long as they stayed on—for a few months or up to a year.

Students of the behaviour of sea otters attach coloured and numbered tags to the flippers, which can be seen a long way off because of the animals' habit of floating with the flippers above the surface (Riedman and Estes 1990). However, for the intensive work on giant otters no such interference was required: their throat patches were sufficient to recognize large numbers of individuals (Staib 2002). Using individual characteristics, I made identity cards for every animal that I met in the Shetland study area, with a number and sometimes with a name as well, if it was a resident. After 5 years I had almost 60 identity cards of otters that had used the Shetland study area.

In field studies throughout the world, workers use digitized data files and images, and digital photography and video can be of great help in behaviour studies. However, in the field this equipment may also get in the way, requiring attention that may be needed to watch the animal and its environment, and at times causing scares. For me, when recording otter behaviour, binoculars and a dictaphone is still the equipment of choice.

Trapping otters for study was a major concern. I disliked it, but it needed to be done, and I tried to do it whilst disrupting the animals as little as possible. Both on the Scottish mainland and in Shetland, I used large wooden box traps, the design based on the traditional stone otter traps, as formerly used by the Shetlanders (see Chapter 14). I left them in place for years, not set of course, but open: this was one of the secrets of their success. The otters were used to them, and could walk through the traps without let or hindrance; several times after I took a trap away, I saw otters inspecting the bare site very suspiciously.

The traps were never baited with food (Eurasian otters do not readily take fish that they have not caught themselves, and bait also attracted other animals such as stoats, mink, cats and hedgehogs), but I often put some fresh spraints inside. When set, the traps were inspected every morning and evening in Shetland, and every morning in mainland Scotland, because otters were active mostly during darkness there. Further details of traps and catching methods for various species of otter are given in Chapter 14.

If an otter was caught, it was usually fairly quiet, as it was kept in the dark. It was immobilized with an injection of ketamine, or a mixture of ketamine and xylazine (Blundell *et al.* 2000; Melquist and Hornocker 1979; Mitchell-Jones *et al.* 1984; Reuther 1991). This was administered with a syringe from a blow-pipe, through an inspection cover in the top of the trap (Parish and Kruuk 1982). The animal would then be weighed and measured, ear-tagged or fitted with a radio transmitter, and put in a box to be released after recovery.

A few otters were followed by radio-tracking in Shetland (e.g. Nolet and Kruuk 1989), but in general it was not a success there. In Britain it was not then legal to use internal transmitters in wild animals (this has changed since 1987), and otters do not easily accept anything attached to their body such as a radio-collar or harness. Mitchell-Jones *et al*. (1984) and Green *et al*. (1984) used a radio-harness that stayed on an animal for a few days or weeks, but we saw otters seriously irritated by such attachments, and it is not possible to affix a collar, as the neck is wider than the head.

Fortunately, it is now permissible in Britain, as in other countries, to use small internal transmitters, implanting them intraperitoneally as pioneered in the USA (Blundell 2001; Melquist and Hornocker 1979, 1983), and this solved the problem for our more recent freshwater studies. Once implanted, the transmitter, which is the size of an AA-cell battery) does not appear to affect the animal. The operation is carried out by an experienced veterinarian, surgeon or physiologist, the transmitter is slipped in when the animal is under anaesthetic, and the otter released at the place of capture.

When following Eurasian otters in freshwater areas, where they cover large ranges usually at night, we moved by car and on foot, to discover with whom they associate, where they fish and which habitats they use. One small extra was to inject each radio-otter with a small dose of a harmless isotope, zinc-65 (Kruuk *et al*. 1980), also used in hospitals for medical treatments. This enabled us to recognize the spraints (faeces) of that individual for about 6 months afterwards, and was useful to study individual food preferences, sprainting habits, range sizes, and the proportion of spraints in areas that came from marked and unmarked otters. In winter it was sometimes possible to follow otters' footprints in the snow (Fig. 3.3), for considerable distances.

Some of the problems I faced during fieldwork had to be solved in a more controlled situation, and studies of captive Eurasian otters played an important role. An example is the interpretation of data from faecal analysis, where there was information on

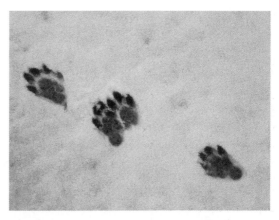

Figure 3.3 Tracks of a Eurasian otter in Scottish snow. Footprints are in groups of four; note five toes and claw marks.

what was in the spraints, and this had to be translated into otter diet. This can be done after feeding captive animals known quantities of different fish species, then assessing their 'output' in the spraints (Carss and Parkinson 1996; Erlinge 1968; see Chapter 7). Equally important were the energetics of foraging; in captive animals my colleagues and I could measure oxygen consumption when they were swimming and diving in their pool, and from that we could estimate the costs of foraging in the wild (Kruuk *et al*. 1994b; see Chapter 9). Our group of captive otters, kept in two connected enclosures with vegetation and swimming pools, were a constant source of information on behaviour, against the backdrop of observations in the wild.

Further details of methodology for otter studies are discussed in the relevant chapters. There is a vast literature on the diet of otters of almost all species, almost invariably based on the contents of their faeces (although in Scottish freshwater lochs and in Shetland I used direct observation). The commonly applied methods for assessment from spraints (Fig. 3.4) are discussed in Chapter 7, and the evaluation of otter distribution and habitat use, also from spraints, in Chapter 4. The use of DNA techniques on otters has provided much needed insight in species relationships, social organization and genetic variability of populations (Dallas and Piertney 1998), and the recent application of these techniques to otter spraints (Dallas *et al*. 2003), and with a 65%

Figure 3.4 Eurasian otter 'spraint' (scat) of typical size on stone along water's edge.

success rate by Hung *et al*. (2004), has opened up new fields of study: see further references in Chapter 11.

Study areas

This book would not have been written if it had not been for my first few years with the otters in Shetland, and I will describe some of this background for many of the results mentioned here. With a 'difficult' species such as the Eurasian otter, which is few in numbers, persecuted and shy, nocturnal and intelligent, one needs a good deal of field experience to be able to watch it, catch it, and to ask the right questions. In an average, low-density area this experience is hard to come by, as all of the problems related to the animals' retiring nature are exacerbated. Shetland provided a superb medicine for such predicaments: otters are common, they live along open, easily accessible, tree-less coasts and, perhaps most important of all, they are active in daytime, not at night.

Some people have criticized my use of this study area for Eurasian otters, because it is so different from many other places where the animals occur. Shetland is said to be 'atypical', not representative: one cannot extrapolate results to elsewhere because the otters swim in sea-water, in daytime, they have only one bank instead of two, the vegetation is different, and so on. My argument is that, obviously, one has to be careful in extrapolating, but that is true for every other individual study area as well.

Shetland throws up hypotheses for the many questions one needs to ask about otters, and these hypotheses can be tested anywhere. There are good environmental reasons for all the differences in the Shetland otters' ecology and behaviour, and these reasons are investigated. Comparisons between any of the coastal ecosystems and those in rivers and lakes are helpful, and indeed there is the practical point: when watching otters anywhere else, I feel at a great advantage with my experience from those cold, but easy, northern shores.

The general ecology of the islands and of the otter habitats in Shetland have been described by Watson (1978), Herfst (1984), Kruuk *et al*. (1989), Kruuk and Moorhouse (1991) and Johnston (1999). Shetland is a group of more than 100 islands at the northern-most tip of Britain, around 60°N—about the same latitude as the southern tip of Greenland. The total area is some 1400 km^2, of which about 980 km^2 is on the island of Mainland. For otters this is largely irrelevant, however, because they use just the very irregular coastlines, with deep inlets everywhere. Nowhere in Shetland is further than 5 km from the sea.

The landscape of Shetland is rugged, barren, tree-less and gale-ridden, one of the windiest and dampest places in Britain. Large areas are covered in peat, little of which is suitable for agriculture. There is an aura of isolation—one feels a long way from everywhere else. For a biologist, the most important aspect of this isolation is rightly emphasized by Johnston (1999): flora and fauna are the result of recent colonization, well after the ice age. There are huge colonies of seabirds on the cliffs, and otters (dratsies to the Shetlanders) are abundant. Two species of seal are common, the harbour or common seal (*Phoca vitulina*) and the grey seal (*Halichoerus grypus*), both quite important to the otters, and rarely also some other seals, and various whales. The rich populations of fish is described in Chapter 8.

The waters separating the islands, the 'Sounds', are often less than a kilometre wide, with ferocious currents, and many of them must be quite effective barriers, even to an otter. Oil and gas tankers wind their way through these Sounds, past treacherous rocks, past the graves of hundreds of ships that have perished here throughout history. This is the most obvious danger to any marine life in the area, although, until now, accidents with tankers have been relatively minor, including even the 1993 disaster of the 'Braer'. However, one cannot help but fear that a large oil wreck is only a matter of time: Shetland is dominated by oil, by the activities surrounding the oil terminal at Sullom Voe. Tankers will continue to ply these waters for a long time to come, the winds are horrific, and conditions unforgiving.

We established an intensive study area on a peninsula on the north-east coast of Mainland Shetland, the Lunna Ness (see Fig. 5.2). Its total coast was about 30 km, but we used only 16 km intensively. There are a few farms and houses, one of which was our 'field station', close to the shore of one of the large bays, the Boatsroom Voe. I, and the various students and others who helped me, could sit against the house, watching otters, seals and the many birds.

Many of the characteristics that make Shetland such a good place to see otters are applicable also to other suitable sea coasts in Europe, such as the shores of Skye, Mull, Uist and Harris in western Scotland, and of Hitra and other low-lying Norwegian islands. Parts of the Portuguese coast have many Eurasian otters, but access is very rocky and difficult, the sea is rough and, worst of all, the animals are nocturnal (Beja 1996b).

Where North American river otters are active by day, along Alaskan, Canadian and Californian shores of the Pacific, they are just as easy to watch as the Eurasians in Shetland, provided the waters are smooth and the coast has some easy access. Steep, wooded slopes are not conducive. In addition, marine otters in Chile can easily be watched from vantage points (Ostfeld *et al*. 1989), and the diurnal Cape clawless otters are a well known tourist attraction along the coasts of the eastern cape, in Tsitsikama National Park (Arden-Clarke 1983, 1986), as are the spotted-necked otters in Lakes Malawi, Tanganyika and Victoria. The South American giant otters (Fig. 3.5) have become a major attractant for researchers (Fig. 3.6).

Our freshwater study areas in the north-east of Scotland could hardly be more different from those of Shetland. There were various tributaries, sections of main river and lochs in the valleys of the River Dee and River Don, which drain from the Highlands, reaching the North Sea in Aberdeen. Both are

Figure 3.5 Giant otter eating fish in shallow water. © Nicole Duplaix.

Figure 3.6 Otter research in progress. Elke Staib observing giant otters in Peruvian Amazon. © Christof Schenck/ Frankfurt Zoological Society.

fast-flowing rivers, in a region of mixed agriculture, forestry and natural vegetation that is characteristic of a large part of Scotland. Agriculture here is mixed pasture and cereal farming in relatively small fields, with some conifer plantations along the banks, as well as narrow strips of natural woodland here and there, with alder *Alnus glutinosa*, willow *Salix* spp. and other deciduous trees.

The rivers run mostly over gravel and boulders, good habitats for their populations of Atlantic salmon *Salmo salar*, and also for sea trout (migratory brown trout *S. trutta*) and brown trout. The fish fauna is poor, however, with only few other species of fish, eels *Anguilla anguilla*, minnows *Phoxinus phoxinus*, three-spined sticklebacks *Gasterosteus aculeatus*, and some brook lamprey *Lampetra planeri*, pike *Esox lucius* and perch *Perca fluviatilis*. The lochs have almost only eels, perch and pike. To me, the exciting inhabitants are the fish predators, which are common: otter, mink *Mustela vison*, osprey *Pandion haliaetus*, goosander *Merganser merganser*, grey heron *Ardea cinerea*, and a few great crested grebe *Podiceps cristatus* and cormorant *Phalacrocorax carbo*. These river systems are beautiful, rich areas, with a wealth of wildlife, and I am fortunate to have them on my doorstep.

A quiet river in a forest in Thailand, the East African lakes, South African coasts and rivers, many large, slow rivers through wooded country in South America—in many such places I was able to see different otters quite easily. The basic principles of observation technique remained the same as in Shetland: being quiet, unobtrusive and down-wind. In general and anywhere, suitable areas for watching otters have open water, rather than narrow streams or densely vegetated marshes, and, if the water is flowing, it does so sedately. Waves and ripples make life difficult for an otter-watcher (the otters themselves don't seem to have any preference for smooth waters, it is just that they are more difficult to spot). However, otters of all species do prefer shallow waters rather than deep ones, which I discuss in Chapter 9; it is a matter of energy requirement and prey abundance. In consequence, most artificial reservoirs, with their steep banks, are not very popular with otters. Shallow, rocky sea coasts are popular, as well as rich, shallow, inland lakes.

In addition to direct observation, or where I find it difficult, I have sometimes been able to use otter tracks (see Fig. 3.3) to tell me what I wanted to know. The footprints of North American river otters have been followed through snow over many miles (e.g. Melquist and Hornocker 1983), and in Sweden and Finland Eurasian otter tracks have provided much information on home range sizes and population densities (Erlinge 1967, 1968; Kauhala 1996; Skaren 1990). On sandy river banks in Thailand, I could distinguish tracks of Eurasian, smooth and small-clawed otters (Kruuk *et al.* 1993c), and in Brazil the tracks of neotropical and giant otters provide telltale evidence of den use (Groenendijk *et al.* 2005).

All species of otter are different in their habitat preferences, some more than others, but wherever in the world one is studying or just watching these animals, their environment, by its very mixed and watery nature, is a concentration of diversity of animals and plants. Fieldwork with otters tends to have a magnificent, rich tapestry.

CHAPTER 4

Habitats

Introduction

Where can one find otters—where are they active, where do they rest? Which species use sea coasts? What are the characteristics of areas used or avoided by otters, of the actual waters they swim in, and of the shores they rest on? Can one generalize?

There is no need for an expert to tell us that all live in and near water, but there is a great deal more to what these animals need from their environment, and not all species are the same in their choices, although there are similarities. In fact, a cursory glance at where otters occur suggests that many of them are very tolerant of where they live, in environments ranging from lakes and bogs to rivers and little streams, from sea level up into the highest mountain ranges (e.g. **Eurasian otters** up to 4120 m altitude in Tibet [Mason and Macdonald 1986] or up to 2000 m in Spain [Ruiz-Olmo 1998]), and from waters in pristine forests to the centres of large cities. Otter are found almost anywhere aquatic, as long as there is sufficient food. However, individual species or populations have habitat preferences, and there may be conditions in which such preferences are of vital importance.

In the following sections I discuss some of the ecological requirements in detail. Habitat selection may be fundamental for conservation management, and many aspects of the otters' environment have a direct bearing on the animals' society, behaviour and populations. Conversely, some otter species themselves have a substantial effect on their habitat.

Methods

Assessing otters' environmental requirements is surprisingly difficult. One reason for this is the problem of establishing how much they use different areas. In some kinds of terrain this may be relatively simple, for instance in Shetland, where the Eurasian otters are active in daytime, and have clear individual markings. There it is possible, with some experience, to recognize individuals over stretches of coast of a few kilometres, and we can see what kinds of coast they use. However, those kinds of field conditions are exceptional, and one is lucky with species such as the South American giant otter and the spotted-necked otter in Africa, which are similarly recognizable and active by day (Procter 1963; Schenck 1997). In general, it is only possible to get some idea of the otters' utilization of any one area by radio-tracking them.

As a cheap shortcut, researchers have commonly used the numbers and distribution of otter spraints, or faeces, as an index of use of different parts of their habitat. This is where the problems begin.

I will use the term 'spraints' for otter scats throughout the book, although the term is traditionally attached to faeces of Eurasian otters only. However, it is a nice word, and there is no reason why it should not be used for all species of otter. Spraints are often the only obvious evidence of the animals' presence, at least in the case of **Eurasian**, **river** and **neotropical otters**. They can be quite conspicuous, and frequently are found in convenient spots, for instance under bridges. Almost all surveys of these species are based on presence or absence of spraints, and a standard method uses a 600-m stretch of bank: if no spraints are found, one scores a negative, etc. At the end of a survey, the percentage of 'positive' sites is taken as a measure of the strength of the otter population, or of its use of a particular type of habitat (Mason and Macdonald 1986).

However, there is little information on the relationship between spraint numbers and otter usage. The few studies that have been done to check this are not always encouraging (e.g. Kruuk *et al*. 1986; see also Chapter 8). Clearly, when spraints are present, there are otters and, in general, over large areas where there are many spraints there are likely to be more otters than in regions where spraints are few.

However, the conclusions one draws from the presence or absence of spraints have to be very guarded. Otters frequently defaecate when swimming, as I have seen of several species both in the wild and in captivity. At some times of year **Eurasian otters** are up to ten times more likely to spraint in the water than in other months. In fact, the spraints we find on the bank have a specific social function apart from elimination; they are deposited in specific sites, and some types of vegetation or habitat are almost totally ignored for this purpose (see Chapter 6).

In consequence, absence of spraints does not necessarily mean absence of otters in that particular vegetation or along that bank. In our study in streams and lakes in the north-east of Scotland, where we followed Eurasian otters with radio-tracking, there were several areas where the animals spent a great deal of time (e.g. reed marshes, far away from open water, and along the numerous very small streams; see below), but where despite intensive effort we could not find any spraints (Kruuk et al. 1998). We tried hard, because we wanted to study the otters' food in those sites. Another radio-tracking study of otters in Scotland had similar problems (Green et al. 1984). It is often wrongly assumed, therefore, that otters do not use bogs or very small streams: there are no spraints there.

Spraint numbers may well be broadly indicative of otter numbers when considered over relatively large areas, such as several kilometres of bank (e.g. Ruiz-Olmo et al. 2001), and when seasonality of spraint-ing behaviour is allowed for (see Chapter 11). Large-scale spraint surveys provide useful information for conservation purposes (see Chapter 14). However, if one wants to assess otters' preferences for different habitats, studies based only on the distribution of spraints, numerous as such studies are, cannot provide reliable conclusions, and they have not been used here.

As a final note of caution for conservation management, we have to remember that not all habitat *preferences* are habitat *requirements*. For instance, many otters rest in areas of good cover along the banks—a clear preference—and one might be tempted to conclude that bank cover is an essential prerequisite for otters. But where prey is plentiful, such as along the coasts of Shetland, otters occur in numbers, without any such cover anywhere near: it is not a requirement.

Fresh or salt water?

Otters, of one species or another, may use almost any kind of water body, from the smallest stream to the fringes of the largest oceans. However, not all species use both fresh and salt water: some live exclusively in fresh water, others exclusively in the sea. The two types of habitat differ in a way that is highly relevant to otters. The seas have some abundant food resources that otters can exploit, but the animals have to adapt in various ways to cope with the problems of salt water.

There are four otter species that have not been recorded along sea coasts: the giant otter in South America, the hairy-nosed in south-east Asia, and the spotted-necked and Congo clawless in Africa. Then there are two species that live exclusively in the sea: the sea otter in the northern Pacific, and the sea cat or marine otter in Pacific South America. All of the others will use sea as well as fresh waters, and the evidence suggests that animals in this last group all adopted the same solution to some of the serious difficulties of the marine habitat, the problems of drinking and washing.

If an animal needs to drink, either it should have access to freshwater, or its physiology should be able to cope with drinking sea water. The two exclusively oceanic otters appear to have chosen the latter option, and at least sea otters are able to survive on drinking sea water (Costa 1982). Where any of the other species live in the sea, they depend for drinking on trickles of freshwater between stones on the shores, or they rely on streams, puddles or underground sources of water.

What is probably at least as important a problem to the otters is thermo-insulation in sea water (see Chapter 10). To enable the dense otter fur to fulfil its role as insulator, the animals need to remove accumulating salt. Again, the sea otter and marine otter have their own solutions for this, such as special fur structure and grooming behaviour (see Chapter 10), but species that use the seas as just one of the options are dependent on the habitat to provide the means to wash themselves. They use freshwater pools, or underground water sources inside their holts, or streams or lakes, all close to the sea shore. For those species, access to freshwater for washing is an important limiting factor when they are using the sea shore (see below).

The aquatic part of the habitat

In the sea only a relatively narrow strip of water is used by otters (except for the sea otter in the Pacific), and only occasionally do they move outside it. For the **Eurasian otter** in Shetland, the vast majority of our observations were made close to the line where land meets water. For instance, in a sample of 500 dives that we saw otters make while they swam along the coast, in March and July 1986, we estimated the distance of the dive site from the shore and found that 62% were within 20 metres of land, 84% were within 50 m and 98% within 80 m (Kruuk and Moorhouse 1991). On rare occasions I have seen Eurasian otters cross the larger sounds between the islands, covering distances of 2 km or more. As a rough estimate, therefore, we assumed that foraging took place within a 100-m zone along the shore. However, the figures suggest that effectively this strip was even narrower than that, because the otters' use of the outer part of this zone was far less intensive than the strip immediately along the shore.

The otters' usage of the narrow zone of water was probably determined largely by depth. The deepest dives that I saw Eurasian otters make in Shetland were at 14 m (where I could see them return to the surface with bottom-dwelling eelpout and rockling). In the Outer Hebrides an otter drowned in a lobster pot set 15 m deep (Twelves 1983), but in general the Shetland otters foraged in much shallower waters. Their diving behaviour is discussed in Chapter 9. Along exposed steep, rocky shore-lines with deep water, Eurasian otters foraged almost only close to the cliff face. These were mostly males, which had a habitat preference that differed significantly from that of females (see below).

Elsewhere along European coasts, the Eurasian otter showed similar preferences for shallow water as in Shetland (personal observation in several sites in western Scotland and Norway; P. Beja in Portugal, personal communication), and other species of otter similarly restrict themselves to the littoral fringe when they are foraging in the sea. North American **river otters** dived off the coast of Alaska in depths similar to those of the Eurasian in Shetland (personal observation, and Larsen 1983; G. Blundell, personal communication). In Chile, the **marine otter** fishes up to 50 m offshore (Castilla 1982), and in South Africa the **Cape clawless** obtains its prey within 100 m of the coast (Arden-Clarke 1986), so depth and distance from land may be a general constraint for almost all species.

The one exception is the **sea otter**, which has evolved a more robust approach to its habitat. The deepest dive is recorded as 100 m, from an animal that drowned in a king crab pot (Newby 1975), although in california their foraging activity is generally restricted to water depths of 20 m or less (Estes 1980). In Alaska, they commonly forage at depths of 40 m or more (Estes 1980, p 5). Sea otters range much further offshore than do the others, though 'rarely more than 1–2 km', but occasionally they are found many miles out at sea. 'In some areas, especially portions of Alaska, water shallow enough for sea otter foraging may extend many miles offshore, and in such areas large numbers of otters may be distributed accordingly' (Riedman and Estes 1990, p 22). Laidre *et al.* (2001) estimated optimum carrying capacity for sea otters along rocky and sandy shores of California as up to 40 m deep.

Within the narrow aquatic zone used by **Eurasian otters** in Shetland, there are distinct preferences for certain substrates and vegetations. Open sandy or muddy bottoms appear to exert little attraction, but most of the time in water is spent diving over rocky areas, especially over bottoms covered in relatively small rocks. The shore itself in Shetland is almost invariably narrow, because there is such a small tide (Fig. 4.1); the vertical tidal range is usually only about 1 metre. This means that the intertidal range of the shore is less than 20 m—usually only a few metres wide. The water is often very clear, with a visibility much better than anywhere else along the British North Sea shores.

Almost everywhere below the high tide line where the otters are foraging is covered in dense algae, down to some 4–6 metres deep, in a distinct zonation of different wracks such as the knotted wrack *Ascophyllum nodosum*, bladder wrack *Fucus vesiculosis*, and in the deepest parts *Laminaria* kelps. The kelps in particular form dense underwater stands, often 2–4 m tall, like waving forests that are almost impenetrable to a diving mammal.

There is much variation in the submerged vegetation of coasts with different exposure to waves. The sheltered bays, for instance, are dominated by

Figure 4.1 Eurasian otter 'holt' (den) in peat, in typical 'good' otter habitat in Shetland with shallow shores. Note oval entrances and large quantity of 'spraints' (faeces) near entrance (conspicuous because of crab remains).

knotted wrack and the kelp *Laminaria saccharina*, but more exposed, rough waters are characterized by *L. hyperborea* and others. In fact, the ratio of *L. hyperborea* to *L. saccharina* is used by scientists as an index for 'exposure' to wave action (Kain 1979; Kitching 1941), and we have employed it in the description of the habitats for the different prey species of otters (Kruuk *et al*. 1987). Fish are abundant, both in species and in numbers. Otters use these different aspects of the habitat in different seasons, a timing that appears to be driven by the presence of the main prey species (see Chapter 8).

Characteristic of sea habitat use by Eurasian otters is their repeated foraging in exactly the same sites, month after month, which we called 'patch fishing'. The properties of these feeding patches are an important aspect of the habitat, and are described in Chapter 9.

Little is known about the underwater sea-habitat details in other otter species, with the exception of the North American ones. In Prince William Sound in Alaska, the river otter forages along shallow coasts (Fig. 4.2), preferably with a bottom of small, rather than large, rocks. In these shallow areas larger algae are almost absent—different from the Shetland shores (Bowyer *et al*. 1995).

The underwater habitat of the **sea otter** is by far the most intensively studied. The important underwater vegetations for this species are the kelp beds, but sea otters also go well outside these underwater 'forests'. In California, the sea otter's favourite areas for foraging have rocky bottoms, especially where the rocks are broken up and varied, rather than flat bed-rock; densities may be as high as five animals per kilometre of coast, compared with 0.8 per kilometre over sandy bottoms (Riedman and Estes 1990). However, in Prince William Sound in Alaska, large numbers of sea otters are found over soft-sediment bottoms, which contain sought-after clams.

Everywhere the animals have a clear preference for waters with extensive kelp canopies, but they are also found in completely open sea. Kelp species, extent and density 'are known to influence distribution patterns as well as territory and home range boundaries' (Riedman and Estes 1990, p 23). The giant kelp *Macrocystis pyrifera* is much preferred over the bull kelp *Nereocystis leutkana*, with territorial males returning to exactly the same site in kelp beds year after year (Riedman and Estes 1990). Clearly these differences in the marine habitat are much more important to the sea otter than to other species, because, as well as feeding there, it also rests, sleeps and has its cubs on the floating fronds of the beds of algae.

Especially intriguing is the well documented evidence that shows the sea otter itself, in many places of the northern Pacific Ocean, as a 'keystone species', as the architect of its own preferred habitat. The giant kelp beds often are a byproduct of the sea otters' activities; in numerous places, where sea otters re-colonized areas from which they had been absent for many

Figure 4.2 Coastal habitat of the North American river otter, in Prince William Sound, Alaska—shallow, stony coasts.

decades, huge kelp beds have grown where previously there were none (Estes and Palmisano 1974, and review in Riedman and Estes 1990). Examples range from the coasts of California (e.g. VanBlaricum 1984) to the Aleutian islands with and without sea otters (Riedman and Estes 1990). The mechanism of this remarkable effect acts through the reduction by the sea otters of populations of sea urchins: the sea urchins, in their large numbers, have a devastating effect on the kelp beds. I discuss this in more detail in Chapter 8.

The underwater characteristics of the otters' freshwater habitats are far less well known than those in the sea. There are several reasons for this; one of them is that otters are more difficult to observe in freshwater, and also underwater habitat differences are less clear-cut than in the sea, and on a smaller scale. In general, the ecology of the freshwater habitat is more tied up with what happens on land (see below), with factors such as stream widths, although lake depths and substrates also play a significant role. Some shallow parts of Scottish lochs are consistently preferred for foraging (Kruuk et al. 1998).

Terrestrial habitat along sea coasts

Most otters (with the exception, again, of the sea otter) spend some three-quarters or more of their lives on land, so the structure of the terrestrial side of their environment is important, especially for conservation management. A large amount of research has been published on this, mostly on the Eurasian otter, the most common currency being the occurrence of otter spraints in different types of vegetation along banks and coasts.

Along sea coasts we assessed the use of different shore types by **Eurasian otters** in Shetland (Kruuk et al. 1989). In this case we established that the distribution of spraints did not reflect usage of different parts of the coast (Kruuk et al. 1986), mostly because many spraints were deposited below the tide-line and were washed away at high tide (Kruuk 1992). Instead, we found that in the open, cover-less landscape of Shetland it was relatively easy to find otter 'holts' (dens), of which there were many (see Chapter 5). There was a good correlation between numbers of holts along a section of coast and the numbers of otters there, as previously suggested by Conroy and French (1987; further details in Chapter 11), and in our intensive study area there were approximately three 'active' holts for every resident female otter. This we used to quantify the use of different coastal types by the animals.

To find where otters occurred in Shetland, we surveyed the coasts of all the large islands and many of the small ones (Kruuk et al. 1989). The main purpose was to estimate the total population of otters (see Chapter 11), but we also obtained information on differences in numbers along the various types of coast. We divided the 1350-km shore-line into 5-km sections, and classified those as, for example, sections dominated by steep cliffs (more than 10 m high), shores along agriculture ('improved' soils), shores along peat layers, or housing and other built-up areas. In addition, there were 38 very small islands (often with peat).

The sections we surveyed (about one-third in all) were chosen randomly from the different types of coast. In each of those we searched for otter holts, by walking, zigzag, through a 100-m wide strip from the shore-line, inspecting all likely sites. It was a relatively easy procedure, as the holts were generally conspicuous, the terrain was not difficult and there was no tall vegetation.

There were large differences in holt density (Fig. 4.3), with highest numbers along peaty shores (see Fig. 4.1), where we found a mean of 13.3 holts per 5 km. On the 12 small islands that we checked there were fewer, whereas cliffs and agricultural coasts had only 1.8 and 1.6 per 5 km, and there were no holts in built-up areas. The agricultural southern part of Mainland had a holt density of only 0.8 per 5 km, the lowest in Shetland. The variation was highly significant, and demonstrated a clear preference of otters for the gently sloping, peaty coasts of Shetland, without agriculture, without cliffs, and without people. The otters' preference for peat and their aversion of cliffs and agricultural areas was also evident when we looked at the coast in greater detail, dividing it into 1-km stretches (Kruuk *et al.* 1989). This is likely to be related to the presence of freshwater (Fig. 4.4), as small above-ground or subterranean freshwater pools and streams (see Chapter 10).

The association between otters, peaty coasts and fresh water that we found in Shetland was useful also for understanding what happens in other areas. In Scotland, otters occur all along the west coast, from Galloway northwards, along the shores of the Western Isles, the Hebrides and on to Orkney and Shetland, but they are virtually absent from the eastern sea boards (Green and Green 1980, and my own observations). Along the Orkney shores there are relatively few otters, in sharp contrast to their abundance in Shetland. To explain this we looked at possible differences in the abundance of prey, but this did not explain the otters' distribution: along the rocky shores of the Scottish east coast and in Orkney there was an abundance of potential prey such as butterfish, rocklings and various other species. In Shetland, too, there were places where otters were few, but where absence of fish was not a likely explanation.

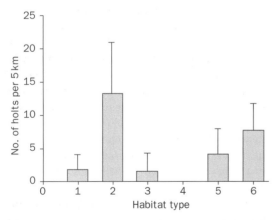

Figure 4.3 Density of holts of Eurasian otters along Shetland coasts; number of holts per 5 km of coast in a 100-m wide strip in different habitats ($F = 28.79$, $P < 0.001$, ANOVA). Error bars denote standard error. 1, Steep cliffs, > 10 m high (89 sections); 2, gentle slopes covered by peat, as in Figure 4.1 (61); 3, improved, agricultural land (40); 4, built-up areas (5); 5, small islands (38); 6, other coasts (47).

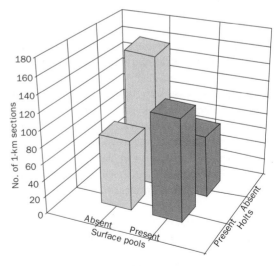

Figure 4.4 Otter holts and surface pools with fresh water in 454 1-km sections of the Shetland coast. Holts and pools are strongly associated ($\chi^2 = 33.5$, $P < 0.001$).

What is suggestive, however, is the association between absence of agriculture and presence of otters. Scotland's east coast, the north-east, Orkney and the south end of Shetland are all areas with few otters, and they are dominated by well drained farmland. Many of the Scottish west coasts have moorland or woods going down close to the shore-line. They also have a higher rainfall than the more intensively agricultural east side of the country, and small freshwater pools are common along these western shores, with many pools showing intensive use by otters (Elmhirst 1938).

A direct comparison between Shetland (many otters) and Orkney (low density) is also instructive. The difference in rainfall is small, but the comparative lack of otters in Orkney may be related partly to human land-use, and partly to a difference in geology: there are mostly non-porous gneisses, granites and sandstone in Shetland (Johnston 1999), and mostly strongly layered flagstone in Orkney (Berry 2000). All this results in relatively little surface freshwater along the sea shores in Orkney, which are well drained, and misery for an otter in search of freshwater in order to rinse off the salt of the sea. My student Paul Yoxon (1999) demonstrated that along the coasts of the Isle of Skye, where the geology is highly varied, otters are closely associated with particular geological formations (especially Torridonian sandstone), whereas they are far less common along, for instance, the Tertiary lava (mostly basalt) coasts. Yoxon found significant differences, in numbers of direct observations of otters and spraints per length of coastline, for these and several other geological formations. Significantly, small coastal freshwater pools are found especially along the impervious sandstone coasts, whereas the very porous basalt has almost none. The freshwater pools on Skye are used intensively by otters (Lovett et al. 1997).

In Portugal, Pedro Beja (1995a, 1996b) found a close association between otter use of the Atlantic coasts and the presence of small freshwater streams. The areas he studied are highly exposed, some with tall cliffs and with few shingle or sandy beaches. Inland there are sand dunes, pine woods and rough pasture, but a number of small permanent streams enter the sea, with low, dense vegetation including many brambles (*Rubus* sp.). Radio-tracked otters spent their resting time along these freshwater streams, although they foraged almost exclusively in the sea.

Everywhere along sea coasts in Europe, otters appear to avoid sandy beaches, as well as large, vertical cliffs and concentrations of human activity, but otherwise most coastal types are potential otter habitat. This appears to hold also for other otter species on other continents, and, as a general rule, freshwater, in small pools or streams, is a key habitat feature for non-obligatory, sea-living otter species along all sea coasts, but not for the two species that occur only in salt water (see Chapter 10).

Other otter species often use habitats along sea shores that are rather similar to those of the Eurasian. Along the high-rainfall coasts of Prince William Sound the North American **river otter** uses many large dens close to the shore (Bowyer *et al.* 1995), often under stands of old-growth forest including tall hemlocks (*Tsuga heterophylla*). There was subterranean freshwater in all eight dens that I visited (unpublished observation), and many small streams. Similarly, the **neotropical otter**, along the Atlantic shores of Brazil, uses rocky, forested coasts, near freshwater streams (Alarcon and Simoes-Lopes 2003).

Along the oceanic coasts of South Africa the **Cape clawless otter** is closely associated with sources of freshwater: virtually all dens, and other field signs such as spraints, are found in the immediate vicinity of freshwater (Arden-Clarke 1983, 1986; Van Niekerk *et al.* 1998; Verwoerd 1987). Where Cape clawless otters occur in brackish mangrove areas in Nigeria, West Africa, there was no evidence for a presence of freshwater sources (Angelici *et al.* 2005). In other species this association with freshwater pools has not yet been investigated, except in the case of the **marine otter**, which lives in a coastal desert climate, and where no evidence was found for any dependence on freshwater (Ebensperger and Castilla 1992; see Chapter 10).

Amongst the obligate salt-water species, the **sea otter** hauls out on sandy or rocky spits into the sea in Alaska, or along rocky beaches on islands, but if there are none of these it will spend almost all of its time in the sea (Kenyon 1969; Riedman and Estes 1990). Probably, these terrestrial habitat features are not essential. The **marine otter** occurs only in the presence of large rocks and rock formations along rough and very exposed oceanic shores (Ebensperger and Botto-Mahan 1997), in contrast to the **southern river otter** in some of the same areas; these animals prefer sheltered coasts with forests. The authors suggest that the difference in habitat between these two species is the main feature that keeps the two apart. My hunch is that southern river otters are dependent on the presence of fresh water.

An interesting aspect of coastal habitat selection is a difference between males and females. In Shetland, male **Eurasian otters** used the more exposed coasts more often than did females, who clearly preferred the sheltered bays, the voes. Within the study area, the difference was consistent throughout the year, except during the mating season in spring (see Chapter 5). Moreover, many of the Shetland males probably lived along more exposed coasts further away from our main study area. The phenomenon of males, which are larger and aggressively dominant, occurring more often along exposed coasts than females is especially interesting as this appeared to be the more inferior habitat in terms of prey availability. One might expect males to occupy the more productive, sheltered bays. There is a similar sex difference in habitat selection in freshwater areas (see below).

Terrestrial habitats along fresh water: banks and bogs

Habitat requirements of otters along rivers, streams, lakes and other freshwater bodies have attracted interest in many areas, and for several different species. For **Eurasian otters** in north-east Scotland, we studied this in the two main river systems west of Aberdeen, the valleys of the Dee and the Don (Durbin 1993, 1998; Kruuk et al. 1993a).

The rivers Dee and Don are both over 100 km long, and the lower sections were somewhere between 30 and 100 m wide; the width of the tributaries was anything between 0.5 and 5 m. Some of the banks were forested, some had rough pasture or scrubby vegetation, some were used for more intensive agriculture. Eurasian otters were and are found in almost every stream or lake in the north-east of Scotland, from the smallest streams in agricultural fields upwards, even in the rivers right in the centre of large towns, such as Aberdeen. However, they obviously do have areas that they prefer or avoid.

Several authors have commented on a striking association between otters and woody vegetation on the banks (e.g. Jenkins and Burrows 1980; Mason and Macdonald 1986), based on observations of spraints. However, the animals that we followed with radio-tracking did not substantiate this: they showed no preference for a particular vegetation, nor were they affected by presence of agricultural land use (Durbin 1998). They just happen to spraint near trees and shrubs.

An important habitat factor was the width of the stream or river. This affected otter use in a way that was surprising, and somewhat counterintuitive. We followed eight Eurasian otters with radio-transmitters almost daily, some for over a year, and noted the length of time for which animals were active in sections of tributaries, or in sections of the main stem of the river, of different mean width. Simultaneously, we could recognize the spraints of those individual otters, because we had given the animals a harmless isotope (zinc-65) that labelled their spraints for up to 6 months. By collecting all spraints from the tributaries and river sections we could estimate, therefore, the proportion of total otter activity in those areas for which our radio-otters were responsible. From that, combined with the radio-tracking data, we could derive the total amount of time that was spent there by all otters, and compare that with the width of the stream (Kruuk et al. 1993a).

The results are shown in Figs 4.5 and 4.6. First, we compared time spent per length of river and stream, of different widths. It appeared that the otters spent more of their time active per length of river section along wider streams. There was quite a large variation in the results, but they did show that the larger the stream, the more it was used by otters—no surprises there.

However, we then expressed the amount of time active otters spent not just per section of stream, but by size of water area along streams of different widths. This showed a quite contrasting result: an exponential decrease in otter use per hectare of water, in larger sizes of river (Fig. 4.6). So, the smaller the better; clearly the tiny, narrow streams, becks or burns—or whatever they are called in different parts of the country—often no more than half a metre

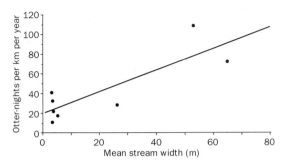

Figure 4.5 Eurasian otters along eight Scottish streams and rivers of different widths; number of nights per kilometre length of stream per year, spent by radio-tracked otters. ($r = 0.83$, $P < 0.02$). (From Kruuk et al. 1993a.)

Figure 4.7 Narrow stream in agricultural area in north-east Scotland—apparently insignificant, but an excellent habitat for Eurasian otters.

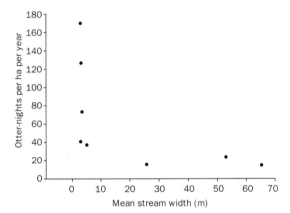

Figure 4.6 Eurasian otter presence per hectare of water, in Scotland. Other details as in Figure 4.5 ($r = 0.86$, $P < 0.01$). (From Kruuk et al. 1993a.)

wide, are hugely important to the otters (Fig. 4.7). This is true even for those small waters running through intensively used farmland, or near villages. The reason for this probably lies in the differences in fish density (see Chapter 8), and in the otters' habit of foraging close to, or under, banks (Durbin 1998).

In these same streams and rivers, Leon Durbin established that Eurasian otters have clear preferences for different substrates. Sections with riffles, and sections with large boulders and/or gravel, are preferred over areas with sandy or muddy bottoms, or with stones. Again, the explanation appears to lie in the foraging behaviour, and may well be related to the fact that our streams were inhabited mostly by salmonid fish. Carss et al. (1990) found that otters catch the larger individual salmon, especially on riffles rather than in deeper waters (see Chapter 9).

Eurasian otters in and around Scottish freshwater lochs showed similar strong preferences in their use of banks and waters. Ten otters that we radio-tracked in and around the Dinnet lochs near the River Dee (see Fig. 13.2), for a total of 462 days, spent 39% of their active time in small streams and bogs, although these areas had less than 5% of the total water surface of the area (Kruuk et al. 1998). In these marshes and little streams there were almost no spraints to show for such usage, but there were lots of frogs and fish to attract the otters.

When the otters were inactive near the lakes, they discriminated strongly in favour of particular types of bank vegetation, especially reed beds. In some places there are wide fringes of reeds (*Phragmites communis*) along the lochs, and otters spent most of their time in those when they were not fishing. As an example, we ranked the use of different, equally sized sections of the shore of Loch Davan by otters with radio-transmitters. We then compared those section scores with (a) a ranking of the numbers of otter paths and couches, (b) the size of the reed bed in the sections, (c) distances from sections where the otters went to forage (places with many eels), and (d) distance from the nearest source of human disturbance, such as houses or main roads (Kruuk et al. 1998).

The results showed that the radio-otters used and slept along the lakes just as we expected, in the places where we found otter paths and couches. The most important variable that attracted the otters was the size of the reed bed. The distance from feeding areas or human disturbance was not correlated with usage, and the animals swam and walked quite long distances to get to and from their preferred places of rest. Clearly, also, they were not easily disturbed by people.

Otter resting sites, of any kind, are not always easy to recognize as such. With radio-tracking we could frequently pinpoint exactly where an otter was sleeping, often for a whole day, but subsequent inspection of such sites, after the otter had left, was often fruitless: we just could not find any couch (Fig. 4.8) or even flattened vegetation (Green *et al.* 1984; Kruuk *et al.* 1998). In 669 days of radio-tracking of four different otters along rivers, the animals slept in couches or other sites in vegetation on 402 days (60% of observations). Some 37% of these were on islands, 22% in grassy vegetation or shrub along banks, 30% under ledges, fallen trees or boulders, and 12% in a trashed car along a stream.

These same four otters spent 267 days inside a holt underground (40% of observations). This was located inside artificial river embankments on 73% of days observed, in a tunnel dug by the animals themselves in 16%, and in human-made field drains on 11% (Kruuk *et al.* 1998). So there were many different types of resting site, often human-made, and our overall impression was that, along these freshwater rivers, streams and lakes, almost any place with some cover would do. This stood in sharp contrast to the sites chosen by otters along sea coasts, where the animals spend much more time underground, and where the presence of freshwater inside their holts appeared a *sine qua non*.

Although there are no radio-tracking data on Eurasian otters in eastern Europe, snow tracking has enabled researchers there to show that otters have a clear affinity to beavers, to beaver ponds and to their dens, lodges and dams (Sidorovich 1997). I noticed that in the large flat, marshy areas of Belarus and Russia beaver lodges are often an outstanding feature, providing otters with dry protection in the damp. Beavers decidedly are facilitators for otters; unfortunately, as yet, there are none in Britain, although reintroductions are planned.

As one final point about freshwater habitat usage of the Eurasian otter is that, even more than along the Shetland coasts, males and females show clear differences (see Chapter 5). Our males in the northeast of Scotland spent more time on the larger rivers, with females on the small tributaries and lochs (Fig. 4.9). Interestingly, a similar sex difference was observed in smooth otters in India (see below), and in boreal Canada, in the river otter.

Reid *et al.* (1994b) noticed that radio-tracked **river otter** males spent their time on the larger rivers and drainages, moving between the females, which stayed mostly on the smaller tributaries and the lakes.

In that area in Alberta, which freezes over and is snow-covered during winter, the river otter used especially bog lakes, with clearly banked shores and

Figure 4.8 'Couch' of Eurasian otter in reed bed, constructed by the otter itself.

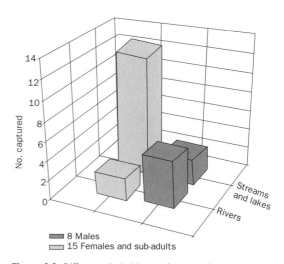

Figure 4.9 Difference in habitat preference of male and female Eurasian otters in a Scottish freshwater area. Males are found more along large rivers, females and sub-adults in lakes and small streams ($P < 0.05$, Fisher's exact test).

many beavers. But in general, and especially during the open-water season, they showed little selectiveness of habitat. In winter, however, they were often confined to the few streams and lakes with open water, which appeared to be critical to the otters' survival. To some extent they could still use ice-covered streams and lakes in winter, if a drop in water level created air space between water and ice, and in this way some otters could move long distances without actually surfacing.

In addition, in many other areas of North America, river otters are closely associated with beaver ponds, and they are frequent inhabitants of beaver lodges (Fig. 4.10). Melquist and Hornocker (1983) established, in over 1280 observations, that 38% of resting sites of their radio-otters were in beaver dens or lodges (18% were inside log jams and 11% just in vegetation). In Maine, Dubuc et al. (1990) found a clear preference of otters for watersheds with beaver ponds and dams; they provided 'an abundant food source in summer, stable water levels, abundant herbaceous cover, den and resting sites, and relatively low levels of human disturbance' (p 598).

These authors also noted that river otters prefer areas with a diverse vegetation, rather than densely wooded river banks. River otters that were radio-tracked by researchers in Idaho used just about all water there was, but with a clear preference for rivers in wider valleys, rather than between the steep mountains. In a clear similarity with our Eurasian otters, they also preferred smaller streams over lakes. Especially during the harsh winters there, river otters moved down from the mountains into the valleys.

In more subtropical climes, river otters in Florida live in highest density in the almost impenetrable swamp forests, and lowest numbers are found in the more open freshwater marshes there, with salt-water marshes intermediate (Humphrey and Zinn 1982).

Freshwater habitat in Latin America, Asia and Africa

Observations on habitat use by the other otter species are more sporadic. Sheila Macdonald and Chris Mason (1992), with long experience of research on the Eurasian otter, found spraints of the **neotropical otter** in Costa Rica, in habitats very similar to those that a Eurasian otter would have used: along rivers, mountain streams, even streams in towns, and all the way down into the coastal mangroves. Others deduced the preference of vegetation types by this species from the distribution of spraints in areas of Brazil. There are the same reservations about conclusions from this as for the Eurasian otter: presence of spraints shows the presence of this species, but a lack of faeces does not mean absence of otters.

Along rivers in a forested area of south-eastern Brazil, Pardini and Trajano (1999) found neotropical otter spraints at many different kinds of shelter: various types of natural cavity, including caves, spaces under tree roots or under banks, and areas of dense grass. Proximity of human disturbance did not appear to affect the otters' use of such shelters, although the researchers could not know whether

Figure 4.10 Snow-slide tracks of a North American river otter near a beaver lodge. Note tracks of jumps between slides, and marks of the legs in slide track. © Angie Berchielli.

otters actually stayed in these shelters, or merely visited them and sprainted (as Eurasian otters often do).

In the Brazilian Pantanal, I observed 12 different neotropical otters foraging along large rivers, where they fished in daytime almost exclusively in shallows immediately along the bank, within 1–2 meters, between branches of overhanging trees and shrubs, and in patches of floating water hyacinths. They may prefer this narrow strip because there are piranhas in the deeper waters of these rivers and oxbows.

The environment of the **giant otter** has been well studied, as these animals are also easy to see in daytime. Nicole Duplaix (1980) surveyed rivers in Surinam, and by recording direct observations she concluded that giant otters prefer slow-flowing creeks and rivers (Fig. 4.11), rarely go into the coastal areas, and avoid mangroves. They occur most often in rivers with low-sloping banks with good cover, and easy access to small forest creeks and swampy areas. In broader rivers they especially use shallow areas, such as rapids, or pools and ponds between sandbars. In an extensive review of literature, Carter and Rosas (1997) agreed with this outline, but also mentioned use of oxbow lakes, reservoirs or even agricultural canals. There is a distinct preference for clear black waters, and giant otters that live in more turbid streams will hunt in isolated areas of clearer water.

Giant otter habitat has been analysed in detail in an intensive study area in Peru, by Christof Schenck (1997). There, the species occurs mostly in large oxbow lakes or 'cochas'; elsewhere in the country, they are (or were) more common in white water rivers. Schenck found that the animals would occupy almost any kind of water body, but they preferred the fish-rich, larger areas. The cochas have generally clearer water; in addition, they are slightly warmer, and fish densities are significantly higher. Cochas with forested banks are more popular with the otters than those without forests. Many of them have large areas with floating vegetation, but otters were seen to

Figure 4.11 Habitat of giant otters in Surinam. © Nicole Duplaix.

hunt almost everywhere, right in the middle of large areas of open water, or in dense aquatic vegetation, or between the submerged, dead branches of fallen trees. A highly significant trend was that the larger the cocha, the more otters it would have. This was true not only for numbers of otters per entire oxbow, but also for otters per hectare of water: the largest cochas had 15 ha per otter, the smallest 20 ha. This is possibly related to fish populations, and provides an interesting difference with the findings in river and Eurasian otters, which show an opposite trend (see above).

Giant otters appear to be less affected by the presence of piranhas than neotropical otters. Perhaps this is because the giants are faster, with their broad, flat tails crossing broad stretches of water in an instant, and they forage in groups, sometimes actually catching piranhas.

The **southern river otter**, on the same continent, is similar to the other freshwater *Lontra* in its choice of abode. It is mostly a freshwater species, occurring in the sea in sheltered areas (see above), and in general appears to use lakes, ponds, rivers and streams in a way similar to, for example, river or Eurasian ottrs (Aued *et al.* 2003; Chéhebar 1985; Medina 1996; Medina-Vogel *et al.* 2003). It also uses dens or holts, or couches, in similar fashion, located in natural cavities, crevices or tall vegetation (Medina 1996).

In Thailand I noticed that, at least in some areas, the **Eurasian otter** confines itself to the upper reaches of rivers and to their small tributaries, whereas the larger **smooth otter** occupies mostly the main stem, and lakes and dams. Having seen habitat use by the Eurasian otter elsewhere, I predicted that the Eurasian otter would also have used these larger and deeper waters if no smooth otters had been present. In our main Thai study area, there was a clear cut-off point where the distribution of one began and the other stopped (Fig. 4.12) (Kruuk *et al.* 1994a).

In Malaysia, smooth otter habitat is found along large, main rivers, reservoirs and lakes, but they also make extensive use of rice paddies (Foster-Turley 1992; Sivasothi and Nor 1994). In India, too, radio-tracking studies with direct observation of the animals showed that large water bodies, such as main rivers, are the important parts of the smooth otters'

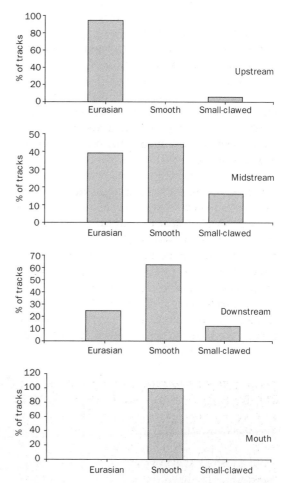

Figure 4.12 Tracks of three species of otter along a river that runs into a lake in Thailand. Eurasian otter is found mostly in upstream, narrower parts; smooth otter in the wider parts and mouth of the river; and small-clawed otter in between. (From data in Kruuk *et al.* 1994a.)

ranges (Hussain and Choudhury 1995, 1997, 1998). These last authors found that, along a 425-km stretch of the Chambal river, smooth otters occurred mostly in the upper reaches, away from cultivation and human activities. The river had rocky banks over about 28% of its length, and the animals spent 78% of their time there, avoiding cultivation, and sandy or clay banks.

Hussain and his colleagues found that dens, dug mostly by otters themselves, were used only during the cool time of the year. Their smooth otters also used backwaters of the river, small lakes with reeds and other vegetation. Interestingly, it was especially the females with cubs that stayed and foraged there—a sex difference in habitat use similar to that shown in other species.

Where smooth otters occur in lakes or dams, in a reserve in southern India, they were seen to spend significantly more of their time along the shallow, gently sloping coasts (Anoop and Hussain 2004). Rather than in deep and steep areas, they foraged especially in the shallows close to outflows of streams, and generally in parts of the lake where there were many dead trees. Each otter used several dens, usually dug by themselves, and often under tree roots or fallen trees, no more than 20 m from the water.

There have been few systematic studies of the other freshwater otter species in Asia. The **small-clawed otter** is often associated with rice paddies, feeding on the many small crabs there (Foster-Turley 1992), and I saw them doing just that in Palawan (Philippines). It also commonly uses many other types of water body and marshes, including mangroves, and in Malaysia thse animals appear to be common along small streams (Nor and Ahmad 1990, cited in Larivière 2003). In Thailand (see Fig. 4.12), I noticed that small-clawed otters were the species that frequently foraged well away from the river banks in damp vegetation, where freshwater crabs would often be active at night (Kruuk *et al.* 1994a); the other otter species there confined themselves to the river for foraging.

The **hairy-nosed otter** is even less well known. In the flats of south Thailand these animals are especially common in dense swamp forests (*Melaleuca*), and also along mountain streams above these wetlands (Kanchanasaka 2004). In Vietnam these otters are to be found in similar swamp forests, low down in the Mekong delta (Nguyen *et al.* 2003).

In Uganda the **spotted-necked otter** is a species of the shores of lakes and large rivers (Baranga 1995), and, similarly, in South Africa it is 'confined to the larger rivers, lakes and swamps which have extensive areas of open water' (Skinner and Smithers 1990, p 452). On the island of Rubondo in Lake Victoria the species was common, and I watched several groups consistently foraging within some 10 m of the shore (usually within 2 m). They fished in the shallows between rocks and branches of trees (Fig. 4.13), in

Figure 4.13 Lake Victoria provides a good habitat for the spotted-necked otter, with rocky coasts and many overhanging branches.

murky waters that were full of small cichlids (Kruuk and Goudswaard 1990). The shores were densely forested and almost impenetrable. Lake Muhazi in Rwanda has similarly large numbers of spotted-necked otters fishing muddy waters. There, Anne Lejeune noted that the animals spent 71% of their foraging time in, or next to, the fringe of dense reeds and papyrus of the lake, and only 29% in the much larger water body further away from the shore (Lejeune 1989).

John Procter (1963) frequently saw spotted-necked otters use holts in between tumbled boulders along the western shore of Lake Victoria, with entrances sometimes underwater. Similarly, I found their sleeping sites in Lake Malawi under and between large boulders, often accessed by the otters from underwater (personal observation, 2003). The waters in Lake Malawi were very clear, in contrast to those of Lake Victoria, and (perhaps as a consequence) otter density was only a fraction.

South African spotted-necked otters, radio-tracked in Natal, used slow-flowing, well vegetated, small rivers, dams and fish ponds, and occasionally swamps (Perrin and Carranza 2000b). They avoided river banks with grasslands, but concentrated their activities significantly in rocky habitats. Their holts also were somewhat different, often amongst tree roots along banks, and frequently the animals slept in tall vegetation.

On several occasions I watched **Cape clawless otters** foraging in partly floating aquatic vegetation in South Africa, clearly favouring the dense tangle of grass and various weeds above open water, and frequently emerging with a crab in their mouth. The species is known to occur along rivers and small streams, along lake shores, in swamps and in small farm dams. The animals often wander far from water in search of new sites, much more often than the spotted-necked. The riparian vegetation of their habitat can vary from forest to desert to mangrove, from township to agricultural, and they range from streams in high mountains down into the ocean (Skinner and Smithers 1990).

Radio-tracking these Cape clawless otters, Michael Somers followed them at night along small rivers, into towns such as Stellenbosch, sometimes in rivers with high pollution. However, they did show preferences: there were significant associations with places such as reed beds, or boulder sites, or overhanging vegetation (Somers 2001; Somers and Nel 2004). Some would spend almost all of their active time, every night, in fish farms stocked with trout. All of these sites were places with high densities of the otters' main food, mostly freshwater crabs, or fish. Somers also noted that in areas with much disturbance, such as towns, the otters spent much time foraging, but did not go on land to spraint; a spraint survey, therefore, would have made wrong deductions.

He observed that, in larger rivers, dams and lakes, the otters rarely foraged in water further than 8 m from the banks, which made it difficult to estimate the intensity of use of different size rivers. But, generally, the animals were far less choosy in their habitat than the spotted-necked otters—a function, perhaps, of the habitat selection of their main prey.

The very similar **Congo clawless**, or **swamp otter**, is confined to the tropical rainforests of interior central Africa. It can be readily observed in large and small rivers, and not only in swamps as one of its names suggests. It is often seen, foraging in mud, in swampy forest clearings ('bais') (Jacques *et al.* in press).

Otter habitat: some general comments

If one is familiar with any one species of otter in its natural habitat, then there is little problem, wherever one is in the world, in picking out which part of the landscape any other otter would use, of any species. There is much in common between the chosen environments of all of them—of all, except the sea otter.

The most common method for establishing which parts of the natural habitat otters use is the survey for faeces (spraints). Unfortunately, as we have seen, it is unreliable for several reasons, and at present solid conclusions can be drawn only from observations using radio-tracking, or from systematic visual contact with the animals. Commonly, also, otter habitat utilization is expressed as otter (or spraint) numbers per length of coast or stream. For many reasons it would often be better and more consistent, to express it per area of water, especially if one needs to refer to prey numbers and productivity.

Eleven of thirteen species use freshwater habitat, and seven of them are known also to use the sea. At least some of these seven, possibly all, need access to fresh water to cope with marine existence. Two species live only in the sea. This scenario suggests that the ancestor of recent otters may have been a freshwater species, from which those now living in the sea have evolved.

Despite the many similarities, the previous sections show distinct differences between species in their use of water bodies. For example, some prefer smaller and others larger streams or lakes, some like it rough and others sheltered, some prefer shallow coasts and others deeper waters. This emerges especially where more than one species use the same waters, such as a river in Thailand (Eurasian, smooth and short-clawed) or Brazil (giant and neotropical), or a coast in Alaska (river and sea otter) or southern Chile (southern river and marine otter). In African lakes and rivers two species (spotted-necked and either of the two clawless otters) also differ in habitat preference, and they use different prey.

Such differences in preference, however, cannot conceal the large, overall similarities in habitat use between species. Almost all of them use any kind of water body that holds sufficient prey, generally preferring shallow and well vegetated waters and banks. Holts (dens) are more often used in some areas than in others, and frequently otters sleep or rest out in the open, even in cold climates. Any otter living along sea coasts will use freshwater puddles, streams or underground sources to wash salt out of its fur, and only the obligate sea-living species appear to achieve this by specially structured fur and/or unique grooming behaviour.

In several solitary-living otter species one finds clear differences in habitat selection between males and females, with males going for the larger, deeper rivers and lakes, or the more exposed sea coasts. Where males are larger and more aggressive than females, it is interesting that they seem to use apparently inferior habitat, a phenomenon found also in some herbivores such as red deer (Clutton-Brock *et al*. 1982).

At least one otter species, the sea otter, has a large, often dominating, effect on the structure of its own habitat (see Chapter 8). There is no evidence that such substantial effects are also exercised by other species, which could be because they have been studied less thoroughly. However, the sheer density and biomass of sea otters on many Pacific coasts vastly exceeded that of other otter species anywhere else, so perhaps it is unique also in its effects on habitat.

The implications for conservation management, of any species' use of its environment, are far reaching, because habitat management is often our main tool. However, we should keep in mind a reservation that I mentioned earlier in this chapter: the animals' habitat *preferences*, in any one area, are not necessarily habitat *requirements*. In general, all otters are shown to be highly tolerant of differences in landscape, vegetation or human land-use, of human disturbance, and sometimes even of industrialization or urbanization. The otters' priorities therefore lie elsewhere, for instance with prey availability, and sometimes focusing otter conservation effort on habitat factors such as vegetation or disturbance may be chasing a red herring. There are situations, however, where habitat factors may be demonstrably critical to otter populations; the availability of freshwater pools or streams along sea coasts is a case in point.

CHAPTER 5

Groups and loners: social organization

Introduction

Amongst the Carnivora, more than in most other mammalian orders, there is a large variety of spacing patterns, with many different social systems. However, there also are many similarities between species. For instance, almost all carnivores are territorial, defending an area against others of the same species (Ewer 1973), even if only against particular categories of other individuals. Usually such an area has a clear and well defined border, and the 'owner' backs its territorial claims with aggression, and with scent marks of some kind. In many species each territory is inhabited by a solitary individual, but there are also group-living carnivores, living in small, tight packs or in widely dispersed groups, defending the borders of their own group territory against other packs, clans, prides, or whatever their group system is called. Within the territorial system, we recognize a simple distinction between group-living and solitary species (Sandell 1989), although we now know this to be an oversimplification.

Knowledge of such spacing patterns and organization is important for various purposes. Gene flow should be a significant consideration in conservation management of populations, and under some conditions restricted gene flow may be a consequence of population bottlenecks, or of constraints due to social organization (Pimm *et al.* 1989). We will see that these are important considerations in otter populations (Dallas *et al.* 2002).

In addition, information on social organization is required in order to model the relationship between, on the one hand, numbers and densities of the predator, and on the other hand the dispersion of its food, shelter and other requirements. It is likely that spacing patterns are strongly affected by resource availability and utilization (Macdonald 1983). To understand these relationships, social organization needs to be quantified, as well as underlying mechanisms such as competition, cooperation or reproductive suppression (Creel and Creel 1991)—all important aspects of the regulation of animal numbers.

These are not only theoretical concerns, but there is a practical need to know the extent of a species' variability in spacing, dispersal patterns and underlying mechanisms. For conservation purposes we want to know these, as well as the overall limiting factors, if ever we are going to manage the animals effectively. The need is urgent, especially for conservation priorities such as otters.

In many carnivores there is considerable discrepancy between the observed numbers and the *effective population size*, that is, that proportion of the population that participates in reproduction (Creel and Creel 1998). This discrepancy is evident especially in the highly gregarious species, and effective population size may be as low as 0.2 of the observed. (e.g. African wild dogs). It is generally assumed in conservation management that, to maintain genetic diversity, the effective population size needs to be at least 500 individuals (Franklin 1980). Actual population sizes will then be several times that figure, also for otters. It should be an important objective to determine effective population size, and the analysis of social systems is a prerequisite for this.

The organization of animal societies into groups is particularly fascinating, especially where we can study how members of such communities cooperate in one way or another, for example in hunting (Kruuk 1975b), rearing offspring (Macdonald and Moehlman 1983), defence against predation (Rood 1986), or in other ways (Sandell 1989), and where we can compare with others that behave in solitary fashion. The otters are of particular interest for their organization and behaviour, because of large social differences between species that are closely related, and that have remarkably similar ecological requirements. Some live in tightly organized groups, others appear to be almost totally solitary, and one occurs in large aggregations.

In the more social species, the division of areas into group territories is highly relevant. Frequently in the Carnivora there is little or no correlation between the size of a social group and the area it occupies; one finds that individuals of the same species may live in groups of variable size, and they may be in small territories or in relatively huge ones, as for instance in badgers, mongooses, meerkats, kinkajous, hyaenas, wolves or foxes (review and references in Kruuk 2002). There have been several studies to explain such relationships, asking why some individuals in some species disperse whereas others stay in their natal range and add to the group, what it is in the environment that sets limits to groups and to the size of their territories, and how the structure of these societies is adapted to their environment. Such questions also need to be asked about the different species of otter.

Much of the variation in spatial organization within a species can be explained by the distribution and richness of the resources. This has been formalized for carnivores by David Macdonald, in his 'resource dispersion hypothesis' (1983) (see also Kruuk 1989, and Chapter 13). One could expect the relation between resources and territory size to be relatively simple in the 'solitary' species (Sandell 1989), uncluttered as they are by variations in group size. However, there are not many studies of solitary carnivores in which the dispersion of resources has been documented. One of the aims of our Eurasian otter project in Shetland was to do this for just such a solitary animal: to describe the distribution of prey in relation to that of the otters.

Unexpectedly, however, this objective was frustrated because the supposed simplicity of a solitary otter existence turned out to be a front for a highly complicated social organization (Kruuk and Moorhouse 1991). What we thought to be a solitary species (Chanin 1985; Erlinge 1967; Mason and Macdonald 1986) was not. In fact, as I will show, the distinction between 'solitary' and 'social' has little meaning in cases such as that of many of the otters. In Shetland, for instance, Eurasian otters neither live singly nor do they exist in clear groups, and no single term suffices to describe the degree of their gregariousness.

In the following sections, I will first discuss the spatial organization of Eurasian otters in Shetland, where I know it best, then describe what we know of the sociality of the same species elsewhere. I will compare this with the very different set-up in other species of otter, such as the river otter in North America, which show remarkable variations that allow for important generalizations.

Shetland: organization of Eurasian otters in a marine habitat

When I started our intensive project with **Eurasian otters** in Shetland, there was considerable frustration at being unable to see any pattern in the movements of the otters. Previous studies by others were of little help, because they rarely involved actual watching of the animals themselves. We were also disadvantaged in being able to do only limited radio-tracking, for various reasons. And, although we knew many of the otters individually by sight, it took several years along those Shetland coasts before we recognized some social relationships and the way in which the animals used the area (Kruuk and Moorhouse 1991), in a pattern quite different from what was known for any other carnivore.

The animals were shy, but we learned how to follow them around the coasts whilst watching through binoculars or telescope, recognizing individuals from the characteristic patches on their throats and coloured ear-tags, and we did use a few radio-transmitters. We came to realize that some females were always there, in the same area for at least a couple of years, whereas others were only transitory. Individual males (Fig. 5.1) were much more erratic than females, and they used much larger areas of coast than the females. We found that the females who were 'residents' did not go beyond particular points along the coast. They were always alone or with their own cubs, but, at the same time, they did share their ranges with other female otters. There were no clear, simple territories.

On average there was about one adult otter for each kilometre of coast (see Chapter 11), but every otter used a stretch of shore several kilometres long. We divided them into three distinct categories: resident females, resident males and transient animals. The most important category, around which the entire system revolved, was that of the resident females.

Figure 5.1 Eurasian otter (male) swimming conspicuously along the Shetland coast.

All otters confined themselves largely to a narrow strip of water and land along the coast (see Chapter 4). Occasionally they moved outside this, in long distances over land or across the waters, but the vast majority of our observations were made close to the line where land meets water. By following and recording individual animals, we established otter ranges in terms of length of coast used, not size of area as one does for most other mammals, and previous authors have done the same (e.g. Erlinge 1967, 1968; Green et al. 1984; Melquist and Hornocker 1983). For some purposes this is somewhat misleading, because the width of the strip of coast used by otters varied with the slope of the coast, and also in fresh water the width of streams appeared to be very important (see Chapter 4). However, for the purpose of describing spatial organization, I will maintain the use of that much easier and more relevant measure: the length of coast or bank. In recent years methods of analysis have been developed that are amenable to probabilistic modelling of home ranges (e.g. 'adaptive kernel estimates'; Blundell et al. 2001; Sauer et al. 1999), but their application would not have altered the pattern that emerged in Shetland.

Over 5 years we recorded movements and observations of many different females through the numbered sections of the coast of Lunna Ness, and for fifteen of them there was sufficient information to estimate home range size (Kruuk and Moorhouse 1991). They did not all use the area at the same time, and simultaneously with them there were other otters present for which we did not get enough observations, especially in the beginning of our study. For the adult females we found three boundary areas, therefore four distinct ranges, each being used by several animals (although we obtained almost no data on the animals in the most southern range).

Despite the fact that the final picture was not complete, it was obvious that several females living along a particular coast recognized the same boundaries, also during consecutive years. The exact location of these boundaries was found by marking turning-around points when otters were swimming along the coast, as well as from mapped sightings.

Boundaries were not conspicuous; one was at a stretch of rocky beach, another was a small stone dyke sticking out into the sea, and a third was a little stream entering the sea.

We knew the complete length of only two of the ranges with some certainty: one was about 4.7 km and the other 6.4 km long. The next range, stretching along cliffs and with deep water, was about 14 km long, but some females within that area probably had ranges smaller than that. It was much longer (and narrower) than that of the others, with otters staying very closely inshore and next to the steep rocks. The boundaries between the female ranges were stable, at least over the 5 years that we were there, with a succession of different Eurasian otters involved.

We called the areas group ranges, because at any one time there were several otters living in each of them; one was used by two females, the other by up to four females simultaneously, and the longest range by at least five. These females were resident as adults for up to 2 years, then they disappeared and probably died (see Chapter 12), and others took over.

Within each home range we also defined, for each otter, a *core area*: that 350-m section of coast where the animal was seen most often, plus the adjoining sections that would constitute the smallest part of the range where the otter spent at least 50% of its time (Kruuk and Moorhouse 1991). This definition followed the study of core areas in Eurasian badgers by Latour (1988), after a method developed by Samuel *et al.* (1985). Within the home range this core area would often be apparent with far fewer observations than we needed for a total home range, within days or a few weeks. The core areas varied in size between 0.5 and 1.6 km of shore.

An example of core areas in two home ranges is summarized in Fig. 5.2. It shows that, despite the fact that several females have identical home ranges, their core areas were quite separate. Interestingly, the core areas of some females changed from one year to the next, although the home range remained the same. For instance, one mother and daughter had exactly the same core area during the daughter's first independent year of life, meeting frequently and often foraging together. But a year later the daughter moved to another part of the range, from where the

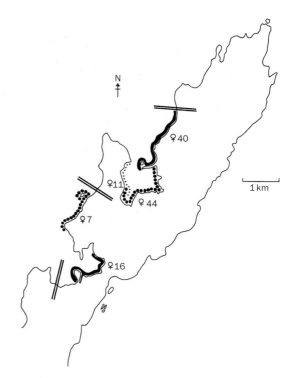

Figure 5.2 Group home ranges and individual core areas of five female Eurasian otters along Lunna peninsula, in Shetland. Each female used the entire group territory (section of coast between double lines), but spent more than 50% of her time in the core area.

previous female occupant had disappeared; mother and daughter still occasionally visited each other's sections, and both produced a litter of cubs (Fig. 5.3). The mother lost hers, however, and she herself disappeared soon after. Elsewhere we observed a female that had cubs in one core area one year, then changed to another core area for the next year where she had cubs again; the previous occupant disappeared.

In general, adult female Eurasian otters living within the same coastal group territory in Shetland avoided one another, and core areas were more or less exclusive. However, there was also a great deal of mutual tolerance between group members, quite different from relationships with females from outside the group home range: within the group they avoided, rather than join or confront each other. This was demonstrated on the few occasions when I saw two families meet that belonged to the same group territory. These were females with their cubs,

Figure 5.3 Family of Eurasian otters resting ashore in Shetland: adult female and four large cubs, on exposed algae at low tide.

which was interesting because the cubs do not have the same scruples as do adults. The following is an account of two of such observations.

October 1985. On a small rock skerry two small cubs are eating a big rockling, whilst the mother is fishing nearby. They are in the centre of their core area, in the group range. Suddenly, seemingly from nowhere, another female from the same group (but different core area), arrives with her big cub which enthusiastically throws itself at the fish and the two smaller cubs. Some loud wickering and quick dashes back and forth, then somehow all three cubs eat, right next to each other. The mothers, however, are having more serious problems. The second female sits downwind from the party of cubs, well out of the way and slightly crouched, not coming any nearer. The first, in the meantime, has caught another rockling and approaches with the red glittering prize, landing upwind of the cubs four metres away. Normally she would have taken the fish directly to the cubs; now she looks at the eating party, then slides back into the water fish and all, off to another rock out of sight. After a couple of minutes of status quo, the second female's cub looks up, then walks to the water's edge, and both swim off. Ten minutes later the two families are several hundred metres apart again.

Early November 1985. Following one female, I see her with her single cub, quietly swimming between some small rocky islands in her own core area. She lands on one of them, obviously unaware that the other female from the same group range, and her two offspring, are curled up between boulders on the other side. When she comes downwind of the other party, she crouches slightly, hesitates, then slowly walks up. The two adult females sniff each other fleetingly, sniff anal regions, both in a low posture, circling warily but without overt aggression. In the meantime the cubs are running about in a mêlée of brown bodies and tails, then two walk up to their respective mothers and try, unsuccessfully, to suckle. The females both spraint, close to each other. This chaos only lasts about 40 seconds, then one slides into the sea with her cubs close behind her. For a short time the other female, with her cub, continues to sniff the site of confrontation, then she spraints again and also leaves, fast, in the opposite direction.

There was no real aggression between these Eurasian females here. They were probably closely related and their ranges overlapped totally. But they kept out of each other's way, using their own preferred areas within the group range and away from the others. The system appears to be based on each otter having an interest in staying away from others, within their joint group range. Only the cubs did not seem to bother one way or the other (Kruuk and Moorhouse 1991). However, interactions with females from outside the home range were aggressive, and animals were effectively excluded. The Eurasian otter's female group ranges were *de facto* intrasexual group territories, that is, defended areas.

Male otters are quite considerably larger than females, as in many other mustelids (Powell 1979). In Shetland, for instance, the mean weight of males more than 2 years old was 7.35 kg (95% confidence limit $\pm$ 0.46 kg; $n = 31$), and that of females 5.05 kg ($\pm$ 0.29 kg; $n = 42$). These bodyweights, incidentally, are much lower than those of otters from the British mainland, where an average weight of 10.1 kg is recorded for males, 7.0 kg for females (Chanin 1991). Therefore, Shetland males were almost 1.5 times heavier than females, and one could expect *a priori* larger home ranges on that basis alone, because home range size is generally correlated with bodyweight, at least across species (Clutton-Brock and Harvey 1977).

We saw male otters less often than females in our Shetland study area, and when we followed them they moved over much larger distances. Their different behaviour meant that it was difficult to obtain sufficient observations on individual males to recognize their total ranges. In 1985 and 1986, in a sample of 887 observations of adults when we could be certain of the sex of the otter, only 34% were males, 66% females. In addition, males occurred more often in rough water along the more exposed coasts, where they could be overlooked more easily, rather than in the sheltered bays, the voes. Along wild, turbulent and rocky shores, 39% of the adult otter observations were of males, but in the bays as few as 28%, a significant difference (Kruuk and Moorhouse 1991). There was a strong seasonal trend (Fig. 5.4): significantly, during April and May (the mating season) 61% of 887 otter observations in the study area were of males, versus only 7% in December. When males were present, their ranges were relatively large (up to 19.3 km), but there was

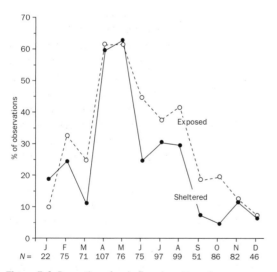

Figure 5.4 Proportion of male Eurasian otters along sheltered and exposed sections of coast of study area, Lunna, Shetland, at different times of year. N denotes numbers of observations. There were significantly more males during mating time in later spring ($P < 0.05$, runs test) and along exposed sections of coast ($\chi^2 = 11.4, P < 0.001$), although overall females predominated.

not one single male for which we knew the full extent of his movements, and we were not able to do any radio-tracking on them.

The boundaries between female ranges in the study area appeared to be observed also by at least several of the males, which turned back at the same point along the coast. But at least one female boundary was ignored by males, and they always travelled on there without let or hindrance. The ranges of individual males, therefore, overlapped with at least two groups of females, and probably often with more. Each male range was used by several male otters simultaneously, but the pattern of overlap between them was unclear. Whatever the range-sharing pattern between males, they were almost invariably aggressive to one another, much more so than females (see Chapter 6), and there appeared to be some aggressive, intrasexual territorial system underlying the dispersion, as in many other mustelids (Powell 1979). We never saw groups of several males together (nor was this ever reported by any other observer), as is commonly found in other otter species.

Eurasian otters on the move—transient animals—were easier to study in Shetland than were the resident male otters, because for some reason they were easier to catch in our box traps (perhaps because they were less suspicious). We were able to get some good radio-tracking data on four of them, three females and one male. One of the female transients was followed over 28 km of coast, another moved over at least 40 km before we lost her. One female disappeared immediately after release, and she was discovered 3 months later on the island of Whalsay, a distance of only 15 km as the crow flies, but much further when measured along the winding shore line, which would have been her route.

Following these transient otters along the shores, I noticed that they were often doing something new or unexpected, behaving as if exploring, quite differently from our regular residents. An otter was not recorded as a transient until it had been followed for a few days, when we realized that it did not stay in one particular area but just kept on enlarging its range, moving further and further afield. Characteristically, such transients used rather inferior types of holt, away from the usual otter haunts—often rabbit burrows with nothing to show from the outside that there was an otter in it: only the radio transmitter told the story. Their food, too, was often inferior compared with that of the residents; for instance, one female transient specialized in catching rabbits in their burrows, and some others would eat mostly crabs (see Chapter 7).

Observations on the spatial organization of Eurasian otters in other marine areas were made along the shores of Mull, on the Scottish west coast, incidental to research on the energetics of otter foraging. Johnny Watt's study suggested a social organization similar to that in Shetland: there were several females sharing relatively small ranges in sheltered waters, with males ranging far more widely and overlapping with several female ranges (Watt 1991). In addtion, along the coast of the Ardnish peninsula, close to the Isle of Skye in western Scotland, Ray Hewson and I found several different individual Eurasian otters exploiting the same coasts (Kruuk and Hewson 1978).

Organization of Eurasian otters in freshwater areas

In an early, classical study of Eurasian otters in freshwater, Sam Erlinge (1967, 1968) provided some excellent insights into the social life of the animals along rivers and small lakes in southern Sweden, based on snow tracking. The populations he studied were relatively dense, with on average one adult otter for every 4–6 km of stream. Females lived in more or less exclusive areas, no more than 6 km of river, with some small overlap between neighbours. There were no group territories. Male territories were much larger, including the ranges of the females, and there was evidence of aggression between males on the borders of their ranges. Erlinge's otters, therefore, conformed to the typical mustelid social organization (Powell 1979). By its nature, his study was confined to winter, when the lakes were frozen over, and even streams were often covered in ice and snow. That may have been one reason for the relatively small home range sizes he found; possibly, the otters ranged more widely in summer.

Until recently, few radio-tracking studies were available of Eurasian otters in freshwater areas, and the earlier ones were unable to focus on numbers of animals within the same social system. The first was carried out in central Scotland by Green et al. (1984), providing insight into the large sizes of the areas covered by these animals: three individuals occupied stretches of 16–39 km of river. There was evidence of other otters simultaneously using the same areas as the focal animals—a similarity with our Shetland results. Later, when the law changed, my colleagues and I were able to implant a number of radio-transmitters in otters in freshwater areas of north-east Scotland, and some interesting results emerged. For instance, just like otters in Shetland, males used different types of areas compared with females.

Adult male otters spent most of their time on the main stem of the rivers, with frequent excursions up the tributaries. For instance, in the Dee and Don valleys, of eight large adult males that we trapped and/or radio-tracked, five were most often along the main stem of the river (see Fig. 4.9), whereas 13 of 15 adult females or independent sub-adult otters were mostly along the small tributaries or in lakes. Just as along the coasts of Shetland, the difference in habitat of males using the main stem and females the tributaries was by no means absolute, but a statistically significant preference.

The size of otter home ranges in freshwater habitats is quite staggering, especially that of the males. However, this is only because we express this size in a somewhat unusual way: as length of stream or river. Green et al. (1984) found the length of range for one male Eurasian otter to be 39 km, and one of Leon Durbin's (1998) males in north-east Scotland moved around in 84 km of stream. The mean length of river and stream used by all otters in the Scottish studies was 38.8 ± 23.4 km for adult males ($n = 6$) and 18.7 ± 3.5 km for adult females ($n = 10$) (Durbin 1998; Green et al. 1984; Kruuk et al. 1993a; and unpublished data). The figure for adult males shows a large variance, and does not include observations on two animals with which we lost contact. This was probably because of their very large range size, and therefore the true figure for male range size is larger still.

In terms of area of water used by otters (based on widths and lengths of streams and lakes), the figures are of an order of magnitude that we would expect for carnivores of this size. Calculated for adult males this was 63 hectares, and for females 34 hectares, of water (Kruuk et al. 1993a), comparable to range sizes of Eurasian badgers (Kruuk 1989) or red foxes (Macdonald 1980b).

Within each home range we frequently saw or heard encounters between the animal that we were radio-tracking and other otters, anywhere in their range, and not just near boundaries. In the case of the adult males, we knew that they were seeing several females in the different tributaries and on the lochs. However, we had no good evidence that neighbouring home ranges of females on the tributaries overlapped; there was only the fact that there were other otters of unknown sex and status using the same area as the target otters in our radio-tracking studies (Kruuk et al. 1993a).

On lakes, such as the Dinnet Lochs in the Dee valley, I often saw several females in daytime, using the same waters at the same time, sometimes close together. There never was overt aggression between them, even when they were foraging with litters of cubs, as observed also by Jenkins (1980). Several times I saw six or more otters, including cubs, simultaneously: they more or less ignored one another—just happened to be there at the same time. We were in the dark about the relationships between animals that shared a range, but I assumed that these were similar to what we had found in Shetland; they were possibly closely related. The observations were consistent with a scenario of female group ranges in tributaries and lakes, probably with boundaries that were less well defined than those in Shetland, and these ranges visited by males that were usually based along the main stems of the rivers.

More recently, part of this general picture has been confirmed for Eurasian otters on Kinmen, one of the islands of the coast of China, with DNA extraction from spraints (Hung et al. 2004). About half of the otters encountered were resident in the study areas (using much smaller ranges than in Scotland), and the rest were mere visitors. One river had two group ranges of mean 0.6 km long, of three and two females and one single male; the male ranged over 0.9 km. The other river also had two group ranges, one occupied by two females and two males, the other by two females and one male, with females ranging over a mean of 1.2 km and males over 2.3 km. It is likely that the home ranges of these groups were larger, though, because otters may have used areas that were not sampled for spraints (e.g. sea coasts), or the animals may not have sprainted everywhere. On each of the two rivers the otters were more related to one another than to animals from the other river, and within groups the females were either mother–offspring or siblings.

Spatial organization of Eurasian otters: some generalizations

Given certain habitat preferences, for instance for cover when resting or for narrow streams when foraging, the spatial organization of Eurasian otters in freshwater appeared to fit into the same general pattern as that along the sea coast. We found a system of home ranges and, at least under some conditions, a large overlap of female ranges. There may be a similar system of core areas and intragroup tolerance in fresh water as in the sea, but this needs further study. In freshwater and in the sea, males live in larger ranges, in a somewhat different and more exposed habitat (large rivers or rocky coasts), and they frequently visit the female ranges. Overlap between neighbouring male ranges, and the relationship between males within the same range, are unknowns, but observations do show that males of this species never go about in packs. There are considerable numbers of vagrant animals (in some areas as many as half of the population), which may not be reproducing.

The sum total of our knowledge of the Eurasian otter's social organization is meagre, mostly because of methodological problems. The glimpses we have suggest that there is considerable fluidity, with few strict boundaries and little permanence, which does not make it easy to generalize. Another complicating factor is the short lifespan of otters (see Chapter 12), causing a fast turnover within groups and populations.

The social system of North American river otters

Much more is known about the social life of the North American river otter (Fig. 5.5) than that of its Eurasian counterpart. In both cases the social system is highly variable, but especially for the river otter

Figure 5.5 River otter. © Thomas Serfass.

this has been documented in some excellent studies. Despite the variability, one general trend is obvious: the river otter is decidedly more gregarious. *Lontra canadensis* is predominantly a freshwater species (Melquist and Dronkert 1987), and along the coasts they live in ecological conditions very similar to those of coastal *L. lutra* in Europe.

A classic study is that of Wayne Melquist and Maurice Hornocker (1983), using radio-tracking and direct observations. They showed that, in a system of rivers in Idaho, home ranges of adult females, with or without cubs, averaged 38 ± 9 km, and an adult male regularly covered 81 km of river. Two or more female river otters could occupy the same range, with separate 'activity centres' (places where they spent more than 10% of their time). Some sites, such as fish (kokanee, *Oncorhynchus nerka*) spawning areas were used simultaneously by several females. All this was not too different from our observations on the Eurasian otter, but where the river otter varied strikingly was that frequently, and for periods of weeks or longer, animals joined other otters in 'packs'. Family groups (single females with cubs) were often accompanied by an additional adult otter, most often a female. Sometimes, in addition, unrelated lone juveniles joined. One group consisted of three females, of which two were offspring of the third. Siblings often remained together for months after separating from their mother, and the authors mention one report of groups of 15 to 30 river otters. There was no mention of groups of only males (but these may have escaped observation). The general comment was that 'river otters appear to be far more sociable and tolerant of conspecifics than previously thought' (p 52).

An interesting contrast with this research is a similar, more recent, study by Reid *et al.* (1994b) in boreal Canada. They found home ranges considerably larger than those in Idaho, between 34 and 249 km of river and lake shore: females averaged 58 ± 33 km, males 182 ± 50 km. Unlike the Idaho study, there was little overlap between female ranges, but much between males and females, and also between males. Remarkably, the males often associated in packs; they did not join family groups, but packs of three or more males were a consistent feature in this study, mostly in spring and summer.

Several observers have commented on these groups of male *Lontra canadensis*, which clearly is also a common phenomenon along the Pacific seaboard of North America. There, as many as 18 individuals may be involved, moving around as a group, and fishing simultaneously in the same sites (Blundell *et al.* 2000; Rock *et al.* 1994). Along the coast of northern California, Scott Shannon (1989) monitored a group of six to eight males, which remained gregarious throughout the year, sharing dens, grooming, playing and foraging together, even sharing food. They formed smaller, temporary associations amongst themselves, or at times were solitary. At the same time, a female with her cubs sometimes tolerated males nearby or chased them; cubs of previous years' litters helped with provisioning the cubs, playing and guarding against males.

Large overlaps in the coastal ranges of female river otters have been reported (Larsen 1983; Noll 1988; Woolington 1984), some of those observed by radio-tracking from aeroplanes. Bowyer et al. (1995) found female home ranges of 8–20 km ($n = 7$), for males 21–45 km ($n = 15$), along Pacific coasts in Alaska. However, precise boundaries and core areas were not documented until a major study by Gail Blundell, also in Prince William Sound, Alaska. In her analyses, using 'adaptive kernel methods' (Sauer et al. 1999), she found mean home ranges for females of 15 km of coast ($n = 9$) and for males 60 km ($n = 20$). These animals used 'core areas' (i.e. 50% of observations) of 4 and 10 km respectively (Blundell et al. 2000).

Interestingly, Blundell found that the core areas of female river otters in Prince William Sound usually did not overlap one another, although the full home ranges did. However, in male river otters there was extensive overlap even of core areas, with females as well as males, and also between groups of males as well as between solitary individuals.

In conclusion, the pattern of social organization of North American river otters throughout their geographical range in freshwater and in the sea shows an underlying common theme of exclusive female core areas but with usually overlapping home ranges. Males have larger core areas and ranges, each of them overlapping with those of other males as well as females. This simple, solitary system, as found in most Mustelids (Powell 1979), is then developed somewhat further, in some sites by mature female river otters joining other family groups; perhaps they are older offspring, and at times help with rearing the cubs (Rock et al. 1994; Shannon 1989). As a separate phenomenon, males may club together and cooperate with other males in sizeable groups, especially outside the breeding season.

Thus, *Lontra canadensis* has gone several steps further towards gregariousness than *Lutra lutra*, in a social system that shows similarities with that of the honey badger *Mellivora capensis* (very large male territories encompassing several females, with at times males associating in all-male groups; Begg et al. 2003).

As to the biological function of the large groupings of males in the river otter, there still is much uncertainty. Gail Blundell suggested that males did not derive any reproductive advantage from grouping, because solitary males had as many offspring as gregarious ones; nor were gregrarious males more closely related to each other than to others (Blundell et al. 2004). But she noticed that their diets were different: the profile of isotopes, indicative of prey species in the river otters' spraints, varied between social categories, and she suggested that the pack-living males preyed more on schooling fish, in cooperative hunts (Blundell et al. 2002). However, the isotope profiles of the local schooling fish were not known, and there are no direct observations of river otters hunting schooling fish. Possibly, the male packs may be an anti-predator strategy against killer whales, pumas or wolves, but why then are so many of the males solitary? Perhaps they balance increased predation risk against advantages gained from solitary foraging or mating strategies. Clearly there is a need for further study.

Social organization of the sea otter

At a glance, the social organization of the sea otter *Enhydra lutra* is totally different from that of other otters. For one thing, sea otters are more gregarious, they occur in much higher densities, and they may join in aggregations—'rafts' (Fig. 5.6)—of hundreds, even up to 2000, animals (Estes 1980). However, there is large variability, as well as distinct similarities with the social life of other otter species. Several excellent research projects have provided us with insights.

Almost all sea otter life is spent at sea, in a social system perhaps best understood as a complicated result of two conflicting tendencies. On the one hand is territoriality, as in most other carnivores and other sea mammals; on the other a strong

Figure 5.6 Part of a 'raft' of sea otters, California. © Richard Bucich.

gregariousness, as also found in many sea mammals. It is the latter that is most conspicuous in the field.

One of the first things that biologists noticed in this gregariousness is that the sexes are quite strongly segregated. There are distinct male and female areas, where the animals rest when they are not foraging (Garshelis *et al.* 1984; Kenyon 1969; review in Riedman and Estes 1990). Male areas, with often rough seas, are along more exposed parts of the coast than those of females. Along the coast of California (between Los Angeles and San Francisco), male areas are at the ends of the main distribution of the species, more than 150 km apart, and on either side of the female area. They are permanently occupied by mature and immature males, often in groups, and outside the breeding season of September and October most of the fully adult, breeding males are based there (Garshelis *et al.* 1984; Jameson 1989; Riedman and Estes 1990).

In the Aleutian Islands, Alaska, male areas were less far apart, and rather scattered throughout the female areas, but in Prince William Sound the fronts of the expanding geographical range of the population were male areas (Garshelis 1983; Garshelis and Garshelis 1984; Garshelis *et al.* 1984). Within these Alaskan male areas, the animals occupied home ranges of 11 ± 0.7 km of coast, shared with many other males. Apparently, when unoccupied habitat in the Prince William Sound was colonized by sea otters, the initial arrivals were old, white-headed males and only many months later did groups of younger males move in, followed by females and pups after one or more years (Garshelis and Garshelis 1984). Males moved between areas much more than females did.

One measure of the size of areas covered by sea otters is the large distance between the outermost points of the observed ranges (Ralls *et al.* 1996). Conveniently, this is also a measure that is comparable to the linear home range used for other otter species. In California, where these authors measured the activities of 40 sea otters with radio-transmitters, the linear ranges averaged 98 km for adult males (54–181 km), 128 km for immature males (44–228 km), 24 km for adult females (5–86 km) and 47 km for immature females (13–120 km). The distances covered by sea otters in Prince William Sound were not exactly comparable, because of the more convoluted shape of coast, but they were fairly similar (Garshelis *et al.* 1984; Ralls *et al.* 1996).

Within the male areas animals regularly rested together in daytime, especially mid-morning, in large 'rafts'. Males were more gregarious than females, and their rafts averaged 70 to 100 animals in Prince William Sound, compared with only two to five for females (Garshelis *et al.* 1984). The researchers suggested that this gregariousness had evolved in response to predation pressure, or to facilitate food-finding, or to promote social interactions. Of these possibilities I believe predation pressure to be the most important, as predation on sea otters can be severe (see Chapter 12), and the other possibilities are hardly supported by direct observations.

Female sea otters, whether alone or with a pup, tend to stay in more or less the same area all the time, although there are many observations of long-distance movements (Riedman and Estes 1990). Excluding such long-distance trips, the females' year-round home ranges are smaller than those of males:, on average 4.8 ± 0.9 km in Alaska ($n = 7$; Garshelis and Garshelis 1984) and 18 km of coast in California ($n = 22$, R. J. Jameson, personal communication, in Riedman and Estes 1990). They have recognizable 'activity centres' or core areas, but these do not appear to be exclusive and have not been measured. The home ranges of females may overlap completely with those of any number of others, and they are in no way exclusive: females are decidedly non-territorial.

Then, especially during the summer breeding season, some males leave the male areas, move to the female ranges and set up territories. These are the older, often white-headed, males (at least 6 years old; Garshelis et al. 1984), who defend a small area of sea against others: for instance, about 35 hectares in California (Loughlin 1980) and about 23 hectares in Prince William Sound (Garshelis et al. 1984). It is a territory that during the breeding season is rather smaller than the ranges of the female sea otters. Thomas Loughlin described the activities of these males:

The territory formed an exclusive resting location for the territorial male; no other male was allowed to rest within the defined area. Females were allowed to rest within the territory and were subject to attempts at copulation by the resident male. Other sea otters were allowed to move unmolested through the territory on their way to or from a feeding site. However, males and females that foraged within a territory were subject to having their food stolen by the territorial male. Females with pups often were accosted as they passed from one territory to the next. (Loughlin 1980, p 580).

The male territory is where copulations take place, and female otters go round and select. Garshelis et al. (1984) ranked territories in terms of 'quality', that is, size, degree of enclosure by coast and other territories, accessibility for females and food availability. As expected, they found that 'good' territories scored more copulations than 'bad' ones.

Riedman and Estes (1990) have suggested that the territorial behaviour of male sea otters is driving the social system of these animals, for example causing the segregation into female and male areas, a system quite unique amongst carnivores. The organization of male territories in areas with large concentrations of females clearly has similarities with the social organization of other sea mammals such as seals, sea lions and walrus. These social similarities with sea mammals are more obvious than those with the systems of immediate relatives, such as all the other otters. As mentioned above, other otter species, for instance, have female territories, and they do not show substantial aggregations. All this suggests that *Enhydra*'s organization has been formed in response to strong environmental selection pressures from the oceanic habitat, irrespective of phylogeny.

Social systems of Latin American otters

Vying with the sea otter for the position of most charismatic otter in the world is the **giant otter**. Not many of them are left, but they are conspicuous, living in noisy groups in dramatic settings of rainforest, active only by day and endearingly inquisitive. Individual identity is easily established because of the animals' strikingly patchy throats, so there is less need to resort to radio-tracking equipment to study their behaviour and ecology. Several PhD theses have been written about giant otters, including the excellent studies by Nicole Duplaix, Christopher Schenck and Elke Staib, and enthusiastic research teams are dedicated to ensuring the species' survival.

Giant otters, weighing up to 32 kg, live almost exclusively in tight packs, with only a few male or female loners in between. The groups consist of extended families (Fig. 5.7) with the odd immigrant; there is a central pair, a female and a male, and the offspring from the last one or more years. Normally there are three to nine animals involved, with occasionally groups joining or congregating at a prolific feeding site, and numbers within a group

Groups and loners: social organization 67

Figure 5.7 Family of giant otters (dominant male in centre). © Nicole Duplaix.

reaching 20 (review in Carter and Rosas 1997; Duplaix 1980; Staib 2002).

The inclusion of a permanent male partner to the leading female of the group is remarkably different from the organization of other otters and, for that matter, of other mustelids. Usually, male mustelids keep territories separate from, but overlapping with, those of females (Powell 1979; another exception to this is the group-living Eurasian badger *Meles meles*, Kruuk 1989). Observations of giant otters generally suggest that it is, indeed, the lead male in each group that fathers the cubs in that group, although until now no observations have been published of copulations in which the identity of both partners was known (Staib 2002). So there is yet the outside possibility that cubs are fathered by outsiders (nomads and immigrants), as is the case within groups of brown hyenas (Mills 1990). However, observations of social behaviour such as the association between the lead male and female, and grooming, suggest that the male 'partner' of the lead female is also the father of her cubs.

The members of the group usually move around the home range together, and within the pack the males tend to associate with one another (Duplaix 1980). All group members help with looking after the latest batch of cubs, and when cubs are small they 'babysit' when the rest go out foraging, whilst any of them may carry fish back to the cubs. However, there is very little cooperative feeding (see Chapter 9)—usually it is each one for itself. When there is a spot of bother, such as strange otters or people, males are the first to confront it, but usually any group initiative, such as for foraging trips, or visits to scent-marking sites, comes from the lead female (Carter and Rosas 1997; Duplaix 1980; Staib 2002).

Each group has its own home range, usually a large oxbow along a river, or a lake, or more rarely the river itself—at least that is the case during the dry season. During the wet, groups disappear into flooded swamp forests, and in the absence of radio-tracking nobody has been able to follow them there. The dry season home ranges are well over 12 km of wide stretches of water long, with group core areas of 2–6 km (mean 3.5 km; Duplaix 1980). There is some overlap in home ranges between neighbouring groups, but little or none in core areas; it all functions largely without overt aggression, but through clear mutual avoidance. Interestingly, apart from the mixed-sex family groups in this species, in a high-density giant otter population in an artificial lake in the Amazon area, all-male groups have been observed, consisting of fairly young animals (N. Duplaix, personal communication).

What the advantage is to the giant otters of living in such conspicuous, tight groups is far from clear. It does not appear to function as a means of foraging; possibly, it helps to protect against predation (e.g., by piranha, caiman, python or jaguar; see Chapter 12).

Far less is known about the other Latin American otters. The **neotropical otter** also had not been studied with radio-tracking at the time of writing. My own observations in Brazil, as well as those in many different parts of its geographical range, from Mexico down to Argentina and including marine as well as freshwater areas, all report that it is a solitary species. There are no records of groups other than a single female with her cubs, or the occasional two-some (Duplaix 1978; Parrera 1993; refs in Larivière 1999b; M. Muanis, personal communication). Recently, radio-tracking of the **southern river otter**, in a riverine habitat in southern Chile, showed this species also to be solitary (Sepulveda *et al.* 2004). Despite brief studies of this otter in quite a few different sites, including marine ones, there have been no observations of groups, apart from one-parent families (Parrera 1996, in Larivière 1999a).

The **marine otter**, along the Chilean and Peruvian coast, may have a more complicated social system than the previous two, but there is much uncertainty. Ostfeld *et al.* (1989) found several dens being used by more than one adult otter, each bringing in food for cubs. Various stretches of coast were shared by several different otters (see also Postanowicz 2004). The researchers could not distinguish sex, however. Several other observers have watched these otters foraging, and groupings have been seen of two, three or four animals together (19%, 7% and 1% of observations, with all others solitary; Ebensperger and Castilla 1991, in Larivière 1998). But in these cases, too, it was not known whether these were family groups or other associations. A great deal remains to be done to understand the social organizations in these last three species.

Social organization of otters in Asia

The **Eurasian otter** has had little attention in Asia (apart from recent DNA research on an island off the coast of China; Hung *et al.* 2004), although it occurs in most countries throughout the continent. We must assume that its social systems there are little different from those of the same species in Europe, and where I have seen it in rivers in Thailand the animals were solitary, both males and females. Its nearest relative is the **hairy-nosed otter** in south-east Asia, which appears similar in all aspects (sometimes it is treated as a subspecies of *Lutra lutra*). Almost nothing is known about its social life, apart from the fact that it appears to be a solitary species like the Eurasian otter, living in swamp forests and along mountain streams (Kanchanasaka 2004; Nguyen 2001).

The **smooth otter**, in waters from the Middle East to Vietnam, shows an organization rather like that of the South American giant otter. It is also a rather large, diurnal otter of often wide rivers, and it usually lives in groups of up to nine or more individuals (Fig. 5.8), with a mean group size of 4.6. A crucial difference is that here the groups do not consist of a central 'pair' and offspring, but of a single female and offspring of her last one or more litters, and such groups often associate with a resident male (Hussain 1996). There also are some solitary individuals, but a pack is the norm (Foster-Turley 1992; Hussain 1996; and personal observation).

Radio-tracking studies of the smooth otter along the Chambal river in India indicated home range sizes of 17 km for an adult male and 5.5–7 km for females or independent sub-adults, with 657 and 213 ha of water (Hussain and Choudhury 1995). Within these home ranges the researchers found the otters concentrating their activities in core areas: females spent about 72% of their time in only 1.2 km of river.

Figure 5.8 Extended family of smooth otters, India. © S. A. Hussain.

A possible adaptive value of this group existence could lie in quite spectacular cooperative foraging, which I saw in Thailand and others have observed elsewhere (see Chapter 9). Another possible advantage is protection against the many predators in the smooth otters' habitat, such as several different crocodiles, python and many carnivores. Groups are also known to approach human predators aggressively, for instance farmers in rice paddies (Foster-Turley 1992).

Adult male smooth otters are apparently well tolerated within groups, although observations suggest that they are not permanent members (Hussain 1996). Interestingly, in captivity female smooth otters fully tolerate males around their cubs, and males help with provisioning and nest-building. This is in contrast to captive Eurasian otters, where males are not tolerated anywhere near the cubs (Wayre 1974; and see Chapter 6). There are no published observations of groups of male smooth otters.

The last south-east Asian species, the **small-clawed otter**, is gregarious like the smooth otter, but less than half its size. It takes very different prey along small rivers and streams, in marshes, rice paddies and mangroves (see Chapter 7). Little has been published about the behaviour of this animal in the wild (though more from captivity). Observed group sizes vary from four to eight (Furuya 1976), 12 or 13 (Timmis 1971) or up to 15 (Foster-Turley 1992), and several authors comment on larger group sizes in coastal habitats than in freshwater (summary in Sivasothi and Nor 1994). There is no good evidence about the composition of such packs. Presumably they involve at least one adult female and offspring, but whether the group is centred on an adult pair or on just the female has not been established. Observations in captivity suggest permanent inclusion of an adult male (Foster-Turley and Engfer 1988), but such evidence is somewhat dependent on keepers' preferences. There is no published information on the size of home ranges.

Otter social systems in Africa

The two clawless otters south of the Sahara are closely related to the Asian small-clawed otter, and of the same genus. However, the two African ones are really large, as otters go, and the social system of one of them has been studied to some extent. North of the Sahara one finds the Eurasian otter, of which in Africa the social organization has not been studied, but more is known of another *Lutra* further south, which shares its geographical range with the two large *Aonyx*.

The **Cape clawless otter** is the best known in Africa, and the subject of some intensive studies, several involving the use of radio-tracking. Charlie Arden-Clarke conducted the first of these, along the south coast (1983, 1986); apart from radio-tracking, he also detected the distribution of spraints from study animals that had been injected with the isotope zinc-65. With this combination of techniques he found that there were male group ranges, each inhabited by four or more male Cape clawless otters. There was a sharp boundary between male ranges, so it was likely that they were exclusive 'territories' in the sense of defended areas. Arden-Clarke obtained less information about females, but it appeared that female ranges did not coincide with those of males. Two male group ranges extended over 13 and 19 km of coast.

Within the group ranges males were usually solitary (64% of observations), but in a third of encounters the males were in twos or threes (Fig. 5.9), and there were a few sightings of groups of four or five males. Subsequent observations of other researchers were similar, for both freshwater and otters in the sea (Somers 2001; Somers and Nel 2004).

Figure 5.9 All-male group of Cape clawless otters, along coast of South Africa.

Occasionally otter males were seen sharing a large prey with one another (Arden-Clarke 1983), but there was no evidence of males being attached to families or assisting with family care.

In freshwater, Cape clawless otters have much longer home ranges, comparable in size to those of the Eurasian otter. Mike Somers and Jan Nel (2004) found a mean home range for three males of 42 km (maximum 54 km) of river, with a 'core area' (> 50% of all use) of 7 km. The mean area of water involved was 459 ha (core 91 ha), and they suggested that the size of these ranges was determined by the distribution of reed beds (favourite foraging areas) along the river. Female home ranges in rivers were much smaller, with a mean for three females of 17 km (maximum 19 km), involving an area of water of 111 ha. Female core areas averaged only 3 km. The researchers found that the core areas of females did not overlap, but those of males did, and males also overlapped extensively with females. There was no information about the numbers of otters, male or female, sharing a range.

For Cape clawless otters a general picture emerges of group ranges, for males and females independently, although the issue for females has not yet been resolved adequately. Within those ranges there are temporary groupings ('fission–fusion'), with only small numbers involved, and with occasional food sharing, but group members do not cooperate in foraging. There is the possibility that the function of groups lies in predator detection or protection; Arden-Clarke (1983) mentions the presence of sharks, and along African rivers there are crocodiles and various large carnivores.

A close relative is the **Congo clawless otter**, or swamp otter. Almost nothing is known of its social system except that it is usually seen singly, with occasional family parties of mother and up to three offspring (Jacques *et al.* in press; Kingdon 1997; Larivière 2001a).

Throughout the sub-Saharan region, and only in freshwater, one finds the **spotted-necked otter**, of the same genus as the Eurasian otter but with a different social structure. It is only about a quarter of the size of *Aonyx*, which occurs in the same areas. When I studied them on Rubondo island, in Lake Victoria, an early morning would find me on some rocky promontory, watching sometimes seven or more of them foraging in the dense scrubby vegetation along the watery edge of the forest. Usually, they were all scattered within about 100 m and working independently, with great speed winding between branches and rocks, frequently emerging with some small *Haplochromis* fish from Lake Victoria's murky waters. Then, for some reason, all of them would move on, coming together into a tight pack and swimming across the wide forest-lined bay for about half a kilometre before spreading out again into the vegetation along the shore, and starting another fishing bout. Such observations neatly confirmed what John Procter had seen of these animals many years earlier: a compromise between existence in a pack and solitary foraging (Procter 1963).

In both of our studies the otters were diurnal, and usually in small packs. On Rubondo they had a mean group size of 3.2. Occasionally they were alone, and I counted one pack of ten otters (Kruuk and Goudswaard 1990). The local National Parks rangers reported packs of around 20 animals. Some 30 years earlier, along the western shores of Lake Victoria, John Procter had also watched these animals over a much longer period, and he reported some marvellous observations (Procter 1963). He saw only a few solitary spotted-necked otters, but many large packs, which he called 'schools', of up to 21 individuals. The most common school size was four to six otters. Interestingly, Procter noticed that the large groups were usually all males, and the small groups (up to five otters) were females, or a female with offspring.

In nearby Rwanda, Anne Lejeune observed, in the much smaller Lake Muhazi, that spotted-necked otters usually occurred in groups of two or three animals, but there could be as many as 11 (Lejeune 1989). So the groups she watched were smaller; significantly perhaps, there were no crocodiles in that lake (Lejeune and Frank 1989), whereas Lake Victoria has many and large ones, as potentially important predators. Procter's and my observation of otters in pack formation when crossing large stretches of open water could suggest an antipredator function of this grouping behaviour. None of these studies could produce any estimates of the size of home ranges, or the pattern of overlaps between individual ranges.

In Natal, South Africa, spotted-necked otters behave quite differently. There they are nocturnal, and are found in streams and ponds in the absence of crocodiles, usually singly but in groups of up to five (Rowe-Rowe 1992). A radio-tracking study by Ilaria Carranza in the Drakensberg area, Natal, showed that ranges of several males were almost identical and these males often moved around together (Perrin et al. 2000). Ranges of females were not completely known, but they were smaller, and either quite separate from one another (overlapped by male ranges), or a range might be almost the same as that of another female, but the two would always avoid each other (presumably using different core areas). However, females were often seen together with other, unidentified, animals.

In summary, the social organization of *Lutra maculicollis* is obviously quite variable, and there may be a dichotomy between populations from eastern and southern Africa. A common denominator is the existence of male territories and male groups, with sometimes quite large numbers involved, and there is a suggestion of female territories with exclusive core areas in southern Africa, and of female groups in eastern Africa. There is no evidence of cooperative foraging or food sharing, but the presence of crocodiles may be a decisive factor in the formation of large groups.

Social systems of otters: general comments

For a monophyletic group of mammals (derived from a single ancestor), of species with usually remarkably similar ecological niches, there is a surprising amount of variation in their social organizations, from a solitary existence to organized groups, to 'rafts' of hundreds of individuals, and variations on the theme of territoriality thrown in (Fig. 5.10). There is a great deal missing in our knowledge of their sociality, but some generalizations are possible.

If we compare what little we know of social systems, most of the variation between species appears to be unrelated to their taxonomic positions. Socially, species of *Lutra* or *Aonyx* have less in common with others of the same genus than they have with quite different otter species, and species that are closely related (Bininda-Emonds et al. 1999; Koepfli and Wayne 1998; Van Zyll de Jong 1987; see Chapter 3) may show large social differences. But both the giant otter and the sea otter, which are taxonomically distant, also each have a quite unique social system, with packs centred around a female plus a male, and large aggregations, respectively.

All otter social systems are variations on a theme of female territories and independent, larger male territories, the classical mustelid system (Powell 1979). This is different from, for instance, the canids, where social organizations are derived from a system of pair territories (Macdonald and Sillero-Zubiri 2004). Moreover, there is considerable similarity in the size of otters' home ranges, in the lengths of shore-line that they occupy, which are mostly of the same order of magnitude. Where social systems vary is in individual cooperation, in the formation of family groups or packs, in the retention of offspring from previous litters, in the presence or absence of packs of males, and in features such as 'rafts'. This variation, sometimes within species, sometimes between closely related species with at first glance rather similar ecology, suggests either that their niches are more dissimilar than appears, or that food resources are not as important as we think, in shaping these aspects of social organization.

One variable of which we do not fully understand the biological function is grouping, or pack formation, especially of males and in diurnal species. In these packs, cooperative foraging or cooperative territorial defence is rarely important, so why do animals join? I suggest that the most likely function of these groupings is anti-predator protection, against crocodiles, sharks, killer whales or terrestrial carnivores (see Chapter 12). Grouping of potential prey animals has a confusing effect on a predator (Kruuk 2002), and enables effective alarm signals between participating individuals. It is difficult to demonstrate in animals such as otters, but this hypothesis would explain, amongst others, the absence of pack formation in otters in Europe, and the occurrence of packs in dangerous situations.

When we try to understand the reasons for the differences in social organization of various otter species, we have to evaluate many aspects of the biology of these animals. Sociality is affected by the kind of resources on which they are dependent, by the advantages of cooperation and disadvantages of

72 Otters: ecology, behaviour and conservation

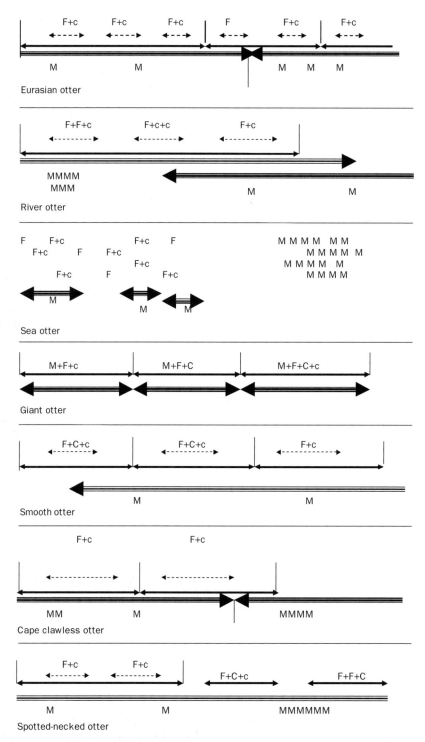

Figure 5.10 Summaries of spatial organizations of seven species of otter, shown as solid lines (home ranges of females, with or without cubs), triple lines (home ranges of males) and dotted lines (core areas of females), and individuals grouped or solitary. F, female; M, male; C, large cub; c, small cub.

competition. One has to consider the advantages and disadvantages to individuals of postponing dispersion, the occurrence and effects of inbreeding, and of reproductive suppression. I discuss these issues in connection with sociality in Chapter 13, after having considered data on resources, exploitation and populations.

The social organization of otters is important, not only because of the scientific interest in its evolution, causes and effects. When designing conservation management strategies, such as the areas of waters to be protected, the proximity of fish farms to wild populations of otters, numbers and status of otters in reintroductions, as well as other aspects, detailed and quantitative knowledge of otter society is vital (see Chapter 14).

Otter dens, or 'holts'

Even in otter societies where several adults share a home range, it rarely, if ever, happens that adult individuals simultaneously use a 'holt' (den), unless they are moving around together in a group. They clearly avoid having to share a bed, even if the holt is a sizeable one, and their scent marking behaviour helps in this (see Chapter 6). In group ranges, holts are used on a time-share basis.

There are large differences between the resting sites of otters of various species. They vary from extensive, complex tunnel systems with bedded chambers and even a 'bathroom' of sorts, to simple holes, or above-ground couches of vegetation. The sea otter does without a den altogether. We made some detailed observations on the denning habits of Eurasian otters in Shetland.

Within the system of home ranges along the Shetland shores, the holts of **Eurasian otters** play an obviously important role, much more so than in other, freshwater areas (see below). Otters use them to sleep in at night, but they are essential also for other purposes. Little was known about these holts along sea shores, apart from a few casual observations (Harvie-Brown and Buckley 1892; Kruuk and Hewson 1978), until Andrew Moorhouse did a detailed analysis of their structure, location, spacing and use in our Shetland study area (Kruuk 1995; Moorhouse 1988). He found that otters usually constructed their own holts, digging extensive systems of tunnels and chambers, and providing them with bedding, although sometimes they also used existing caves and tunnels, or rabbit warrens.

Moorhouse excavated nine holts in the Shetland peat layers, reinstating them again afterwards, and he measured and mapped the system of tunnels and chambers (Fig. 5.11). Some of these holts were used by the otters as natal holts, where females gave birth to cubs (see Chapter 6), but Moorhouse excavated them only at least 1 month after the family had left in order not to disturb them. Once a natal holt had been abandoned, the otters rarely returned until the following year. Natal holts frequently were far from the sea, sometimes 1 km or more, and difficult to find, because in contrast to ordinary holts they show no spraints, hardly any clear pathway in, and a very unobtrusive entrance. This seclusion may be related to male cannibalism of cubs (see Chapter 6).

Main entrances to all Eurasian otter holts have a characteristic shape, being quite large and wider

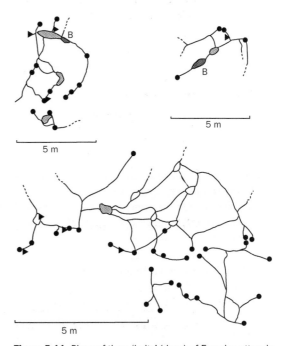

Figure 5.11 Plans of three 'holts' (dens) of Eurasian otters in Shetland. ●, Entrance; ▼, underground latrine; shaded areas, sleeping chambers with bedding; B, bath chamber. The largest of the three holts was abandoned at the time of excavation, and no fresh water was present. (After Moorhouse 1988.)

than high (on average 27 cm wide, 17 cm high). This is different from the burrows of rabbits, which are also common in Shetland; they have a circular entrance. Regularly used otter holts also show a characteristic smoothly worn entrance, and often one finds tracks and/or faeces ('spraints'; see Fig. 4.1).

The total length of tunnelling inside the holt may be considerable—in Moorhouse's sample up to 51 m—but several holts we saw must have been much larger still. The usual length of tunnel was between 10 and 20 m, extending horizontally at a depth of commonly about 0.5 m under the surface (see Fig. 5.11). The chambers could be just widened parts of a tunnel, or at the end of special side-tunnels, and they were lined with bedding, often masses of it. Otters used heather (*Calluna vulgaris*) for this, or grasses, and holts close to the shore-line were often lined with fresh sea weeds, especially knotted wrack (*Ascophyllum nodosum*). Several times we watched otters taking bedding in, after first biting off the vegetation, then carrying it into the holt in their mouth; on one occasion Moorhouse observed a male dragging bedding in by taking it under the chin and on his forelegs, walking backwards into the holt entrance—exactly like the Eurasian badger *Meles meles* (Kruuk 1989; Neal 1986). Many of the chambers in excavated holts had large sheets of plastic as bedding, such as the fertilizer bags that litter the Shetland shores, and I saw otters drag them in. We may object to this horrible evidence of civilization along the beaches, but otters took to the plastic culture with gusto.

All the holts excavated by Moorhouse that were used by otters at the time had pools of freshwater inside, often quite far from the entrance and invisible from the outside. These pools could be connected with small underground streams, or they could contain water that collected in a hollow in an impermeable layer below the peat, and they were very important as built-in 'bathrooms' (see Chapter 10). Otters abandoned holts in which the water dried up.

Most holts were found close to the sea coast; in the intensive study area, 77% of 112 holts were within 100 m of the shore-line, and only 13% were more than 200 m inland. The inland holts were more difficult to find, so Moorhouse focused on holts within 100 m of the sea. There was a significant difference from random in the distribution, the holts were clearly clustered, and in almost all areas they were at or near sources of freshwater—often clearly 'wet', with muddy entrances, or with water visible within. In a large otter survey of the whole of Shetland we found that 61% of 1-km coastal sections in which there was an otter holt also contained open freshwater, compared with only 34% for sections without otter holts—a significant difference (see Fig. 4.4; Kruuk *et al.* 1989). There was a close association between holts and wet, peaty coasts (see Fig. 4.3); small islands, too, were inhabited by otters, but only if sources of freshwater were present. It appeared that this association of holts with freshwater (on the surface, or underground) was an important reason why we found so few signs of otters near agricultural areas.

In every home range of otter females there were many holts that were frequently used by the animals, and we often saw them go from one holt to another, especially when they were accompanied by cubs. There was a clear correlation between the number of resident females inhabiting an area and the number of 'active' holts there; using observations from the Lunna study area as well as several small islands (Moorhouse 1988), we found a good linear relationship (Fig. 5.12). In any one stretch of Shetland coast, there were about three times as many active holts as there were resident female otters.

Andrew Moorhouse distinguished 'main holts', 'subsidiary holts' and an intermediate category,

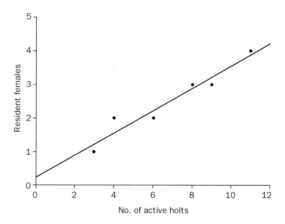

Figure 5.12 Number of active dens in group ranges, compared with numbers of resident female Eurasian otters present in the ranges, in Shetland. There were about three dens per female ($r = 0.97$, $P < 0.001$). (From Kruuk *et al.* 1989).

based on frequency of usage. Main holts were in use on at least 60% of his visits during the one and a half year study, subsidiary holts on fewer than 30% of occasions. The main holts were significantly further away from the shore (median distance 90 m) than subsidiaries (median distance 30 m); we did not have an obvious explanation for the difference. There was a striking difference in usage by different social categories of otters: independent, sub-adult otters and non-resident, subordinate otters were never seen using main holts—they always stayed in inferior places (see Chapter 6).

In freshwater areas the animals behave differently. While Eurasian otters along the Shetland coasts appeared to be dependent on elaborate, specially constructed holts, their freshwater conspecifics were much more casual about it. There are many descriptions of otter holts along streams, often between tree roots, and with underwater entrances (e.g., Chanin 1985; Mason and Macdonald 1986), duly provided with spraints by the animals. But, although otters often visit such places and spraint there, they rarely sleep or rest in them; at least, the animals that we radio-tracked did not.

In fact, on loch shores Eurasian otters almost never rested or slept underground, but always in the reeds or thick vegetation. They slept on 'couches' (see Fig. 4.8) or just curled up somewhere, even in mid-winter, regardless of the frequent misery of the Scottish climate. Reeds appeared to be the favourite bank habitat for otters, but here in Scotland also rhododendron, which occurs as a weed in many areas, is frequently used. Couches have been described in detail by Hewson (1969). Measuring 0.3–1 m across, couches are made by otters biting or pulling off vegetation, and then dragging it over some distance before flattening it to curl up in. The main purpose of the couch appears to be to provide a dry base. As a considerable refinement, we found that, on the Dinnet lochs in north-east Scotland, breeding females made special covered couches, consisting of a large ball of reeds, somewhere on a dry spot in a large reed bed but often close to water, with one or two entrances at the sides, and a lining of soft grass—comparable to nests of birds such as magpies (Taylor and Kruuk 1990) and first described by Buxton (1946). One of our radio-otters gave birth to two cubs in such a natal couch, which was about 1 m in diameter, 0.5 m high and about 2 km away from the nearest open water where the other otters were living.

Along rivers and streams otters were more inclined to sleep in holes, proper dens, than they were at the lochs. Even so, these holts were less important in that habitat than on the sea coasts and, especially if there were small islands in the river, otters appeared to prefer sleeping above ground on those islands rather than in a bank-side hole. It also struck us that, when otters did sleep underground, this was often in some human construction, a heap of boulders or an artificial road embankment. Upturned roots of windfall trees, often quite far from water, were popular daytime sleeping sites, and one male otter frequently slept in the wreck of an old car (a Vauxhall) that was half submerged along a very small stream.

The fact that, along the rivers, proper holts were few and far between was probably largely because the otters had little use for them, not because they could not make them. Shetland otters had shown us that they have no problem in digging quite large holts themselves, and along rivers potentially otters could have enlarged one of the many rabbit warrens. However, otters often prefer to rest above ground, some individuals more than others, as a purely individual preference. For instance, observations on two adult males that we had followed along the River Dee in 1990 showed that one spent 23% of 84 days in holts, and the other 37% of 214 days; the rest of the time they were above ground. This was a significant difference (Kruuk 1995). The observations on both otters were spread over different seasons, and both were more likely to be in holts during the winter (24% and 45%) than the summer (17% and 8%).

Wherever there are beaver lodges (holes in the bank, or large constructions of sticks and branches) and beaver dams available, especially in central, northern and eastern Europe, otters are highly likely to use these as holts, irrespective of the presence or absence of the original owners. In Belarus and Estonia I found otter spraints in and around many beaver lodges and, although no radio-tracking data are available, at least during the winter otters were sleeping inside (see also below for the North American river otter).

Holts along stream banks were relatively simple affairs compared with the Shetland holts. In freshwater areas holts usually had just a single entrance, within a few metres of the water, and more often than not we found them when radio-tracking in places that otherwise we would have overlooked. Holts could be in open, grassy banks or in woodland or scrub; there was little obvious selection of habitat. We never found a holt with an underwater entrance, which has so often been described as 'typical' in the literature.

The use of dens by the North American **river otter** *Lontra canadensis* is very similar to that of *L. lutra*: dens were not important in freshwater habitats. Also, dens were never dug by the otters themselves (Liers 1951; Melquist and Hornocker 1983). This in contrast to dens of the same species along the Pacific coast, which are often enormous, with tunnels much larger than those of *L. lutra*, deep and with many entrances, clearly at least partly constructed by the otters themselves (personal observation; Larsen 1983; Noll 1988; Woolington 1984). The actively used otter dens that I found along the coasts of Prince William Sound were in old coastal forest, usually on steep slopes within 50 m of the sea, and with access to freshwater inside. The otters left many spraints in and between the entrances, and radio-tracking demonstrated that these were the main, probably the only, sleeping sites (G. Blundell, personal communication).

In freshwater areas of boreal Canada, radio-tracking of river otters established that beaver lodges provided ideal dens that were frequently used by otters (Reid *et al.* 1994b), confirming observations by Melquist and Hornocker (1983). These last authors found as many as 88 different holts and resting sites along rivers, used by one single otter over a 16-month period. Their animals used beaver dens in banks in 32% of their observations, logjams and beaver lodges in 24%, and dense vegetation in 11%. Liers (1951) found river otters denning in burrows of woodchuck (*Marmota monax*). Just like the Eurasian otter, the river otter may also construct large, covered 'nests' of vegetation with a sideways entrance, used as natal dens (Johnson and Berkley 1999).

In Latin America **giant otters** have several different dens in use (Fig. 5.13) within their home range, in which they frequently spend the night. The **neotropical otter** also uses many different shelters, including caves and cavities between rocks, and tunnels which they dug themselves; along 7 km

Figure 5.13 Adult male giant otter at the entrance of a den in the river bank. © Nicole Duplaix

of river, 108 such places were identified. These places were frequently used by the otters, and they left faeces, as described above for the Eurasian otter (Pardini and Trajano 1999). This species also constructs covered couches (Harris 1968). The **marine otter**, along Chilean coasts, uses dens between large boulders and in clefts of rocks, some of them with an underwater entrance (Ostfeld *et al.* 1989).

The **giant otter** does not always rest and sleep in dens overnight; it also uses the 'campsites' for this, but much more rarely. Dens are large, immediately along the edge of water, with an entrance up to 1.5 m wide, leading into tunnels 40–50 cm wide and 30–40 cm high, typically some 3 m long and entering a chamber of some 1.5 m in diameter, without bedding. Some tunnels may enter from under water, and some dens have an additional rear entrance, leading into the forest. Outside and close to the entrance is a latrine, and it is often the smell of the latrine that gives the den away. There were at least 21 dens that had been used at one time or other along the 12-km study creek of Nicole Duplaix (1980).

Similar to Eurasian otter, in Africa the **Cape clawless otter** has many different holts along coasts, each used by several males. In one 13-km stretch, Arden-Clarke (1983) found 15 holts, all used by the same animals. These, however, were not undergound structures but tunnel systems in very dense and bushy vegetation, and always close to a source of freshwater (personal communication). Along rivers, the Cape clawless may make its den inside large accumulations of debris, or in dense, impenetrable vegetation. I found the **spotted-necked otter** based deep inside large rock formations along the shores of Lakes Malawi and Victoria, but was unable to assess the number of such holts. In India, the **smooth otter** mostly dens in crevices between rocks, usually on islands in rivers, but sometimes it also digs its own holt. A family group moved between nine such holts over almost 2 years, but only during the cold season; the rest of the year the animals rested above ground (Hussain and Choudhury 1995).

In summary, the use of many different holts was evident in most species of otter, and they also spend a lot (if not most) of their time resting in the open, in couches or somewhere in vegetation. Their behaviour shows that they are not dependent on just one or two individual holts. This demonstrates a remarkable difference with, for instance, the Eurasian badger, which almost always sleeps in the single same den throughout the year (Kruuk 1989). In badgers this combines with a strategy of 'central place foraging', a strategy that is suggested as optimal when dealing with unpredictable resource availability. In otters the availability of prey populations is likely to be different (see Chapter 8), and the animals are therefore less dependent on one central, focal resting site.

The observations on otter holts have significance also in a conservation–management context. In Britain, attempts have been made to facilitate the existence of Eurasian otters along rivers by building artificial holts, made of logs or concrete; the above observations suggest, however, that the animals are unlikely to be in need of such human-made structures.

CHAPTER 6

Scent marking and interactions: social behaviour

Introduction

One of Niko Tinbergen's important contributions to biology (1963; see also Kruuk 2003) was his insistence on the analysis of animal behaviour in terms of four main frameworks. These are *causation* (immediate internal and external influences that produce the behaviour), *ontogeny* (its development during an animal's lifetime, through learning, environmental and genetic effects), *function* (the consequences for survival) and *evolution* (its evolution in a species). These frameworks pose different kinds of question for the same behaviour (Tinbergen's 'four whys'), and they should be developed in concert. Spatial organization, as discussed in the previous chapter, can be approached similarly, as a behavioural strategy. Here I will be concerned with its causal aspects—some of the mechanisms and behaviour patterns that are important in the maintenance of social organization.

For each of the species we should be asking what strategies are used to maintain its social system. How do the otters react to one another, how do they disperse, what actually happens in the field? Second, one wants to know about variability of behaviour, and whether otter behaviour is optimally adapted to maintenance of the spatial system, and to the pattern of exploitation of resources.

Only a few species have been studied in any depth, and in the solitary ones it is a fairly rare occurrence to see otters face to face with one another (except for mothers and their offspring), even in areas such as Shetland where several Eurasian otters use the same stretch of coast. In my observations the animals appeared to be very adept at avoiding confrontation, a phenomenon in which the otters' system of scent communication must play a dominant role. Over the years in Shetland, I saw almost 100 interactions apart from those within family groups, and they gave at least some idea of how complicated and subtle the relationships are. In other, more gregarious, species more is known about social behaviour in the field, but mostly about parental and within-group relationships. Finally, there are several helpful studies on the behaviour of different otters in captivity. In the following pages I will describe and quantify some observations that are related to the otters' pattern of spacing, including aggression and territoriality, and their sexual and parental behaviour.

One immediate difficulty that we encounter is that much of the communication between these animals is by olfaction, by scent and smell. There are remarkably few visual displays and calls, and the same was found in other, more social, mustelids such as Eurasian badgers (Kruuk 1989; Neal 1986), or all the more solitary ones (Sandell 1989). Even the highly gregarious South American giant otter uses few visual signals, but 'their vocal and olfactory communication systems are highly complex and sophisticated, which probably compensates for the lack of visual clues' (Duplaix 1980, p 548). The most visually ostentatious carnivores are probably the canids (e.g. Leyhausen 1956; Tembrock 1957), but in general the Carnivora, and especially also the Mustelidae, make little use of displays (in the widest sense of the word) compared with, for instance, ungulates (e.g. Clutton-Brock *et al.* 1982; Estes 1967, 1969; Jarman 1974) or birds (e.g. Tinbergen 1960).

Possibly, the disadvantages of conspicuous displays are greater in Carnivora than in other groups. In addition, the nature of their food resources and the way in which these are exploited may enable carnivores to get by without elaborate displays (see below). Nevertheless, there are carnivores such as the spotted hyena that are extremely noisy and have many very obvious displays (Kruuk 1972). Somehow, otters manage to do with far fewer, and it is likely that, instead, olfactory communication plays the key role in the spatial organization of otters.

Some of this communication by scent is direct, between individuals. When, in the field, one otter approaches another it will frequently make a detour, then do so upwind, as do many other carnivores, and this strongly suggests that it receives information. As yet, however, we have no idea what it learns from this. When a Eurasian otter, or a river otter (Melquist and Hornocker 1983) is seriously frightened by something—by a person or when it is caught in a trap—it will expel contents of the anal glands, and a strong smell of otter fills the air. There must be many other odours used in social interactions that are wasted on us but carry important information for the animals themselves.

Otters also have several different methods of scent marking; one of these scent communication systems is what in Europe is called *sprainting*, of which the functions could be several. Spraints (scats) are usually the first, best known, and often only sign to naturalists in the field that there are otters around—almost all scent-mark this way, but in some species spraints are more conspicuous than in others.

Sprainting behaviour and other scent communication

An otter spraint is usually tiny, no more than a small dollop of faeces. Both **river otters** and **Eurasian otters** usually produce spraints that are grey or black and tarry, with fish bones, shapeless, sometimes very liquid but, if solid, generally less than 1–2 cm across. Even a human nose can smell the insignificant, fishy-reeking objects several metres away, at least during the first hour after an animal has been there. Spraints are incontrovertible evidence of the passage of an otter, and are all that most people will ever see of the elusive, shy, nocturnal animal. In other species of otter, spraints are often somewhat different, as explained below, yet clearly recognizable.

Spraints consist mostly of food remains, such as fish bones, to which are added the fairly inconspicuous secretions of two anal glands, situated along the gut close to the anus; they produce the 'ottery' smell. Often confused with these anal gland secretions is a jelly-like substance, slimy brown or greenish, which is secreted somewhere in the intestine itself. Sometimes a spraint consists of nothing but this jelly. Captive Eurasian otters produce this jelly when they have not fed for a day or more (Carss and Parkinson 1996), and I assume that at such times the animal is motivated to spraint not for purposes of elimination, but for other reasons.

Eurasian otters (and probably all others, except the sea otter) tend to spraint on vantage points or other striking places, for instance on top of prominent rocks along the water's edge (see Figs 3.4 and 11.2), under bridges and near trees, or at the junctions of tributaries. Along sea coasts many spraints are found near small freshwater pools (see Fig. 10.12), near holts, or on promontories. In or near the entrances of holts there may be accumulations of spraints in large, smelly heaps (see Fig. 4.1), except near natal holts with newly born cubs, which from the outside usually show no sign of otter occupation (this is true also for the river otter; Melquist and Hornocker 1983).

After landing, a Eurasian otter will walk up to a spraint site and sniff it for many seconds, for up to half a minute. It will then spraint and/or urinate, curling its tail slightly upwards (Fig. 6.1), and sometimes sniff again and repeat the process. When we see the animal urinating, we can often confirm whether it is male or female from the direction of the stream or urine: forward if male, backwards if female. Unexpectedly, spraints of males in the wild are usually much smaller than those of females, often no more than tiny droplets of faeces. The spraints of cubs are the largest of all.

The spraints that we find along the shores are not the otters' only defaecations: the animals also excrete frequently when swimming (see below). On land, the same spraint sites are often used for many years, and this may produce piles of up to 15 cm high, especially along sea coasts (called otter 'seats'; Elmhirst 1938). Because of consequently high nitrogen concentrations, many such sites on rocks attract a characteristic flora of green algae, or elsewhere of nitrophilous grasses such as Yorkshire fog (*Holcus lanatus*). The dark green colour of the grass on and around a spraint site makes it stand out, especially in winter.

Because of their location in the landscape, spraints and spraint sites are often conspicuous, and otters go well out of their way to deposit a spraint on just such

Figure 6.1 Female Eurasian otter with two large cubs, sprainting on seaweed at low tide in Shetland.

a prominent place. A Eurasian otter spraint is so small (especially that of a male) that an otter will have to produce many of them each day (dozens) to excrete all its food remains. Together, these different observations imply that spraints have a further purpose: communication by scent (Erlinge 1968; Kruuk 1992; Kruuk and Hewson 1978; Mason and Macdonald 1986).

All species of otter conform to this picture (except the sea otter, which seems to leave a scat on land more or less by accident—it does not scent-mark). Not surprisingly, the spraints of different species show quite a range of variation. The scats of North American **river otters** are very similar to those of the Eurasian species, but somewhat larger and smelling less fishy, more faecal. Especially along the Alaska coast, river otters leave enormous sprainting sites, with large heaps of scats over large areas, outside holts and next to freshwater pools. The vegetation on and around such sites, including the spruce, is clearly affected by the additional nitrogen provided by the river otters, as demonstrated by the presence of stable isotopes from the marine environment; the otters are fertilizing the beach-fringe forest (Ben-David *et al*. 1998).

The **neotropical otter** spraints in somewhat similar fashion to the Eurasian otter, also more frequently in winter than in summer (Parrera 1993, in Larivière 1999b), in and on similar places (Macdonald and Mason 1992), and to me the scats smell similarly fishy. Its scats are often associated with scratch marks against banks of rivers or with 'sandcastles'; most of them are deposited as single markings, and large 'sprait sites' or latrines are unknown in this species (Groenendijk *et al*. 2005). The **southern river otter** uses latrines (sprait sites) near holt entrances and inside holts like *Lutra lutra*, as well as elsewhere some 3–6 m from the shore and 50–80 m apart (Chéhebar 1982, in Larivière 1999a).

Sprainting by **giant otters** is of a different order of magnitude, resulting in large accumulations of faeces: 'their haunts are easily known by a strong and disagreeable smell, in some instances so strong that we increased by all means in our power the speed of the canoes to get out of its precincts' (Schomburgk 1840, in Duplaix 1980, p 581). Apart from sprainting in ones and twos on rocks and vegetation along the water's edges (Fig. 6.2), as the other otters do, groups of this species also establish a number of 'campsites' throughout their ranges (Fig. 6.3) (Duplaix 1980;

Scent marking and interactions: social behaviour 81

Figure 6.2 Giant otter sprainting. © André Bärtschi.

Staib 2002). These are large activity centres, several metres across, on which they rest and spend many hours on end, performing different types of scent marking. There is at least one large latrine on each of these campsites, and through rolling, rubbing and paddling with front and hind feet a porridge of mud, urine and spraint is created. Whilst sprainting, giant otters move their feet about (which may also result in scent marking with the pedal glands), and they wave their tail up and down and sideways, which may function in dispersing scent (Duplaix 1980).

The animals go through bouts of 3–12 minutes of very vigorous scent marking (Duplaix 1980). As a

Figure 6.3 Giant otter scent-marking station—a conspicuous and strongly smelling 'campsite' on which spraints, urine, and secretions from interdigital and other skin glands are deposited. © Nicole Duplaix.

result of these activities, otters and area will share a strong smell—a unique group smell (Staib 2002). Another gregarious mustelid, the Eurasian badger, shows a parallel group smell phenomenon, as members of badger clans mutually scent mark one another by pressing their anal regions together (Kruuk 1989). As another scent-marking activity, giant otters on the campsites will also reach up, standing on their hind legs, and clasp saplings between their front legs, which they then drag underneath their body, often urinating at the same time (Duplaix 1980). For none of these activities do we have any suggestions as to the messages that are being communicated.

Giant otter campsites are scattered along the home range, and are often situated in places where otter paths lead inland, towards distant pools or other resources likely to be group specific. The overall result of the vigorous and intensive marking and scratching activities on the campsites is that to a visiting otter they must be very obvious indeed, even over several hundred metres across wide rivers. Interestingly, spraint sites of giant otters are frequented also by the much smaller neotropical otters that live in the same rivers, and whose spraints are recognizably much smaller (Duplaix 1980).

In other regions where there are several different otter species living in the same waters, such as southeast Asia or most of Africa, one can recognize them from their spraints (Kruuk et al. 1993a). In Africa, I found spraint sites of the **spotted-necked otter** on large flat rocks along Lakes Victoria and Malawi, usually on small promontories and just out of sight from the water. They smell much more strongly than Eurasian spraint sites, often detectable to us more than 10 m away. The spraints themselves are similar to those of Eurasians, but more scattered all over the site, never in heaps (Kruuk and Goudswaard 1990).

Along the same lake shores and river banks (and sometimes on the same spraint sites) one may also find the scats of **Cape clawless otters**, which are more distinctly scat shaped, full of bits of crab and, by otter standards, very large (3–4 cm in diameter). Because of the difference in size of scats, there are far fewer of these in the otters' habitat than we find, for example, in the Eurasian otters' range, so perhaps their scent-marking function is less important. Yet the animals do have conspicuous spraint sites, both in fresh water and along coasts, using them in similar manner as coastal *Lutra* in Shetland, even to the extent of also sprainting at fresh-water pools along the shores of the Indian Ocean (Arden-Clarke 1986; Van Niekerk et al. 1998; and personal observation). But numbers of spraints are far fewer than for *Lutra* (as is the case for the small-clawed).

I collected some interesting observations in Thailand, where **Eurasian**, **smooth** and **small-clawed otter** occurred together, along the river Huay Kha Khaeng (the upper ranges of the 'River Kwai') and its tributaries (Kruuk et al. 1993c). Their main spraint sites were somewhat different from one another, as we discovered from tracks and direct observations, with the last two species usually sprainting high up, well above the water, on large, flat rocks (but not always), and the smooth otter in more prominent sites than the common and the small-clawed. The Eurasian otter sprainted lower down near the water, with fewer spraints per site than the others produced, but in the same type of place as this species does in Europe.

The larger spraints of smooth otters, smelling strongly of rotting fish, were found on boulders large or small, on flat rocks or sandbanks, often on their own or with just two or three of them. But some of their spraint sites were large and reminiscent of the giant otters' campsites, with quantities of faeces characteristically flattened and spread out by smooth otters rolling and rubbing in them. The other two species did not do this. The small-clawed spraint sites were conspicuous because the scats were large (despite the relatively small size of the animal), and because of the many crab remains in their faeces. Especially interesting was that, although there were species differences in the kinds of spraint site they established, they did visit and sniff each other's, as we determined from tracks. They also had a clear overlap in diet and, if these spraints do have a function of preventing competition (see below), then such interspecific interest in spraints is to be expected.

Spraints are of interest to us for several different reasons. First, as in many other carnivores, scent marking with scats is likely to feature prominently in the maintenance of territories. A great deal has been written about this (summaries in Gorman and Trowbridge 1989; Gosling 1982; Macdonald 1980a, 1985), and for species such as spotted hyaenas and badgers much was learned about their territorial systems just from studying the distribution of faecal scent marks (Kruuk 1972, 1989).

The second reason for an interest in spraints is more pragmatic: many research workers have used them to monitor otter populations, to establish trends in numbers and habitat preferences (e.g. Chanin 2003; Crawford 2003; Green and Green 1980, 1987; Jones and Jones 2004; Lenton *et al.* 1980; summary in Mason and Macdonald 1986). There are some reservations about this technique, because of the assumption that more spraints means more otters. If we could verify this, spraint surveys would be a powerful tool with which to assess otter numbers and activity, but more research needs to be done to support this methodology (Kruuk and Conroy 1987; Kruuk *et al.* 1986; see Chapters 4 and 11). Finally, of course, spraints provide the basics from which one can assess the diet of otters, and they provide genetic material enabling individual identification. In research on otter ecology, spraints they are hugely important.

To provide insight into the biological function of sprainting by **Eurasian otters**, I spent some time in Shetland observing actual sprainting behaviour (Kruuk 1992). If spraint contains messages from one otter to others, then who does the signalling, when and where? And to get clues about what makes otters spraint, I asked what do otters do just before and just after they spraint? If it were important in sexual behaviour, one might expect differences in sprainting between males and females, and it would be seasonal in areas where reproduction is seasonal, such as Shetland. If sprainting played a role in group territorial or core area defence ('keep out!'), boundaries might show concentrations of spraints (Gorman and Mills 1984; Gorman and Trowbridge 1989).

Some relevant work had already been done before we started in the Shetland study area, especially by Jim Conroy. He found that numbers of spraints that he collected along stretches of coast of the Yell Sound in Shetland varied widely between his bi-monthly visits, and the use of sprainting sites along different sections of coast often fluctuated quite independently. There was also a huge seasonal variation in spraint numbers, and about ten times more spraints were found per visit in winter than in summer (Conroy and French 1987). A similar seasonality has been found elsewhere, in English rivers (Macdonald and Mason 1987), and confirmed in central Europe (Kranz 1996). Whilst reproduction in Shetland otters was highly seasonal (Kruuk *et al.* 1987; and Chapter 12), in English and continental European fresh water it is not, with otters giving birth at almost any time of the year (Harris 1968; Kranz 1996). If sprainting had anything to do with reproduction, then seasonality of sprainting might have been expected in Shetland, but not along fresh water.

The seasonal differences in numbers of spraints along the shores in Shetland were caused not by Eurasian otters spending more or less time there during winter or summer, but by individual animals actually sprainting more or less often on land in winter (Kruuk 1992). I calculated how often I saw an otter deposit a spraint (see Fig. 6.1) for every hour that it was observed, and also for every time I saw an animal land. On both scores there was a striking seasonality, including a large difference between the summer months and the months when winter merged into spring (Fig. 6.4). Sprainting frequency in March was 12 times higher than in June. This was not because otters produced fewer scats in summer, but because they defaecate when swimming (often

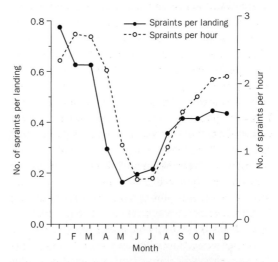

Figure 6.4 Observations of sprainting behaviour of adult Eurasian otters in Shetland at different times of year; seasonality of spraints over time and per landing. (Total observations, 292; $\chi^2 = 117.3$, 9 d.f., $P < 0.001$; curves smoothed as three-point moving averages). (From Kruuk 1992.)

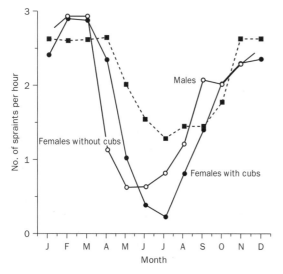

Figure 6.5 Frequency of sprainting of male and female Eurasian otters in Shetland, with or without cubs, at different times of year. There were no significant differences between the three categories; curves smoothed as three-point moving averages. (From Kruuk 1992.)

easy to see, as they lift their tail out of the water when they do so).

All categories of otter—males, solitary females and females with cubs (see Fig. 6.1)—behaved quite similarly, and I found no significant overall difference between them in rates of sprainting (Fig. 6.5). Their seasonal patterns were somewhat different, with males apparently sprainting more than females in summer and less in winter, but these differences were not significant.

Sprainting was therefore unlikely to be related to sexual behaviour or reproduction. Was it some simple territorial behaviour? Most probably, it was not. Three territorial boundaries were accurately known, and the sprainting rates in the sections of coast at either side of these boundaries proved to be no different from sprainting rates anywhere else. In the 350-m sections near boundaries, I saw 5.2 spraints deposited per section, whereas overall along the coast the average was 6.3 (Kruuk 1992). Expressing it differently: near borders otters deposited an average of 0.26 spraints each time they landed, elsewhere 0.37. None of these differences was statistically significant. I was surprised about this pattern of faecal scent marking in the Eurasian otters, because it was very different from what I had seen in other carnivores, such as Eurasian badgers or spotted hyaenas, which cram their latrines along the borderlines of the clans (Kruuk 1972, 1978, 1989; Mills 1990).

Whilst badgers, hyaenas and many other carnivores are highly aggressive against intruders, and back this up by scent marking along borders and elsewhere, Eurasian otters are far less aggressively active in their defence. It has been argued convincingly by Gosling (1982) that, in general, scent marks are effective in territorial defence especially at the time of aggressive encounters between intruders and territory owners. During an encounter, an intruder can identify an owner after matching the latter's scent to the smell at the borders and within the territory. This would render escalation of such an aggressive encounter into a full-scale fight less likely: the intruder desists after assessing the extent of the owner's 'investment' in the territory.

Gosling's 'match hypothesis' appears plausible, and explains many features of scent marking. However, judging from the distribution of spraints in the case of Eurasian otters, it may well be that either the match hypothesis does not apply to otter sprainting or, more likely, that the spraints have another and possibly more important function.

An alternative (and more likely) function of sprainting by Eurasian otters is that the scent marks serve as signals not between, but within otter group territories—in fact between any otters fishing along the same coast (Kruuk 1992). With spraints anywhere along the water-line, otters may simply signal to others that they are, or have been, feeding in a particular site or stretch of coast or stream. For such a mechanism to be effective, it is not necessary for otters to be able to recognize individuals from their spraints, although it is known from experiments in captivity that they can do so (Gorman and Trowbridge 1989).

Individual Eurasian otters are creatures of habit; they come back to the same feeding sites (see Chapter 9) or to the same small washing pools or holts (see Chapter 10) day after day. It will be advantageous for a second otter, arriving later, to go elsewhere, because the site will have been used, or is in use, by another otter who has the advantage of prior knowledge. If this second otter does go somewhere else, the first otter will benefit because it can then forage without competition, or can return to that site before long (often the next day) and should find the resource in better shape if no other otter has been there in the meantime. The same argument could apply to sprainting next to a small freshwater pool along the coast, or fresh water in a holt (see Chapter 10): if many otters were to wash in it, the water could become too salty or dirty. An otter advertises by sprainting that it has been bathing there, which might induce others to keep out, to mutual benefit.

The observation that all Eurasian otters—males, females and juveniles—spraint at more or less the same rate, and that the animals would spraint anywhere in their range where their resources could be, is consistent with the idea that sprainting means advertising the use of a resource. Also consistent is the observation that in Shetland many spraints, about 30%, are deposited in the intertidal area, so they can function for only a few hours at the most before the tide sweeps over them.

If this comparatively simple explanation, of spraints preventing competition, holds true, one could expect that otters would spraint (a) near places where they were fishing (i.e. before, during or after feeding bouts) and (b) near other resources where competition between individuals could arise. A logical consequence of this hypothesis would be that (c) sprainting should occur especially at times when resources were scarce. In addition, one would expect to find that (d) otters exploited repeatedly a system of patchy, replenishing resources in their home range, and derived some advantage from persuading other otters not to use, for instance, a food patch after they had fed there themselves.

To test these ideas, I noted what otters were doing immediately before and after they sprainted (Fig. 6.6). There was no doubt that sprainting was closely associated with fishing or eating. In 66% of the 331 occasions in Shetland that we saw otters sprainting, it was either preceded or followed immediately by a feeding sequence. An otter's feeding bout, that is, a single foraging session in any one area, was always preceded by sprainting, but this occurred somewhat less often during or after a feeding bout (Fig. 6.7) (Kruuk 1992).

In support of the hypothesis, I found that the striking seasonality in sprainting behaviour in Shetland (see Fig. 6.4) coincided with the annual fluctuation

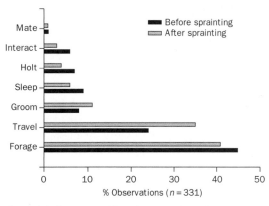

Figure 6.6 Behaviour of Eurasian otters immediately before and after sprainting in Shetland. A close association with foraging can be seen. (After Kruuk 1992.)

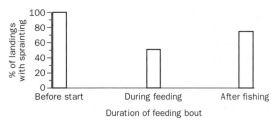

Figure 6.7 Sprainting of Eurasian otters during foraging in Shetland. The otters always sprainted before the start of a spell of fishing along the coast; they sprainted intermittently at landings during and after fishing. (After Kruuk 1992.)

in food availability (Kruuk et al. 1988). There was a peak in numbers of potential prey in mid-summer (when there are fewest spraints), and a trough in winter and spring (the times of most spraints). The same appears to be true in freshwater ecosystems, where productivity in fish populations is related directly to water temperature, and fish biomass is, therefore, highest at the end of the summer and lowest in spring (Kruuk et al. 1993a; and see Chapter 8). Spraint densities fluctuated correspondingly (Erlinge 1967, 1968; Jenkins and Burrows 1980; Mason and Macdonald 1987; Murphy and Fairley 1985).

For the hypothesis on the relationship between spraints and resource utilization to hold, we needed to demonstrate that otters exploit their resources in such a way that it would be beneficial for them to signal this use to others. I have no doubt that they know every nook and cranny in their regular range.

We have shown that otters use exactly the same feeding sites again and again (see Chapter 9), and that coastal sites that were cleared of fish were repopulated with potential prey within 24 hours (see Chapter 8). It would make sense, therefore, for an otter to signal to others to keep out because it would be detrimental for the otter to arrive at a known food patch after someone else has just depleted it. Just as important, it would make sense for the receiver of the signal to go elsewhere, because the chances are that another otter has just denuded the place of its resources.

With this system of signalling with spraints, otters possess a mechanism whereby they can partition resource utilization within a group territory without direct confrontation. It is a system that is well adapted to the use of any resources that a carnivore exploits with some prior knowledge and experience, and the repeated utilization of the same prey sites is a prime candidate. It may well be that many other species of carnivore use scent marking in a similar strategy.

However, we should keep in mind that, apart from the role in resource partitioning, sprainting have the potential for carrying many other messages. We may have to admit that their exact information content will not be known to us for a long time to come.

Apart from sprainting, all otters (except the sea otter) also use other means of scent marking. One of these involves urination, which must be very important for signalling. The animals often urinate on sprainting sites, but, again, we know little about the communication significance. Quite likely it will transfer information between the sexes, at least sometimes; in dogs, for instance, urine from females has long been known to flag oestrus status (Beach and Gilmore 1949), and both cats and dogs detect gender from urine (Dunbar 1977; Verberne and de Boer 1976).

Eurasian otters also rub their cheeks against stones, they roll on special rolling sites, males scrape up small heaps of sand ('sandcastles') or vegetation at, or close to, sprainting sites, or anywhere along sandy river banks, and they often sprain on top. Scraping probably leaves secretions of inter-digital scent glands. All such behaviour patterns are means of scent communication, as in other carnivores (Gorman and Trowbridge 1989; Macdonald 1980a).

Different from the Eurasian species, the male North American **river otter** has clearly distinguishable pedal glands on the pads of the inner toes of the hind paws (unpublished observations). As they occur only in one sex, the function is likely to be sexual, and secretions will be left wherever the animal walks on land, or on the striking 'moss castles' that they make near the holts and sprainting sites. The nature of the message is as yet a mystery.

River otters leave their scats in quite similar places as do the Eurasians, next to their dens, near freshwater sites along sea coasts, at junctions of streams, near trees or other conspicuous obstacles. It is likely that their message is similar, functioning to prevent competition for resources of various kinds. Researchers of captive river otters have suggested that sprints communicate he social status of males, but these observations were done out of natural context, and with old scats (Rostain et al. 2004).

Aggressive behaviour

Any system of exclusive home ranges of animals is likely to be based on some kind of aggressive interactions between the inhabitants. This has been demonstrated for simple territories of solitary animals (Sandell 1989), for clan territories of spotted hyenas or badgers (Kruuk 1972, 1989), and for many others. I had expected it also for female group territories of Eurasian otters. However, at least in **Eurasian otters** in Shetland, such aggressive interactions were remarkably rare, and despite many years of observation our information on aggressive otter behaviour in the wild is still poor. When clashes do occur, however, the consequences may be serious.

Really hard fights I saw only between males, and, conversely, whenever males met they almost invariably fought (eight of nine observations) (Fig. 6.8). The following is an account of such an interaction:

January 1986 in Shetland, in the fading light of a dull afternoon, about four o'clock. Following a large male otter that swims along the coast, I see him land on a small rocky island, just out of sight between some large boulders. Within minutes, another otter arrives, from the same direction as the first—it must have been following, without me being aware of it. The other one is also a large male, swimming about five metres off-shore, characteristically with his tail showing conspicuously along the surface. He passes the place where the first one landed, clearly gets wind of the other's presence, suddenly switches direction, and lands. A fleeting sniff of each other's faces, then a loud, high-pitched screaming and 'wickering', from both animals. A lunge, followed by an incredibly fast chase, off the rocks, into water, on to land, up the slope directly in front of me. Screams, the fleeing otter is overtaken, bitten hard in the rump. A turn-around, again a chase in that curious lolloping gait, though very fast. This all takes some seconds only, then the totally preoccupied animals run straight up to me where I sit against a peat bank. They only notice me at almost touching distance, the fleeing otter exploding out of the way and back into the sea. The pursuer jumps back, stands about eight metres away, watching the other go, occasionally glancing towards me, uttering a soft wickering noise. Then he, too, runs down and into the sea again, and both disappear.

There is evidence that such aggression between individual Eurasian otters plays an important role in their ecology, and is the cause of significant mortality. In post-mortem analyses of otters found dead by researchers and the general public, Kruuk and Conroy (1991) reported from Shetland that 4% of 113 deaths were caused by bite wounds. In a more detailed study in south-west England on 198 carcasses (mainly traffic victims), Simpson (1997) and Simpson and Coxon (2000) found that 23% of males and 13% of females suffered bite wounds from other otters, and altogether, just as in Shetland, 4% of these animals had died as a result of such injuries.

Clearly, however, deaths from intraspecific aggression were seriously underrepresented in both of these samples, because otters that die in traffic are much more likely to be found and collected than animals that curl up somewhere quiet. Simpson and Coxon (2000) argued that the bite wounds they found on the bodies of Eurasian otters were likely to be product of intraspecific fighting, because the placing of bites on the body is different from that in other carnivores such as dogs or badgers: on the feet, face (cheeks, jaws), scrotum, around the anus, occasionally on the shoulder or hock. In addition, the wounds were never infected with *Pasteurella multocida*, which is characteristic of dog bites, but with several species of *Streptococcus*.

In the field I noticed that, between Eurasian otter females from the same group range, there sometimes was some animosity, but rarely more than a few vocalizations, which resulted in mutual avoidance (see Fig. 6.8). Only once did I see an encounter between known females from different group ranges in Shetland, and that was a much more aggressive affair, with loud screams and a chase. Possibly, therefore, females from strange territories are as belligerent towards one another as males are, but they usually just avoid rather than confront each other.

Even between male and female there can be some aggression, especially from the female, but I never saw actual fights. On mainland Scotland in 1991, a wild, large male Eurasian otter broke from the outside into the enclosure of our captive old female at the Institute of Terrestrial Ecology in Banchory, by climbing over the high wall. In this set-up, where the female could not avoid confrontation, the two had a vicious fight in which both were badly injured on legs and face, especially the old female.

Despite the numbers of observed injuries, however, in our field observations aggressive encounters

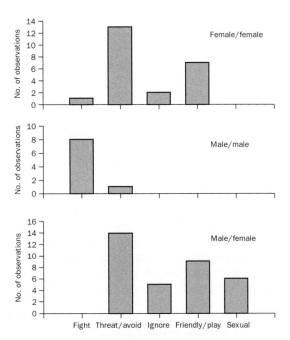

Figure 6.8 Interactions between Eurasian otter males, females, and males and females, in Shetland. Males were significantly more aggressive to one another (eight observations) than to females (34 observations), and more than females to one another (23 observations) ($P < 0.05$, Fisher's exact test), but even between males and females there is some aggression. (Data from Kruuk and Moorhouse 1991.)

between Eurasian otters were relatively rare (except within families), and this was not for want of numbers of animals along rivers and coasts. Somehow, potentially hostile otters manage to stay out of each others' way without actually meeting, quite possibly a result of their sprainting habits.

Some threat postures have been described for otters held in captivity, and they are more or less common to all species (Duplaix 1982). One rarely sees such behaviour in the wild. There was one type of behaviour of Eurasian otters that I saw fairly often in the sea as well as in freshwater areas, and which could be classified as a visual signal directed at other otters. This was the, probably quite aggressive, posture in which large resident males 'patrol' the coasts of their territory. I termed it 'strutting', although the animals were, of course, swimming. They made themselves quite conspicuous, keeping parallel to the shore some 5–10 m out, and exposing more of their body above the surface than otters normally do, especially their tails. Usually when a male swam like that he would not be feeding, but would cover quite long distances, landing here and there for a spraint. **Sea otters** behave similarly: 'Territorial males made highly visible patrols around their territories, involving vigorous grooming and splashing, presumably to deter intruders' (Garshelis *et al.* 1984, p 2651).

For **Eurasian otters**, it is during aggressive interactions that one most often hears various vocalizations. If an animal is cornered by another during a fight, it will produce a very cat-like *caterwauling*, but before things get as bad as that there are calls better described as *wickering* or *chittering*, which are high-pitched, rattling sounds, often grading into a high wailing. Those are the sounds of a young intruder that has been surprised by a resident, or of a rough playfight between cubs. But none of these calls is as loud as those that a noisy cat or a fox can produce. In general this species is much more silent than the South American **giant otter**, which was studied in detail in Surinam by Nicole Duplaix (1980).

Pteronura avoids encounters with 'strange' giant otters, and Duplaix did not see any fighting in hundreds of hours of observation. There, the animals manage to avoid strangers not only by use of scent marks, but by responding to the sounds of screams, wails, barks and explosive snorts that accompany the troops of giant otters everywhere on their forays along rivers in the rainforests. However, giant otters will fight during territorial clashes in high-density areas (Schweizer 1992, in Carter and Rosas 1997). Usually, after hearing another group approach, giant otters will make way before visual contact is made.

Vocal communication in this noisiest of otter species consists of an elaborate system of messages, with many intermediates between the various calls. The system allows for an almost infinitely detailed communication between the animals. Most often, one hears the sharp *Hah!* sound, a low-level alarm signal: '. . . it usually stimulates alertness in other group members who will momentarily stop their activities to look for the source of alarm or uneasiness. While surfacing an otter may Hah! before submerging, or a parent may give it when re-entering a den . . . if the level of alarm increases a *snort* (alarm), a *long-call* (contact), a *hum* (reassurance) or a *growl* (threat) may be given . . .' (Duplaix 1980, p 552). This author described many other loud vocalizations used by the giant otters, including the *wavering scream* during bluff charges towards people, the *hum* which is a contact sound between animals travelling together, the *coo* as an excited friendly greeting, a loud *whistle* between groups, and many intermediates.

Most remarkably, giant otters also communicate under water with low-frequency sounds, as demonstrated in captive animals (Schenck *et al.* 1995). When under water, they can also hear the calls from group members on the surface. Such elaborate communication may well be an essential prerequisite in an environment with an unusual variety of hazards.

There was no such detailed articulation in the Eurasian otters in Shetland, nor in any other species of otter studied so far. The comparison shows that, in otters, sound communication is more elaborate in a species that lives in cohesive groups than in the more solitary ones. Interestingly, a parallel evolution of systems of calls was observed in hyenas, where the gregarious spotted species is by far the most vocal (Kruuk 1972, 1975a; Mills 1990).

Sexual behaviour

In all the years with **Eurasian otters** in Shetland, I only saw five copulations of otters, all of them between the end of February and the end of May.

Mating always takes place in the water, and I think that it is only in captivity that otters may be more or less forced to copulate on land, giving rise to the notion that they may do it anywhere (Estes 1989; Pechlaner and Thaler 1983).

Early May 1986, along the Yell Sound, Shetland. In the late afternoon I meet one of our otter females, in her own core area, and clearly recognizable from the bottle-shaped throat patch. I find her diving about ten metres out from the rocks, catching the odd eelpout and stickleback; she lands leisurely at a group of boulders. Suddenly she is all alert and tense, plunging back into the water—but now followed by one of our large males, with a yellow ear-tag. He catches up immediately, and for several minutes the two roll, splash and tumble about in an eruption of white bubbles. Yellow-tag grabs her neck and holds on under water, clasping her chest and curling around her. She briefly utters the whickering call, then rolls around so much that only small bits of otter show on the surface. The two quieten down with the male holding on, and occasionally they lift their heads for air. Seven minutes later they split up and both land between the boulders, briefly out of sight. Then I see Yellow-tag again, he spraints, lies down and grooms, next to the largest of the rocks. Ten minutes after that she swims off, again catching the occasional eelpout or stickleback.

Twice females were seen to copulate when still accompanied by cubs, quite grown up by then, but still a tight family bunch. The cubs milled around the mating pair or hung about on land nearby, obviously rather apprehensive, perhaps just somewhat scared of the male.

The mating sequences that I saw, such as the one above, gave the appearance of chance meetings. However, there was more to it than that: often, and perhaps usually, the two animals knew each other well. In the winter before, I had twice seen Yellow-tag and that particular female meet, and there must have been several more occasions. These meetings were brief, with the male approaching, sniffing, and the female usually somewhat defensive or even aggressive, then the male swimming off again. Another large old male, who often came through the range of one of our females, sometimes played with her, and once I saw him catch a rockling whilst she was near, which he took up and dropped for her to eat. Such behaviour often occurs in birds as 'courtship feeding', but I was hesitant to read much into it in this case; I only saw it once and it could have been a chance occurrence. Later in that same year one of my students saw these two animals mate, in March.

In Shetland I saw meetings between Eurasian males and females on 34 occasions (see Fig. 6.8), but only in spring did this result in copulation. Sometimes meetings between male and female are highly aggressive, as was also suggested by the observations of Green *et al.* (1984), and in the previous section. But mating itself is a far less aggressive affair in this species than, for instance, in sea otters (see below). In mink, for example, females may be left with large wounds in the neck after copulation (Dunstone 1993); I have never seen such love bites in Eurasian otters.

Generalizing from this small amount of information, our Eurasian otter appears to be polygynous as well as polyandrous: males overlap with ranges of several females and probably mate with them, and for each female range there are several regular male residents. In some form or other, this appears to be the common pattern for mustelids (Kruuk 1989; Powell 1979; Sandell 1989). If males mate especially with female otters that are familiar, this might well make it more difficult for a complete outsider to mate successfully.

Within groups of **giant otters** one assumes that it is the dominant female and her male partner that copulate, but this has not yet been confirmed. They are more often in close proximity, and groom each other more frequently than others in the group (Duplaix 1980; Staib 2002). Elke Staib reported observations on several copulations in her study group, but she could not be certain of the identity of the male, nor did the female give birth at the appointed time afterwards—it happened much later than expected. As in all otters, mating takes place in water, probably always, with the male biting and holding the female in the scruff, and the entire procedure lasting some 20 minutes.

Sea otters are, like the Eurasian otter, polygynous as well as polyandrous. The female is often consorted by just one male during her oestrus, with a clear 'pair bond' for about 3 days (rarely up to 10 days), but occasionally a female may mate with several males in succession (see review in Riedman and Estes 1990). After the mating period, the female usually leaves the territory of the male, although he may attempt to prevent this. During the consorting period the male may play, nuzzle the female and gently bite her; even the pup may get involved in this. Although the male may attempt copulation during this time, this is rarely successful (Garshelis *et al.* 1984; Kenyon 1969). Possibly, in sea otters and in the Eurasian species, consorting may serve to reduce female apprehension of specific males, at the time of oestrus.

Such apprehension on the part of the female sea otter would be quite appropriate, given the extreme male violence that accompanies copulation, to an extent unknown for any other species of otter. Coitus lasts for about 30 minutes, and during this time the male bites the female's nose or face, with the pair vigorously spinning around in the water (Kenyon 1969).

Nose scars . . . indicate that a female is sexually mature . . . certain territorial males in the Monterey area tended to be consistently rougher on females' noses than other males were during mating. These females often sustained extremely serious nose and facial lacerations. In two cases, most of the nose was removed . . . complications from severe nose or facial injuries inflicted during copulation may cause death in some females . . . some of the youngest and oldest females may drown during vigorous mating bouts (Riedman and Estes 1990, p 62)

It is difficult to speculate on why mating in this species should be so horrendously violent.

Family life and parental behaviour

With mating ends the male **Eurasian otter**'s contribution to the next generation. He has nothing more to do with provisioning or protecting the female and the cubs, and it all falls on the female. This is what happens in almost all mustelids (Powell 1979); only in species such as the **giant otter** (Duplaix 1980; Staib 2002) do males stay in regular contact with the cubs, although they are not actively involved in providing and protecting, and although there is no certainty of paternity.

One often reads of a 'pair' of otters inhabiting a stretch of water, calling up scenes of happy family life. But this is an anthropomorphism, and there is nothing in it as far as otters are concerned. There is anecdotal evidence of a male Eurasian otter feeding young (Macaskill 1992), and especially along sea coasts one may sometimes encounter a male and a female wandering around together. In all cases that I saw, and where I knew the animals individually, these were a mother and grown-up son, a cub larger than the mother, but still partly dependent.

Female **Eurasian otters** may be aggressive to strange males when they come too close to young cubs (also in captivity; Hillegaart *et al.* 1981), even if the male happens to be the cubs' father (as far as we know). The same observations have been made in other species such as the **river otter** (Melquist and Hornocker 1983) and in captive **smooth otters** (Desai 1974). The aggression of the female is probably part of a strategy to protect the young, because male otters are a potential danger to cubs: males of many different species of carnivore have infanticidal tendencies (Kruuk 1972; Packer and Pusey 1984). Remains of small otter cubs were found in the stomach of a **Eurasian otter** male (Simpson and Coxon 2000), and in the Pantanal of Brazil a male **giant otter** that did not belong to the local group was seen taking a giant otter cub from a den along the water, and eating it (Mourão and Carvalho 2001).

It seems likely that the avoidance of danger posed by cannibalistic otter males is the main biological advantage of the secretive behaviour of **Eurasian otter** females about their cubs. 'Natal dens', especially in high-density otter areas, are very difficult to find, and after foraging the female is extremely covert in the way she slips back to the natal den. She is highly secretive then, especially nearby the holt, and when swimming along the coast nearest to the holt she hugs the shore closely, hardly showing herself. During the first months of the cubs' life she spraints almost only in water, not on land, and we found that this applies in freshwater areas as well as along sea coasts (unpublished observations on radio-tracked otters injected with zinc-65, which was detected in spraints; see also Jenkins and Burrows 1980).

Over the years we gradually pieced together the sequence of events when a female Eurasian otter produces a litter in Shetland. The 'natal holt' is different from regular holts (see Chapter 5), and looks a very unlikely den for an otter, often far from the sea shore. For about 2 months the cubs stay there—they are never seen outside. The mother is with them all the time except for one single foraging trip every day. Only at the end of the 2 months will she occasionally take a small fish to the cubs. In freshwater areas, natal holts or 'natal couches' of Eurasian otters may also be far from open water, as much as 1–2 km (see Chapter 5; Taylor and Kruuk 1990). However, Leon Durbin (1996) found one natal holt in which one of his radio-tagged otters gave birth, in a slope of boulders, close to a small tributary of a main river.

One day, 8–10 weeks after their birth, the Eurasian otter cubs follow the mother down to the water, or the mother sometimes carries them all or part of the way (Fig. 6.9). From then on the family will live mostly in holts close to the shore, frequently moving house, with the cubs regularly accompanying the female on her fishing trips (Fig. 6.10). It is likely that many of the natal holts reported in the literature (e.g. Harper 1981; review in Harris 1968) are, in fact, occupied by a family only after their first move from the natal holt proper.

In the beginning of their life in the open, the cubs mostly wait on the shore whilst the mother dives and then brings fish to them; sometimes she 'teaches' them to fish. In our Shetland observations, the first immersion of a cub in the water was often forced, with the mother diving whilst carrying the cub by the scruff, but after a few days, the cubs seemed as much at ease in water as on land. They are still rather clumsy and, being fat and fluffy, they are much more buoyant than the adults. All of this activity is made conspicuous by frequent *whistling* between mother and cubs (see below). I have often thought that, if there were predators around such as white-tailed eagles, wolves or dogs, the young otters would be very vulnerable. In fact, one of our Shetland cubs was killed in this early period by a farm dog.

Throughout the Shetland winter otter families stay together, the cubs often playing, and gradually beginning to catch some fish themselves. For otters on the west coast of Scotland this happens when they are 4–5 months old (Watt 1993), but it may be somewhat earlier in Shetland. Usually when they are about 10 or 11 months old the family breaks up; some cubs leave earlier, and some stay much longer with the mother. At least, this was the case in Shetland (Kruuk *et al.* 1991), but on the Scottish west coast cubs stay longer with the mother, until they are 12–14 months (Watt 1993). In North America, families of **river otters** disperse when the cubs are between 8 and 12 months old (Melquist and Hornocker 1983). I had the impression in Shetland that it was mostly the **Eurasian otter** mother who took the initiative in the break-up; suddenly she was off, and if there were several siblings they stayed

Figure 6.9 Female Eurasian otter carrying one of her cubs from the natal holt to the water's edge on its first outing.

Figure 6.10 Two grown-up cubs (about 9 months old) following the mother along the water's edge, Shetland. Most of their food is provided by her.

together somewhat longer, in the same area where they were reared. Some cubs show streaks of independence at an early age, often fishing apart from the others, and leaving for good well before the rest split up.

When the cubs are small and feeding only by suckling, they are in the holt, often far from the feeding (and danger) areas. Maintaining this distance goes without too much cost to the mother, as she makes only one foraging trip a day. After a couple of months the strategy changes, because (a) the cubs are less vulnerable to intraspecific aggression and predation, and (b) weaning starts, so the mother is then carrying fish to the young, one at a time. With the different style of provisioning, the cubs stay close to the female along the shore-line, and they are fed near to where the prey has been caught.

Several stages in this sequence of events are important in the population ecology of the animals. The first is inside the natal holt, when the cubs are dependent solely on their mother's milk for survival. Unfortunately we know next to nothing about what happens down there, about suckling behaviour or competition between siblings, and about conditions when food is scarce. These are issues that are obviously important, especially after we found the close relationship between food availability and the numbers of cubs (Kruuk *et al.* 1991; and see Chapter 12).

During the first days of the cubs' life outside the natal holt, when they are about 2 months old and before weaning, I have seen on several occasions on which Eurasian otter mothers have deliberately abandoned some cubs, leaving them to starve (see Chapter 11). It is unclear why, if females are going to abandon any of the cubs, they do so at this stage rather than much earlier inside the natal holt. One can only speculate: perhaps the abandonments that we watched were only the tail end of a process that mostly takes place underground. Or perhaps it is easier for the mother to abandon a cub during the course of ferrying them individually over long distances, rather than inside the holt where they are together in one bunch. Or perhaps the female avoids having a decomposing corpse in the natal holt. The abandonments are not 'accidental'; nor are the abandoned youngsters in obviously inferior condition.

By the time that the families are swimming along the Shetland shores, usually during late August and throughout September, it appears that, whatever factor it is that determines the number of cubs, this has already taken its toll. From the families that I observed, only a few cubs disappeared in the following months, and it seems that once a cub is about 3 months old it is relatively safe. There is a steady build-up in the numbers of families in the autumn, and in an average winter, on almost half of the occasions when I saw otters in the study area, they were animals in families (see Fig. 11.7).

During this time the female is in continuous contact with the cubs, and rarely out of sight. The family sleep together, swim together and prey together, though often the cubs wait on land or on the sea weed, whilst their mother is diving for fish. If for some reason contact is temporarily lost, a frantic search will ensue. Generally Eurasian otters are silent animals, but there is one curiously melodious exception to this, the *whistle*—the contact call between mother and offspring. Both cubs and mother will do it, their mouth slightly open. When I first heard it, I could hardly believe that it came from a mammal, and I am sure that many naturalists must have heard that clear single note, and registered it as the call of a meadow pipit or some other bird. It carries over several hundred metres, and it is somewhat shorter and higher pitched if it is made by a small cub, compared with the similar call of its mother. Only the Eurasian otter uses it, and it has not been described for any of the other species.

Another well known sound is the *huff*, which is hardly a 'call' but rather an explosive exhalation of air when an otter is alarmed by a person or a possible predator. Typically in Shetland, I would be working on the shore along a voe, emptying a fish-trap or something, and suddenly I would hear a soft huff behind me, only to see a ripple when I looked round. Females frequently huff in reaction to people when there are cubs.

Once the families are moving about along the shores, they can be seen active at any time of daylight. At night they are in one or other of the larger holts near the shore. In daytime the female is very busy keeping herself and the cubs fed, as in these field observations:

Shetland, early February 1987. One of our young females fishes 80 metres out, whilst two 6–8-month-old cubs, her first litter, clamber between the big boulders on the steep shore. I am within a few metres, peering down on them over a large lump of stone. Whilst I am absorbed by the playing cubs, there is a splash pretty close by. Mother is on her way, with the water churning around her: she has an enormous black fish which struggles wildly. Slowly, and with great effort, she gets the prey to her cubs.

She lands a conger eel, a fish longer than she is herself, and maybe also heavier. She drags it on to the rocks, well away from the water; it has almost stopped struggling. As soon as she releases the fish, one of the cubs starts eating, tearing at it somewhere in the middle. The other cub sits close by and sniffs the mother; both groom and roll around together. Five minutes later, she is off fishing again, whilst one of the cubs eats and the other sits nearby. After 10 minutes, the cub that is not eating enters the water and follows, swimming around where the mother is diving. She comes up with a small fish, an eelpout, and quickly eats it herself on the surface, then dives again. This time when she emerges she has a bullrout: almost 20 cm long, with a conspicuous red, white-spotted belly. This is a plump, large fish and a substantial meal; she heads for the shore again, deep in the water, and with a large bow-wave caused by the fish in her mouth. The cub follows as she lands close to the conger eel. The first cub is still eating there, and the second one gets the bullrout. The mother just sits, grooms herself and rolls over, some 10 metres away, mainly out of sight from the cubs.

November 1985. A female with one 4-month-old cub is fishing just off-shore, she in the water and the cub on land. The mother catches one eelpout after another, all of similar length, between 15 and 20 cm, eating all of them herself in the water, floating back-up between dives, chomping loudly on her prey and ignoring the cub. Then, 20 metres out, she captures a big fat bullrout, probably some 18 cm long, much heavier than any of the eelpouts. Immediately she makes a straight line for the shore, dropping the fish next to her cub. The youngster just looks at it lethargically, totally satiated already. The mother sits and waits, watching the cub, but when nothing happens she eats the bullrout herself, and both curl up on the seaweed.

In the first of these observations the conger eel was an unusually large and heavy prey (in fact, I stole the remains after the otters had finished with it and ate it myself). Both observations demonstrated one of the principles of family life of Eurasian otters: when the female is providing for her cubs, she eats small prey herself and takes the larger fish to her offspring (Kruuk *et al.* 1987) (Fig. 6.11). This makes good sense in energetic terms: the food for the cubs has to be ferried over much longer distances than the food for the female, and she can only take one item at a time.

In my observations on Eurasian otters in Scottish freshwater lochs there was no difference in the size of prey eaten by the cubs and that eaten by the mother, because virtually all prey was relatively large (eels) and was almost always taken ashore. Only rarely did I see otters actually eat fish in fresh water whilst they were swimming (except occasionally very small ones). But when Eurasian otters feed on a mixture of small and large fish in rivers and streams, the same maternal behaviour as in Shetland may well apply; good observations are difficult to come by because of the nocturnal habits of the riverine animals.

When Eurasian cubs are beginning to catch some fish for themselves at the age of about 5 months, they are initially much less efficient at this than their mother, and they continue to improve their fishing

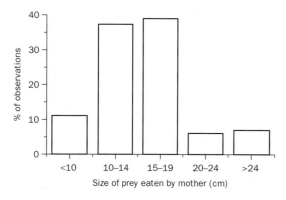

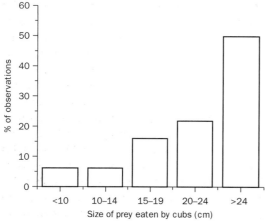

Figure 6.11 Mothers eat the smallest prey: size of 103 fish prey eaten by mother Eurasian otters in Shetland, compared with sizes of 22 fish taken to cubs ($\chi^2 = 49.0$, 2 d.f., $P < 0.001$). (After Kruuk et al. 1987.)

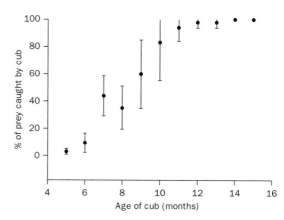

Figure 6.12 Proportion of prey eaten by Eurasian otter cubs of different ages that was caught not by their mother but by the cubs themselves. Cubs get at least some provisioning until 1 year of age. (After Watt 1993; data from Mull, West Scotland.)

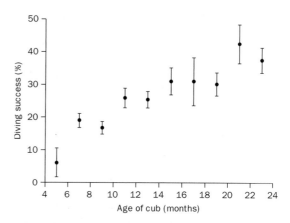

Figure 6.13 Diving success of Eurasian otter cubs of different ages; error bars denote standard errors. Success continues to improve until well after 1 year of age. (After Watt 1993; data from Mull, West Scotland.)

skills well into their second year. This has been studied in detail by Jon Watt, on Mull, off the Scottish west coast (Watt 1991, 1993). He found that even though cubs were self-reliant and independent of their mother after 12–14 months (Fig. 6.12), they were still less adept at catching fish and continued to improve their percentage of successful dives, and also the mean weight of prey they did manage to catch (Fig. 6.13). When some 15–18 months old, they were as efficient divers as the adults, if diving efficiency is expressed in terms of the proportion of time spent under water and on the surface.

The kind of prey that young cubs caught for themselves was often slow and easy, and of rather low quality (low calorific value). For instance, they captured many shore crabs, which have the lowest food value for effort of all types of prey recorded.

Watt found a negative correlation between the age of cubs and the proportion of crabs in their food; in his study area, crab accounted for 32% of the prey captured by dependent cubs, but for independent sub-adults this was 15% and for adults only 3%.

It is tempting to assume that otters have to learn their fishing skills from observing their mother, or by trial and error. Of course, the improvement of performance with age may have nothing to do with actual learning; it could just be maturation. But it certainly has the appearance of learning from the mother, with the cubs watching closely when she is fishing, and they often dive with her from the age of 4–5 months onwards, in beautiful synchrony. Sometimes they may be peering down from the surface of the water, floating along, watching the mother at work on the bottom. Such behaviour is reminiscent of that recorded for fox cubs, when the vixen is catching earthworms (Macdonald 1980b).

However, the most suggestive observations of 'learning' are those where the otter mother appears actually to teach her cubs to catch prey:

September 1986, midday, at low tide along Shetland's Yell Sound. In the kelp nearby an adult female otter is diving and rummaging about. Just below me, no more than 20 metres away, is her single cub, a chubby little animal, about 3/4 of its mother's length and probably about 3 months old. I had just watched a long series of dives by the mother which had lasted more than an hour.

Then she heads for shore, carrying an eelpout. She lands close to the cub, shaking herself whilst still holding the fish in her mouth. With the little one begging alongside her, she walks up to a small rock pool, no more than 20 centimetres deep, and about 2 by 1 metres in size. She drops the eelpout into this; it is an average size fish, less than 20 centimetres long and still well alive, wriggling off into the pool. The cub goes after it clumsily, and misses. The mother comes to the rescue and deftly picks up the eelpout again, releasing it within seconds just in front of the cub. At the second attempt the cub is more successful—perhaps the eelpout has slowed down a bit. The female stands motionless for about half a minute, watching her cub eat.

Before the cub stops feeding the mother returns to sea, back into the huge heaving mass of kelp. She again emerges with an eelpout, a smaller one this time, of less than 10 centimetres. She quickly chews this up herself before getting another, one of about the same length as the one she had taken to the cub. She lands it, takes it to the pool, and the whole sequence repeats itself. The fish is released, the cub chases it and misses, the mother catches the fish and releases it again, then the cub's second chase is successful. After the cub has bolted down the eelpout, mother and cub play, rolling about on the rocks and in and out of the pool; finally both dive into the sea at great speed and swim off next to each other, closely along the shore, into the huge landscape of the Yell Sound.

I have seen this behaviour in different females, and it must be a fairly common occurrence in the life of Eurasian otters. It reminded me of earlier days in the Serengeti, in East Africa, where I had seen cheetah mothers return to their clumsy little cubs and drop a live baby gazelle in front of them, leaving the young ones to chase it, but catching it for them when they could not cope. Would such 'teaching' really contribute to the hunting skills of the next generation? Watt (1993) saw otter cubs release prey close to the shore, then catch it again, sometimes several times with the same fish, and it has been described for mongooses as 'prey capture play' by Rasa (1973; Bekoff 1989; reviews in Fagan 1981).

I saw otters 'playing' with fish a few times in Shetland and once on the Scottish west coast. This usually concerned solitary, sub-adult, independent otters landing and releasing a fish close to the shore, recatching it several times, and mouthing and pawing at it whilst lying or floating on their backs. They rolled with it in the water, sometimes throwing the fish right up into the air. This could go on for up to 10 minutes. Interestingly, in Shetland, the prey with which otters played was usually a flatfish (top-knot *Zeugopterus punctatus*, when I could distinguish species). That was not a very common prey, a less preferred species, one that was sometimes discarded again when an otter caught it after appearing to be more or less satiated. Lumpsuckers *Cyclopterus lumpus* are also used in play.

In freshwater areas, otters sometimes appear to 'play' with eels they have landed on the grass or on ice, rolling them around and flinging them about, even taking them into the water again. This may function to remove slime from the prey: one finds a characteristic patch of slime where it happened, and the eels are always eaten afterwards.

It has been suggested that, amongst the Carnivora, species that are highly specialized hunters are also the ones that take longest to reach sexual maturity and become independent of their mother (Gittleman 1986). In otters this has clearly been taken to extremes, and Jon Watt (1993) noted that the period, of a year or more, for which Eurasian otter cubs are dependent on their mothers is unusually long for a carnivore of this size. Comparison with equally large, sympatric carnivore species bears this out; for example, the fox gains independence at about 4 months (Macdonald 1980b), the Eurasian badger at 4 months (Kruuk 1989; Neal 1986) and the wildcat at 3 months (Corbett 1979). Watt relates this to the long learning period for otters to acquire underwater hunting skills, as catching fish must be unusually difficult for a mammal. It is a theme that impacts on many aspects of otter ecology.

Nor is the Eurasian otter alone in this. Young North American **river otters** along the Californian coast are self-sufficient at feeding at 9–10 months, and are abandoned by their mother at 11 months (Shannon 1998). Along inland rivers they disperse from the family after 12–13 months (Melquist and Hornocker 1983). River otter families are often accompanied by young of previous litters, acting as helpers (Shannon 1998). Young **Cape clawless otters** are still around with their mother when she is copulating during the next spring season (Verwoerd 1987), so their dependency is likely to stretch over at least a year, and similar observations have been reported for the **smooth otter** in India (Hussain 1996).

Giant otter cubs (Fig. 6.14) hunt and travel with the family from as early as 3–4 months of age; they are completely weaned after 5–6 months; quite indistinguishable from the adults in their size and behaviour at 10 months; and leave the group when aged at least 2 years (Duplaix 1980; Laidler 1984). They have been seen to catch their first fish at 2.5 months (Staib 2002). In an interesting difference from the Eurasian otter, the giant's cubs in those early months are provisioned by all the other members of the group, with fish that are much smaller than the fish that are eaten by the adult otters themselves (Staib 2002). In this case the cubs have the advantage of getting small fish that are easier for them to consume, and adults do not have the disadvantage of having to transport small items over large distances.

In these giant otter communities, cubs are played with, assisted, groomed and protected by all members. They benefit from the aggressive group responses to predators; predation on wild giant otters is rare, but of two cubs raised in captivity and released in the wild in southern Guyana, one was soon killed by a black caiman (Duplaix 2004). Group cohesion is maintained with the elaborate system of calls, and even when diving they communicate with

Figure 6.14 Giant otter cubs 1–2 months old. © Nicole Duplaix.

underwater vocalizations (Schenck *et al.* 1995). All of their activities are highly synchronized (Staib 2002), and there is frequent physical contact between group members, such as mutual grooming (which I never saw in the wild in the Eurasian otter). Other sociable species of otter, such as the **smooth** and the **spotted-necked**, also absorb cubs into their group organization, but with less elaborate communication systems (Duplaix 1982; Procter 1963). In captivity, cubs of **small-clawed otters** are cared for by both males and females (N. Duplaix, personal communication) (Fig. 6.15), but what happens in the wild in this species is not known.

Few carnivores take care of their offspring as much as the **sea otter** does. The female rarely has more than one pup, which is born usually at sea whilst she is floating on the surface. From that moment onward she puts an enormous behavioural investment in its survival, swimming on her back she carries it on the chest (Fig. 6.16), often clasping it with both forelegs. The pup is virtually never out of her sight, except when she is diving for food. And that is the time that bald eagles may take young pups (Sherrod *et al.* 1975)—on Amchitka, pup remains at many of the bald eagle nests are testimony.

There is a display of affection between sea otter mother and pup almost like what one sees in primates. The two often clasp each other around the neck, and there are heart-rending stories from hunters and ecologists when they capture either of them. If the pup is netted, the mother often attacks

Figure 6.16 The closest mother–offspring bond: sea otter mother carrying pup. © Bryant Austin.

Figure 6.15 Small-clawed otters—male carrying a newborn cub, the mother watching. © Nicole Duplaix.

the net, with almost human wailing cries, and she closely follows a boat that is taking her pup away. She will also carry the body of her dead pup for many days, even when it is decomposing (Kenyon 1969). Mother and pup communicate with a basic repertoire of ten different vocalizations, many of which are of low frequency and difficult to hear for an observer at a distance. The best known is the scream, uttered by either of them when accidentally apart, and which can be heard at distances of up to 1 km. It has been shown that sea otter screams are individually different, and it is likely that they are useful for recognition, if ever mother and pup get separated when many others are around (McShane et al. 1995).

Initially the pup is suckled about six times per day for about 9 minutes (Sandegren et al. 1973), before gradually the mother begins to provide it with captured prey. The daily time-budget of such females consists of 41% resting on the water, 16% diving and feeding, 13% swimming about, 10% grooming herself, and 20% grooming her pup, with nursing about 8% of the time during other activities (Riedman and Estes 1990). There is no helping by previous offspring.

The sea otter pup's period of dependence is much more variable than that of the river otter, with which it shares some its habitat. It ranges from 5–6 months in Prince William Sound (Garshelis et al. 1984), to 5–9 months along the Californian coast (see review in Riedman and Estes 1990) and to at least 1 year, sometimes longer, in Amchitka, one of the Aleutian islands (Kenyon 1969). These different studies show that in this species the period of dependence is much longer along coasts where sea otters are feeding mostly on fish (Kenyon 1969), compared with places where molluscs and crabs are more important to them (see Chapter 7). This supports the hypothesis that, for otters, the process of learning to catch fish, as a difficult prey, is the factor that determines the very long period of dependence of cubs on their mother.

Interestingly in this context, Jim Estes and his colleagues demonstrated that sea otter females show strong individual differences in prey selection, preferences that are passed on to their pups (Estes et al. 2003b). They stressed the need for specialist 'skilled' foraging methods, and the long period of dependence enables pups to acquire these, presumably by copying from their mothers.

Some generalizations on social behaviour of otters

To anyone watching these animals, several of the otters' social behaviours are uniquely interesting in their own right. I mentioned the giant otter's underwater vocalizations and their group defence against intruders, the river otter's packs of males, the Eurasian otter's abandonment of 'surplus' cubs, and the sea otter's maternal care of its single pup. There is the otters' unusual dependence on scent marking for their communication. But more important are the adaptive links between social behaviour and the overall organization of otter society, and hence with its ecology.

As an example, I referred to several relevant behaviour patterns of the most gregarious otter with a structured group organization, the giant. Compared with the others, the giant otter is highly articulate in its communication: it uses more different and louder vocalizations, and its scent-marking stations carry a host of different messages. In addition, group members groom one another, and they help each other against predators and with rearing cubs. All of this is different in more solitary species. The sea otter, with its gregarious life in large and rather unstructured aggregations, is also more vocal than the *Lutra*, *Lontra* and *Aonyx* otters, though less than the giant, and individual sea otters help each other against predators (and against people with capture nets).

Perhaps more than any other single requirement, it is the long period of dependence of cubs on mothers that is a dominant feature in the social life of all otters. I have presented evidence that this is based on the cub's need to acquire the, for any mammal, intrinsically difficult skill of catching a fish. This has far-reaching repercussions for population dynamics, survival and conservation aspects (see Chapters 9 and 13).

CHAPTER 7

Diet

Introduction

Diet largely defines the ecological niche of an animal. Knowledge of the diet enables us to compare it with what is available, and to evaluate predator–prey relationships. It is a prerequisite for understanding factors that limit populations, for assessing competition and for designing conservation management strategies. Fortunately, we now have fairly good evidence of the otters' daily fare.

The food of otters is the aspect of their ecology that has been studied most thoroughly, and in many different places. For Europe alone, diets of Eurasian otters have been published from well over a hundred localities (Clavero *et al.* 2003). The reason for this, I suspect, is not so much because that is what everyone is most interested in, but because it is more accessible than, say, social behaviour or population ecology. Diet can be studied by analysing the otters' scats, which are not difficult to find on conspicuous sites and in convenient places, such as under bridges (see Chapter 6), and the contents are relatively easy to identify. There are also other methods, as discussed below.

A first finding of the research is that otters are highly specialized animals, compared with other carnivores—just as any naturalist would have predicted. If we compare the data on the prey categories of, for instance, **Eurasian otters** from different areas, with similar data sets from other Carnivora, we can quantify this degree of specialization. We can categorize prey into groups, such as fishes, birds, small, medium-sized and large mammals, amphibians and reptiles, vegetable matter and several others. We then calculate a coefficient of concordance, K, which expresses the agreement in ranking of these prey categories, in diets of the same species from different places. K ranges from 0 to 1 (Siegel and Siegel 1988).

If a species is relatively opportunistic, the rankings tend to vary between sites, and K will be close to zero. Quite early on I found that, taking published data on diets from a few different carnivores species of which many data sets were available, Eurasian otters were highly specialized, with $K = 0.82$ (data from 12 different studies), compared with, for instance, 0.44 for red fox, and 0.67 for European wild cat (Kruuk 1986). Such figures will be modified with more recent data, but the general picture will not change much: Eurasian otters are unusually specialized.

In addition, in most other species of otter the diet is dominated by fish, but not just any species of fish: there are distinct preferences, and some other kinds of prey are also important. It is clearly relevant to establish, not just what otters eat, but what they take from the available fauna, and how they select.

Outstanding early research on otter diet and foraging in fresh water includes the work by Sam Erlinge, in southern Sweden (1967), the PhD study by Margaret Wise in the south-west of England (Wise *et al.* 1981) and the project in western France by Libois and Rosoux (1989). These studies compared the remains of prey in the spraints of otters with prey availability, with numbers of fish caught in nets and with electro-fishing. Their main conclusions were similar: prey other than fish is not important, and fish is taken more or less according to availability, with a preference for the slower moving species, and often for the smaller fish in the populations. The many later studies on Eurasian otter diet have substantiated this (Fig. 7.1; Clavero *et al.* 2003).

For our observations on otter diet on the Scottish mainland, where Eurasian otters were active in streams and lochs, mostly at night, I and my colleagues also used faecal analysis, with all its complications. We analysed spraints from rivers in north-east Scotland, and studied total quantities of food intake, from both field observations and data on captive Eurasian otters (Carss 1995; Carss *et al.* 1990, 1998a,b; Kruuk *et al.* 1993a). These data were used together with information on otter numbers and fish productivity to assess the overall relationship between Eurasian otters and prey populations (see Chapters 5, 8 and 9).

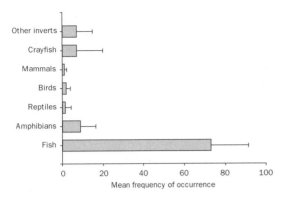

Figure 7.1 Mean frequency of occurrence of prey items in the diet of Eurasian otters, showing them to be primarily fish eaters. Data from 37 diet studies throughout Europe (Clavero et al. 2003); error bars denote standard deviations.

For several other species of otter, the diet and predator–prey relationships have been studied more recently, and are discussed below. None has had as much attention as those of the sea otter, with its overwhelming impact on its habitat.

In Shetland we were in the fortunate position of being able to watch otter predation at work, to see otters in daytime with identifiable prey in their mouths, and this enabled research on many more aspects of foraging than one could undertake with the customary faecal analysis (Kruuk and Moorhouse 1990). The information on diet could be related to availability and productivity of prey populations (see Chapter 8), to differences between individuals, to otter social and foraging behaviour (see Chapters 6 and 9), and we could investigate the role of food as a possible limiting factor in populations (see Chapter 11).

Earlier data from spraint analyses of Eurasian otters along British sea coasts were collected in summer only, and showed consistently that these animals take relatively small fish, mostly bottom-living species, as well as varying numbers of crabs (Herfst 1984; Mason and Macdonald 1980, 1986; Watson 1978). Since then, many more results have become available on Eurasian otters' diet in relation to prey availability along sea coasts, notably from western Scotland (Watt 1991, 1993, 1995), Norway (Heggberget 1993, 1995; Heggberget and Moseid 1994) and Portugal (Beja 1991, 1995a,b, 1997).

In the Shetland study on diet, the starting questions were: (a) which species and what sizes of fish do the otters take; (b) are there differences in the diet with season and coastal type; and (c) do all otters take the same kinds of prey? Results were later compared with availability and with foraging behaviour, and subsequently we did the same (with some modifications) in freshwater areas.

Methods of analysis

The almost universal method for obtaining information on the food of otters is by identifying prey remains in spraints. Many parts of the skeleton of prey are left undigested, and in particular the vertebrae of fish can be identified down to species (methodology and keys in Conroy et al. 1993; Watson 1978; Webb 1975); they can also be used to estimate the size of prey eaten. However, there are some serious problems in the interpretation of the information from spraint analysis, problems often ignored by researchers.

One difficulty is that the otters' digestion of fish bones and other remains depends substantially on the species and size of the prey, and on the activity of the otters themselves; for example, an active otter will leave 60% of salmonid vertebrae undigested and an inactive one 32% (Carss et al. 1998a). In theory, one can estimate correction factors to overcome this, and some attempts have been made by comparing the diet presented to captive otters with the faecal output of these animals (e.g. Erlinge 1968; Jacobsen and Hansen 1992; Wise et al. 1981). However, variation is often large and results so inconsistent that experiments with captive otters are of little help (Carss and Elston 1996; Carss and Nelson 1998; Carss and Parkinson 1996). Another problem of spraint analysis is the impossibility of allocating remains to individual prey animals. In consequence, prey species that are taken frequently will be underestimated, because it is more likely that remains of several individuals will be found together in one spraint.

After a long series of experiments with faecal analysis on captive **Eurasian otters**, David Carss expressed despair: 'It is not possible to quantify otter diet accurately by frequency of occurrence methods . . .' (Carss and Parkinson 1996, p 301). Similar conclusions have been drawn for cormorants

(Zijlstra and Van Eerden 1995) and harbour seals (Harvey 1989). Anne Lejeune demonstrated that the spraints of **spotted-necked otters** do not show any traces for some species of fish (Lejeune 1990). All this means that, under most conditions, from spraint analysis one can draw some careful and approximate conclusions about the rank order of importance of prey species in the diet of otters, but accurate percentage points are out of the question. Results provide a rough approximation only.

In North American **river otters**, Merav Ben-David and her colleagues used stable isotope analysis for an overall impression of diet, by assessing the occurrence of nitrogen and carbon isotopes in groups of prey species, and in the fur of otters (Ben-David *et al.* 1998; Blundell *et al.* 2002). They suggested that the method enables an indication of diet over periods of about 6 months, but of course the results are bound to be coarse, and no critical experimental tests have yet been carried out.

The problems with faecal analyses are many, but alternatives are few. Visual observations on prey capture of feeding otters are possible only in day-active animals, such as sea otters, giant otters and Eurasian otters along Scottish coasts. Furthermore, otters rarely leave any food remains onshore, as they tend to eat the entire fishes. For most situations, therefore, and for most species of otter, one has to make do with faecal analysis for diet assessment.

With our **Eurasian otters** in Shetland we were fortunate in being able to see what they were doing. When an otter was foraging, it would come up with a fish in its mouth, usually of a small manageable size, which it could eat then and there, floating on the surface, pointing its snout upwards and occasionally guiding the fish with a forepaw (see Fig. 9.1). Otters do not swallow prey under water: both in captivity and in the wild I saw them actually catching prey underwater on numerous occasions, and it was always eaten on the surface. Larger fish, or prey that was difficult to handle, would be taken ashore (Figs 7.2 and 7.3). We could usually identify which species the otters came up with and estimate the size of the prey, so we knew when and where a fish was caught, and by which otter.

Some experience is need to identifying prey in the mouth of an otter at a distance. Fortunately, we were also operating a concurrent scheme of trapping fish (see Chapter 8), so were handling fish almost daily. There were only a limited number of species that were relevant to the otters, so it was quite feasible to learn to distinguish them even at a distance of a 100 metres or more, with binoculars or a telescope. Of course it often happened that we could not be sure which species of fish an otter was eating, or we could see only that it was 'something eel-shaped' or 'pollack-shaped'. This was a problem especially with small prey less than 10 cm long; nevertheless, about half of what the otters came up with could be clearly identified. We estimated the size class of prey by comparison with the width of an otter's head, which is 8–10 cm near the eyes.

To convince ourselves that we could recognize these fish in the jaws of an otter, we did some experiments. One of us, the observer, would watch the experimenter with a pair of binoculars from about 70 metres, a usual distance for our otter observations in the field. From a

Figure 7.2 Male Eurasian otter with remains of dogfish *Scyliorhinus canicula*, a small shark, along exposed Shetland coast with thongweed *Himanthalia elongata*.

Figure 7.3 Independent young male Eurasian otter with shore crab *Carcinus maenas* on exposed bladder-wrack *Ascophyllum nodosum* in Shetland.

large collection of dead fish, caught in traps in the otters' range, one would be picked up, fingered and partly hidden in one hand, turned over and manipulated, just as an otter would do when eating, for a period of 5 seconds—which was less than an otter would take. Both the observer and the experimenter recorded species and size of fish. On checking, all of us were fairly consistently almost 80% correct in our assessment of species, and with the sizes we were about 12% out, in a total of 112 presentations.

The observation check was only an approximation of otter reality. In the wild we also used a telescope, and the experimental dead fish had often lost much of their original bright colour, so were more difficult to identify. Further, we could not use the characteristic movements that different species make when caught by an otter. When the observer wrongly identified a species in the experiments, it was often in confusion with a similar species. We did not, for instance, mistake a saithe *Pollachius virens* for an eelpout *Zoarces viviparus*, but we might report it as a pollack *Pollachius pollachius*. The results were sufficiently encouraging to continue with the assessment of diet by direct observations, rather than from faecal analysis.

When we saw an animal foraging, we recorded the coastal section where it was, distance from the shore and the time of day. Apart from identity of the individual otter, and behavioural information (other otters present, what happened before and after), we recorded prey species and size class: class 0 was a prey smaller than 5 cm, class 1 was between 5 and 10 cm, class 2 between 10 and 15 cm, and so on. From the relation between length and weight (Nolet and Kruuk 1989), we estimated the weight of the prey. A total of 45 individual otters was involved.

Diet of Eurasian otters in Shetland

Table 7.1 lists all species that I recorded as part of the Shetland otter diet (Kruuk and Moorhouse 1990). This includes only those observations where an animal was actually seen swimming before catching a prey, and not those cases when I came on an otter eating on the shore. This often happened with large prey such as lumpsuckers *Cyclopterus lumpus* (Fig. 7.4), a bright red fish that took a long time to consume, or dogfish *Scyliorhinus canicula* (a small shark) (see Figs 7.2 and 8.10). It was less often that we saw otters catch any of these big fish, and I had to be careful not to distort the diet record towards large fish—as it was, they were an insignificant, though spectacular, part of the list of species caught. Apart from lumpsuckers and dogfish, we also found otters with large cod *Gadus morhua*, ling *Molva molva*, conger eels *Conger conger*, octopus and various others, although more rarely. Lumpsuckers were an interesting prey, because of the sexual discrimination by

Table 7.1 Prey taken by otters foraging in Shetland, all observations[a]

Species	No. of observations	%
Eelpout *Zoarces viviparus*	686	34
Rocklings *Ciliata mustela* and others	343	17
Sea scorpion *Taurulus bubalis*	283	14
Butterfish *Pholis gunnelus*	201	10
Stickleback *Gasterosteus aculeatus*	114	6
Saithe *Pollachius virens*	106	5
Bullrout *Myoxocephalus scorpius*	73	4
Pollack *Pollachius pollachius*	45	2
Saithe or Pollack	9	<1
Flatfish (mostly *Zeugopterus punctatus*)	39	2
Lumpsucker *Cyclopterus lumpus*	18	1
Ling *Molva molva*	16	1
Cod *Gadus morhua*	6	<1
Eel *Anguilla anguilla*	5	<1
Poor cod *Trisopterus minutus*	5	<1
Pipefish *Syngnathus acus*	4	<1
Sea stickleback *Spinachia spinachia*	4	<1
Dogfish *Scyliorhinus caniculus*	2	<1
Wrasse *Labrus mixtus*	2	<1
Conger eel *Conger conger*	1	<1
Yarrel blenny *Chirolophis ascannii*	1	<1
Other fish	3	<1
Shore crab *Carcinus maenas*	61	3
Starfish *Asterias rubens*	1	<1
Rabbit *Oryctolagus cuniculus*	3	<1
Total	2031	100
Unidentified prey	1557	43 (of 3588)

[a] Includes only those in which otters were seen before capture of prey.

Figure 7.4 Male lumpsucker *Cyclopterus lumpus*, 1.2 kg, taken by Eurasian otter in Shetland. Males of this species are more vulnerable than females.

Presumably, lumpsucker males are more vulnerable because they guard the eggs—an interesting disadvantage of parental behaviour (Clutton-Brock 1991).

Figure 7.5 shows the contribution of each prey species to the otter diet throughout the year, in two different ways. In terms of numbers caught, the eelpout *Zoarces viviparus* (see Fig. 8.1) came first and, together eelpout, rockling *Ciliata mustela* (see Fig. 8.2), sea scorpion *Taurulus bubalis* (see Fig. 8.4) and butterfish *Pholis gunnellus* (see Fig. 8.3) made up about three-quarters of the prey. However, in terms of estimated weight consumed by the otters, the picture was different: rocklings were most important. The four species mentioned above accounted for less than half of the intake in terms of weight.

In terms of bulk, the large, less frequently taken, species of prey made quite an important contribution to the diet, with big, slow fishes such as bullrout *Myoxocephalus scorpius*, lumpsucker and ling accounting for about 35% of the diet in weight, but for only about 5% in terms of numbers taken. The median size of fish prey was only 16 cm, or 28 g; weights varied more than lengths.

We also saw otters catch birds (fulmar *Fulmaris glacialis*, and guillemot *Uria aalge*), several more rabbits and some large fish; one conger eel was almost the weight of the otter female that caught it. However, those were rare events, incidental occurrences that probably we noticed only because they concerned such big and spectacular prey. They are not comparable to the other food data.

the otters: they caught almost only males. On only one occasion in many dozens of observations did I see an otter with a female lumpsucker (which is bright green, in contrast to the male which is red).

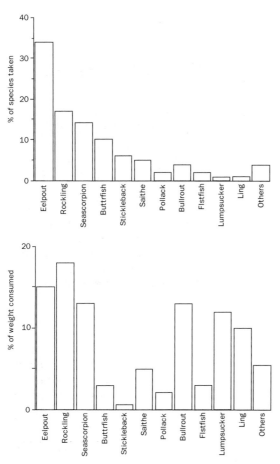

Figure 7.5 Diet of Eurasian otters in Shetland from direct observations of feeding behaviour, as proportion of species seen to be taken and of estimated weight consumed (total of 2031 prey identified; 1557 prey unidentified).

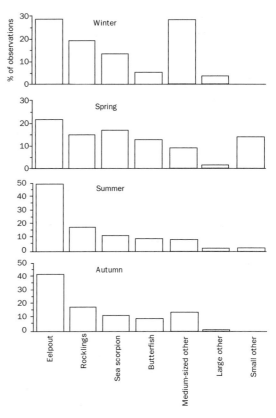

Figure 7.6 Seasonality of Eurasian otter diet in Shetland: frequency of prey species seen to be taken. Spring is the time of small prey, winter of large prey, and summer specialization in eelpout. Numbers of identifications: winter, 361; spring, 754; summer, 434; autumn, 479 ($\chi^2 = 521.7$, 3 d.f., $P < 0.001$).

Otters ate remarkably few crabs during our main study in Shetland (almost always when they did, it was the common shore crab *Carcinus maenas*; see Fig. 7.3), and crab was only a minor constituent of the diet. Crabs were exceedingly common along the coast, but only a few individual otters would eat them (see below). The spraint analysis by Herfst (1984), of material collected in summer, supported this. However, in a quite striking difference, when years later I returned to the area (in 1993 and 2004), otters were catching crabs much more often. Along the coast there were many more spraints with crab remains than we had seen earlier. The change coincided with unusually low fish populations (see Chapter 12).

Seasonal changes in diet were relevant, because of fluctuations in availability of prey species and seasonality of breeding in otters. For simplicity, Figure 7.6 shows just the annual changes in the proportion of numbers of each prey caught. Eelpout were, at all times of the year, more numerous in the food than any of the other species, but more than twice as frequently in the summer diet as in spring. In summer and autumn, they dominated the food. Rocklings and sea scorpions fluctuated little; at all times they were a steady 15–19% and 12–17% of the diet. The other species varied more markedly and, importantly, in spring the small, lightweight and/or low-calorie prey (butterfish, sticklebacks *Spinachia spinachia*, crabs) were prominent. In winter otters caught more of the heavier saithe and pollack, as well as several other heavies such as ling, lumpsucker, bullrout, conger eel and cod (Kruuk and Moorhouse 1990).

Otters in spring had a more varied diet, evenly spread over many species; generally, it was low-quality food. Spring was also the time when we saw them catch birds on the water, and rabbits in their burrows. In summer otters specialized by eating larger numbers mostly of only one species, and of larger sizes than in spring. In winter the diet was more varied than in summer, although not as much as in spring, significantly with more large, heavy fish being taken than at any other time. Mean weights of prey expanded from spring, through summer and autumn to winter, from 46, 49, to 61 and 61 g, a significant increase (median weights were 22 g in spring and 28 g during the rest of the year). This was a strong indication from the otters' food that spring was a hard time.

The food eaten by the animals in different places could well affect the spatial organization. We therefore collected data on both diet and prey dispersion (see Chapter 8), in different parts of the study area.

The coast of our study area is a mixture of different types of shore. At one extreme are the beautifully sheltered bays, in Shetland called 'voes', and often separated from the wild world outside by a sand bar, with a small gap to let the tides through. There are also exposed steep cliffs, with wild waves breaking against them, and in between these extremes there was a gradation of coasts of different exposure. One side of the study peninsula was high, with the bottom of the sea falling off steeply, to about 80 metres deep close inshore. The other side was shallow, intersected by voes, and one had to swim a long way out to get as deep as 20 metres. The otters used these coasts differently, which showed in their foraging behaviour (see Chapter 9).

When we did see an otter fishing along the cliffs, chances were greater than elsewhere that the animal would come up with a really large prey. Somehow it fitted in with the landscape for an otter to part the waves with a large red lumpsucker in its mouth, or with a cod or a flatfish the size of a pan-lid. A mere eelpout would have been out of place. Figure 7.7 shows the otter food along two different coasts of the study area.

There were some clear trends (Kruuk and Moorhouse 1990). In the capture rates of the two most significant prey species in the diet, with increasing exposure of the shore the frequency of eelpout decreased, and that of the heavier five-bearded rockling increased (Figs 7.8 and 7.9). Generally, otters moved into the voes and sheltered coasts in spring and summer, where in spring they fed on any 'small' prey, and later in summer mostly on eelpout. In autumn there was a move to more exposed coasts, where in winter they did most of their foraging, taking many larger prey species (which they did not catch when they visited those exposed coasts at other times). It appeared that otters followed the shifts in prey populations over

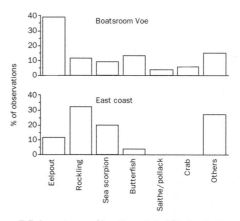

Figure 7.7 Importance of location: diet of Shetland otters along two nearby, contrasting parts of the study area coast (Boatsroom Voe is a sheltered bay, east coast has exposed cliffs). Larger prey species are taken along exposed coasts. Direct observations; numbers of identifications 886 and 25 respectively.

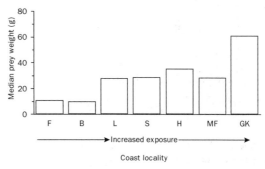

Figure 7.8 Prey weight and exposure: the size of Eurasian otter prey along sections of Shetland coast increased with wave exposure ($\chi^2 = 191.5$, 6 d.f., $P < 0.001$).

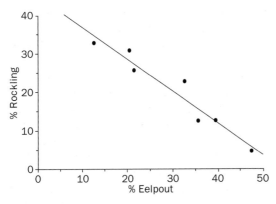

Figure 7.9 Correlation between two main prey species—eelpout and rockling—in the diet of Eurasian otters from different coasts in the Shetland study area ($r = -0.96, P < 0.01$).

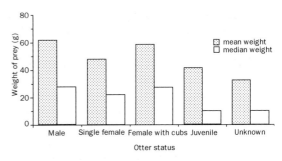

Figure 7.11 The size of fish caught by Eurasian otters in Shetland depends on the otters' sex and status. Numbers of prey: 809 for males, 917 for single females, 946 for females with cubs, 108 for juveniles, and 315 unknown ($\chi^2 = 10.5$, 4 d.f., $P < 0.05$).

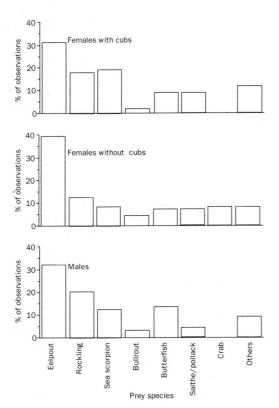

Figure 7.10 Small differences in prey species selection by the sexes (Eurasian otters in Shetland). Numbers of observations: 510 for females with cubs, 660 for females without cubs, and 638 for males. There were no significant differences between the groups.

the seasons, along the various shores within their home ranges (see Chapter 8; Kruuk and Moorhouse 1990).

I compared the main species of prey taken by males, by females on their own, and by females that were provisioning cubs. The differences in species composition of diets were small and not statistically significant (Fig. 7.10). However, there were difference in the sizes of prey: the various categories of otter did select substantially different ones (Kruuk and Moorhouse 1990). For instance, over the whole period of study the largest fish were taken by males, next came females with cubs, then females without cubs, then independent cubs, and the smallest fish were in the diet of the unknown, visiting individuals (Fig. 7.11). The mean weight of prey of males was 13% heavier than that of females; this was due largely to differences between the heaviest fish caught, not to males selecting heavier prey right across the spectrum.

Females with cubs caught larger prey than single females. The prey taken by males was 22% heavier than that of single females, but there was little difference between fish caught by mothers and fish caught by males. Interestingly, the mothers themselves ate the same sizes of fish as did the single females, but they also caught larger prey, which they took to their cubs (Kruuk et al. 1987).

The next question was whether individual otters had their own feeding preferences, their own idiosyncrasies. When different females fished along the same coast during the same season, there was a remarkable similarity in what they brought to the surface, with differences of only a few percentage points. This result was clearly at variance with the behaviour of sea otters (see below), where individuals show large and consistent differences. However, two newly independent sub-adult otters in the same area, a male and a female, showed a striking contrast, although they often operated in exactly the same sites. They fed largely on crabs and three-spined sticklebacks *Gasterosteus aculeatus*, and one took many butterfish (Kruuk and Moorhouse 1990). One sub-adult otter that we radio-tracked in Shetland made a habit of catching rabbits in their burrows. There were genuine individual differences here, in the newly independent animals, and from the other comparisons one might expect that these differences would disappear again when they grew older.

Prey selection by Eurasian otters, from the range of available species, is considered further in Chapter 9. A general characteristic of Shetland otter prey is that most species are rather small, slow moving and bottom living (at least at the time they were caught). We never saw otters with any of the fast pelagic fishes that occurred along our Shetland coasts, such as mackerel *Scomber scombrus*, and saithe and pollack were taken only during winter, when these species swam between the weeds, rather than in open water as they do in summer. During the spring otters ate more varied, and more 'low quality' food, such as crabs, sticklebacks, butterfish and, in general, lightweight prey. I will show later that this occurred in conjunction with other signs of food stress at that time of year.

In summer the Shetland otters had a less varied diet than at other times: a much greater proportion of their food consisted of just one species (eelpout). Other mustelids also have a more specialized diet during times of plenty, and diet from sparse prey populations may be more varied than when prey is available in high densities (Kruuk 1989; Kruuk and Parish 1981). If a similar mechanism operated in the Eurasian otters in Shetland, this would indicate that the summer was a time of plenty, and this was confirmed by fish trapping (see Chapter 8). A diet that lacks in variety suggests that food is plentiful.

The differences in the size of prey in the diet of males and female Eurasian otters in a coastal habitat make intuitive sense. In Scottish rivers we suggested similar variation, with larger fish (especially salmon) taken more often by males (Carss *et al*. 1990). There is a large discrepancy in size between the sexes (see Chapter 5), and similar sex differences in diet are found in many species of carnivore, especially in mustelids. The difference can be as great as in the wolverine (*Gulo gulo*), where males feed mostly on reindeer, females on rodents and birds (Myhre and Myrberget 1975). The subject has been reviewed by Moors (1980) and Ralls and Harvey (1985), who concluded that the biological function of sex difference in diet was not the avoidance of competition, but a byproduct of the size differential between males and females. This size variation would be favoured by evolution, because small females need less energy for daily maintenance, so they channel more energy into reproduction.

This hypothesis is contradicted, however, by results from stomach analyses of Eurasian otters along Norwegian coasts. There, as in Shetland, male Eurasian otters took substantially larger prey than did females, although Thrine Heggberget showed that this was independent of differences in size of the otters: there were no size-related differences in diet within the sexes (Heggberget and Moseid 1994). As discussed in Chapter 4, there is a sex difference in habitat selection by Eurasian otters, and diet is obviously related to this, as either cause or effect.

Eurasian otter diet elsewhere

The Shetland phenomenon of young, independent otters taking relatively low-value prey for several months after independence has been documented in detail elsewhere, by my student Jon Watt (1993). He used direct observations for his diet assessment, as well as spraint analyses, and showed that Eurasian otters along the Scottish west coast decreased the proportion of crabs in their diet, from about 50% of prey caught by cubs of 4 months old, to about 10% when they were 1 year of age. It was almost zero for adults, which took predominantly slow-moving, benthic fishes.

The fish caught were mostly butterfish, which are smaller, with lower calorific value than the staple prey in Shetland, and sea scorpions (Watt 1995). In my study area in Shetland, butterfish were common, but often ignored by otters, except during the difficult times of spring. The comparison begs the question of why Shetland otters, if they were food limited (see Chapter 11), did not take more butterfish. This may be tied up with seasonal availability, as discussed in the next chapter. Other studies of Eurasian otter diet by spraint analysis, along the shores of the Scottish west coast on the Isle of Skye (Yoxon 1999) and elsewhere (e.g. Mason and Macdonald 1980, 1986) suggest that the Shetland conclusions on prey characteristics, such as size and mobility, are generally applicable.

The food of coastal Eurasian otters in Norway is rather similar to that in Shetland—not surprisingly, as the areas are quite close. They also take few crabs, largely the same genera of fishes (fewer eelpout and saithe, but more cod, etc.), and with mean lengths of fishes of about 20 cm (Heggberget 1993; Heggberget and Moseid 1994—using spraint and stomach analyses). In Norway, and in parallel with our Shetland results, stomachs of otter cubs (killed on roads) contained fishes that were substantially larger than those of females that had given birth recently, and males ate fishes that were larger than those taken by females. However, in the Norwegian study independent sub-adult otters were taking prey that was no different from that of fully adult animals.

In Portugal the marine fish fauna is different, but *mutatis mutandis* the prey of otters was rather similar, comprising largely benthic fish. My student, Pedro Beja, found that Eurasian otters took mostly various small blennies, gobies, rocklings and also the larger, and rather slow-moving, wrasses (Beja 1991). The otters in Portugal were nocturnal, unlike their conspecifics along more northern coasts, and all information on diet was derived from spraint analyses.

Droves of researchers (including myself) have analysed tens of thousands of Eurasian otter spraints from inland waters all over Europe and in several Asian countries. Contents have been expressed in decimals of percentage points, with elaborate statistical analyses. However, given what we now know about the difficulties mentioned above of interpretation of results from spraint analyses, one has to admit to a considerable amount of wasted effort, in producing such precise results. David Carss and his students have demonstrated convincingly that safe conclusions from these studies can be drawn only in terms of an approximate rank order of importance of different foods (Carss and Parkinson 1996).

Earlier studies of diet assessment from spraint analyses have been reviewed by Mason and Macdonald (1986); more recent examples are Adrian and Delibes (1987) for Spain, Kyne *et al.* (1989) for Ireland, Carss *et al.* (1990) for Scotland, Libois and Rosoux (1991) for France, Beja (1991) for Portugal, Sulkava (1996) for Finland, Sidorovich *et al.* (1998) for Belarus, Taastrøm and Jacobsen (1999) for Denmark, Gourvelou *et al.* (2000) for Greece, Lanszki and Molnar (2003) for Hungary, and many more.

In general, this large body of information shows that in all countries fish are by far the most common prey for Eurasian otters, with the occasional exception, such as Poland (Brzezinski *et al.* 1993; Jedrzjewska *et al.* 2001), where frogs *Rana* spp. are taken more often. In neighbouring Belarus, too, frogs are more important in otter diet than in western Europe (Sidorovich *et al.* 1998)—perhaps not surprisingly given the vast numbers of these amphibians there. In Scottish lochs frogs and toads *Bufo bufo* are seasonally common prey of otters (Weber 1990).

Another category of prey that sometimes dominates Eurasian otter diet is crayfish (Fig. 7.12), for instance in Ireland the endemic *Austromobius pallipes* (Breathnach and Fairley 1993), or in the Iberian peninsula the exotic American crayfish *Procambarus clarkii* (Beja 1996a). In Sri Lanka rivers, freshwater

Figure 7.12 Freshwater crayfish *Astacus fluviatilis*. This and other species introduced into Europe are locally important prey for Eurasian otters.

crabs *Potamon* spp. are the most frequently taken prey (De Silva 1997). However, generally these food categories are less important to Eurasian otters than are fish.

Apart from these food categories that dominate the scene, Eurasian otters are not averse to taking from other sources of protein, such as mammals (e.g. rabbits and water voles *Arvicola terrestris*), large aquatic insects and (rarely) carrion, of which there is evidence throughout the literature. Amongst those odds and ends, birds occur most often (see review in Cotgreave 1995).

Probably every species of freshwater fish in the area is vulnerable to otter predation, but some more than others. As examples, in scats of Eurasian otters in Scottish rivers, brown trout *Salmo trutta* and Atlantic salmon *Salmo salar* dominate, and the otters also take many eel *Anguilla anguilla*, as well as stickleback *Gasterosteus aculeatus*, minnow *Phoxinus phoxinus* and lamprey *Lampetra planeri* (Carss *et al*. 1998b; Durbin 1997; Jenkins and Harper 1980). This covers almost all fish species in the rivers in north-east Scotland. In contrast, in spraints from the Torgal stream in southern Portugal, Pedro Beja found, apart from crayfish *Procambarus clarkii* remains, eel as next most frequent prey, followed by the cyprinids nase *Chondrostoma lusitanicum* and chub *Leuciscus pyrenaicus*, several other small cyprinids and loach *Cobitis maroccana*. In terms of energy consumed, however, eel was by far the most important (Beja 1996a).

In Greece, spraints collected along Lake Kerkini contained mostly exotics, the American perch-like *Lepomis gibbosus* (pumpkinseed), and amongst cyprinids the goldfish *Carassius auratus* and roach *Rutilus rutilus* (as well as various other cyprinids and small mammals) (Gourvelou *et al*. 2000). In general, slow-moving species are taken most often (Carss 1995), as is evident also from the otters' foraging methods (see Chapter 9).

Eurasian otters prefer the smaller size categories of fish, but not the smallest. Wise *et al*. (1981) found that salmonids taken by Eurasian otters in streams in Devonshire were around 12 cm long (median) and eels 25 cm, weighing 21 and 22 g respectively. In Poland the median length of fish taken by otters (mostly Cyprinids) was less than 10 cm, weighing less than 20 g (Brzezinski *et al*. 1993). Eels taken by otters in the west of France had a median length of 26 cm and a median weight of 25 g (Libois and Rosoux, 1989), and in the Dee valley in north-east Scotland otters took eels of median length 27 cm, weighing 29 g, and salmonids of median length 10 cm, which weighed around 15 g (Jenkins and Harper 1980). Our own results from the River Dee and tributaries showed salmon in the otter diet with a median length of about 8 cm (Kruuk *et al*. 1993a), except in winter (see below).

In Ireland, where freshwater crayfish *Austropotamobius pallipes* made up about half of the otters' diet in terms of frequency, of the available fish species mostly perch *Perca fluviatilis* was taken, with a median length of 10 cm (weight 10 g) (Le Cren 1951). The Irish otters also took salmonids with a median length of 16 cm (49 g), and eels of 35 cm (72 g) (Kyne *et al*. 1989). Danish otters selected eels between 18 and 36 cm in length, and all other fish between 9 and 21 cm (Taastrøm and Jacobsen 1999). For further detailed lists of otter food species and references, see Mason and Macdonald (1986).

These sizes of prey are median values, but along rivers in Scotland, such as the River Dee, otters may take many fish of well over 30 cm in length, mostly in winter (Fig. 7.13). In some salmon-spawning tributaries about a third of all otter spraints, over all seasons, contained remains of salmonids larger than 30 cm (Carss *et al*. 1998b). Of 57 otter-killed salmon found in tributaries of the Dee, the median length was 71 cm and median weight 2.9 kg (Carss *et al*. 1990). At those times, large salmon were the main prey (Fig. 7.14), and otters ate on average 975 g from

Figure 7.13 Atlantic salmon *Salmo salar*, killed and partly eaten in characteristic manner by a Eurasian otter, on shallow riffles in the river Dee, north-east Scotland.

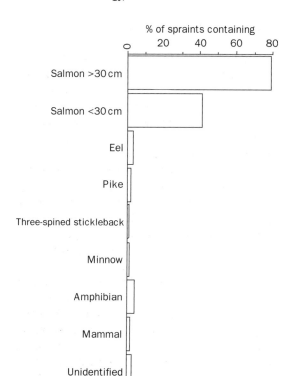

Figure 7.14 Winter prey in a Scottish river, from 324 spraints of Eurasian otters feeding in the River Dee, Aberdeenshire, 1989-1990. Most prey were Salmonid Fish. (Data from Carss et al. 1990.)

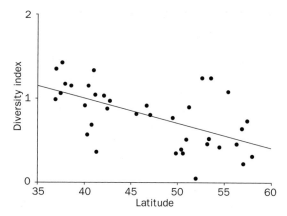

Figure 7.15 Diversity of diet of Eurasian otters, from 37 studies at various latitudes, shows greater food diversity at lower latitudes. Diversity measured with Shannon-Wiener index ($r = -0.57, P < 0.001$). (Adapted from Clavero et al. 2003.)

each fish in one session (range 285–2075 g). There was evidence that it was mainly the large male otters that took the big salmon—a parallel with the observations of otter diet in Shetland.

In summary, the main food of Eurasian otters consists of aquatic vertebrates, mostly fishes with usual sizes of 10–20 cm, or 15–40 g, eels somewhat larger, or frogs and toads with individual mean weights of 20–30 g (Nelson and Kruuk 1997). There are frequent exceptions to these sizes and, where present, large crustaceans also feature prominently.

Recent studies have highlighted some distinct trends in Eurasian otter diet throughout the European continent. Jedrzejewska et al. (2001) reviewed the results of 32 studies of otter diet in 100 localities, and concluded that there were no effects of latitude on the proportions of fish, amphibians or crustaceans in the diet, but there were clear differences between habitat types. The importance of fish in the diet declined from the sea to lakes to rivers to streams, frogs increased, and so did birds and mammals. However, a more sophisticated statistical analysis (Clavero et al. 2003) involved 37 studies and demonstrated that latitude explained 40% of the variation in otter diet throughout freshwater Europe (Fig. 7.15). The higher the latitude, the more prominent was fish in the diet, and the lower the diversity. There was a strong negative correlation between diversity and fish as the otters' main food category: the more important fish was, the fewer species were involved. Obviously, both habitat type and latitude play an important role.

All studies of diet of Eurasian otters that have considered changes over time in freshwater areas have reported a clear seasonality in the occurrence of several types of prey (references above). Eels and crayfish are taken most often in summer, frogs, birds and mammals in winter and spring. In Scottish rivers and lochs, most of the variation in otter diet, especially the occurrence of eels and large salmonids, could be explained by season as well as by locality (lower or higher reaches of rivers, or lakes (Carss et al. 1998b).

Such trends are associated with prey availability (see Chapter 8), and it stands to reason that availability of one kind of prey will also affect the rate at which the alternatives are taken by the predator. Finally, some of the data suggest that, also in freshwater habitats, diversity of diet in winter and spring is greater than that in summer and autumn (Carss et al. 1998b).

Quantities and calories consumed by Eurasian otters

Far less is known about the actual quantities taken by otters, and the variations therein are much more difficult to estimate. One has to rely to a large extent on observations made on captive otters. Two animals kept by Stephens (1957) each ate on average about 1 kg of fish per day over a period of 8 months, which was approximately 15% of their bodyweight per day. Wayre (1979) weighed the food for two captives over 1 week and found daily consumptions of 12.2% and 12.8% of bodyweight. Our own captive otters, which were kept in an open enclosure with a large swimming pool, produced similar estimates. One adult male (9.5 kg) maintained his body mass over 3 months in summer on 1.13 kg fish per day, which was 11.9% of his bodyweight daily. Over the same period a neighbouring group of two females and one male (6.5, 8 and 10.5 kg) also maintained bodyweight by together eating 3.15 (±0.29) kg per day, or 12.6% of their combined body masses.

In the wild, we estimated food intake by one large male along a tributary of the River Dee in the north-east of Scotland, which we could follow and observe intensively in winter by radio-tracking. He ate mostly large salmon and we could weigh the leftovers, measure the length of the fish, from that estimate its original weight, and thus calculate food consumption of the otter. This 8.0-kg animal ate on average 975 g from every fish caught ($n = 39$), and as he took one salmon every night this amounted to 12.2% of his bodyweight per day (Carss et al. 1990). This was an underestimate of daily consumption, as the otter also took some other small prey that we could not assess.

In Shetland we followed almost every move of a radio-tagged lactating female (with one cub), and estimated her total daily intake from the total intake per feeding period (direct observations, with prey sizes estimated) and the length of time spent fishing per day (Nolet and Kruuk 1994). Over a 12-day period, this 5.4-kg animal ate 1.5 (±0.2) kg of fish per day, or 28% of her bodyweight. This was a higher consumption than that of the average otter, as the animal was lactating. Nevertheless, the figure suggests that estimates based on observations in captivity may be too low. This is not unexpected, as energy consumption in captivity is *a priori* likely to be less than that in the wild.

Thrine Heggberget (1995) calculated daily food intake of wild Eurasian otters from the metabolic rates of captive animals, and extrapolated to otters living along sea shores, with certain assumptions. She estimated a daily intake of 1.6 kg of fish by male otters and 1.1 kg by females, or 19% of mean bodyweight. From published data, Pedro Beja (1996a) calculated that an 'average' otter consumes approximately 3150 kJ/day (or 400 kJ per kg bodyweight per day). As a rough approximation, it appears that for otters in nature a daily food intake corresponding to 15% of body mass per day is a conservative estimate, and this is the figure that will be used in calculations in Chapter 9.

The nutritional quality of the different kinds of prey of Eurasian otters varies considerably, and this is likely to have an effect on prey selection. Energetic contents of various species have been estimated using bomb calorimetry. For instance, most of the fish species caught by otters in Shetland (eelpout and all the gadoid fishes rocklings and saithe, etc.) had a calorific content of between 4.2 and 4.5 kJ per g wet weight; eel had 6.1, butterfish 5.0 and sea scorpion 3.8 kJ/g (Nolet and Kruuk 1989). In fresh water in Scotland, these values were more variable: the average eel scored 5.4, trout 5.9, brown frog only 2.9 and common toad 3.4 kJ per g wet weight (Nelson and Kruuk 1997). However, these calorific values are size dependent; for instance, we found a gradient from the smallest eel, which contained 4.6 kJ/g, to the largest with 7.4 kJ/g. Values for fish and crayfish in Portugal were fairly similar to these, for example cyprinid fish 4.8 kJ/g, other fish 4.3 kJ/g and crayfish 4.1 kJ/g (Beja 1996a). Jon Watt (1991) calculated 3.5 kJ/g for the soft parts of crabs, as eaten by otters along the coast, but if one takes into account the large exoskeleton of crabs the figure is only a fraction of this. An average single crab provides less than 50 kJ.

In our freshwater study areas, an average eel (80 g) provided an otter with 430 kJ, compared with an average frog (29 g) that provided only 85 kJ. In Shetland an average size rockling (30 g) provided 120 kJ. The calorific values of fish are therefore greater than those of amphibians and crustaceans, and fish are the premium food.

The New World: diet of the river otter

The American aquatic fauna is different from that of Europe, of course, yet the river otter has the same general preferences as the Eurasian: a *Lutra lutra* transplanted to the USA would not be noted amongst the river otters for its different diet. There have been several detailed summaries of the various kinds of prey taken by *Lontra canadensis* (Hansen 2003; Melquist and Dronkert 1987; Toweill and Tabor 1982), and here I will mention examples. All data are based on spraint (scat) analysis and, although no evaluations of this technique are available for the river otter, I will assume that it is subject to the same reservations as is the case for the Eurasian otter.

A classical project was that by Melquist and Hornocker (1983), in rivers in the mountains of Idaho. They found fish remains in 93–100% of their sample of almost 2000 scats, and far fewer invertebrates such as large beetles, dragonfly larvae and freshwater mussels. Birds, mammals and the occasional snake were also represented. The fishes were mostly suckers *Catostomus macrocheilus*, salmonids of several species (including the migratory kokanee *Oncochynchus nerka* in autumn, and mountain whitefish *Prosopium williamsoni* in winter), and sculpins *Cottus bairdi*, bullheads *Ictalurus nebulosus* and cyprinids. The large majority of fish were sculpins and mountain whitefish. The authors argued that bullheads were underrepresented in the scats, because they have no scales and the sharp spiny remains are often discarded by the river otters.

In Pennsylvania Serfass *et al.* (1990) studied more than 400 spraints of river otters in lakes and rivers. They mentioned the same problem of underrepresentation of bullheads, and also of freshwater mussels, which are eaten by the otters but leave no trace in the scats. Here, also, fish dominated the scene, occurring in 93% of the spraints, followed by crayfish (*Cambarus* spp.) in 44%, and some frogs, insects, molluscs and mammals. The fish were mostly the slow and abundant species, such as perch-like sunfishes or centrarchids, and suckers, as well as various cyprinids, perch and a few pike. There were seasonal fluctuations in the otters' diet, which appeared to be driven by the increase in crayfish abundance in summer.

In rivers and lakes at the much higher latitudes of northern Alberta, Reid *et al.* (1994a) similarly found fish in 92% of the almost 1200 scats analysed. These also were mostly slow ones, such as catastomids (suckers), cyprinids and salmonids, as well as sticklebacks. In winter, when otters had to feed under the ice, their diet was less diverse, and they caught almost only sticklebacks and rather slow cyprinid species. The researchers also found bits of insects (large beetles and dragonfly larvae), snails, birds (ducks), a few frogs and occasionally remains of beaver *Castor canadensis*. It was not certain whether the snail remains in the otter scats were caught by the otters themselves, or by the fish on which the otters fed.

River otters living along sea coasts, such as along Prince William Sound in Alaska, feed mainly on bottom-dwelling fishes (37% of identified food items), but they also take many invertebrates such as marine snails (33%), bivalves (13%) and crabs (9%) (Bowyer *et al.* 1994).

In none of these studies was the size of fish taken by otters measured. Overall, however, the diet of the North American river otter, *mutatis mutandis*, can be described in the same terms as that of the Eurasian otter: the vast majority is fish of virtually all species, with emphasis on the slow-moving and bottom-living ones. These animals also take crayfish and some other invertebrates, and a few other odds and ends such as amphibians and reptiles, birds, insects and small mammals.

Diet of the sea otter

Sea otter diet is of fundamental importance in the story of some fascinating, huge, ecological effects along the Pacific coasts, to which I will return later. The foraging behaviour of this species has enthralled biologists for decades, so that a great deal is known about diet.

Just as for the marine-living Eurasian otters, researchers of sea otters are in the fortunate position of being able to see and identify what the animals bring to the surface when they are feeding. Most information on their food, therefore, comes from direct observations. There are also a few studies of the contents of sea otter scats, left on haul-outs, and

Watt *et al.* (2000) were able to compare results from scat analysis with direct observations. There was an approximate correspondence between these analyses, as for the Eurasian otter.

The first aspect that strikes one, when comparing the many detailed studies of sea otter diet from different areas, is the large variability—greater than for any other species of otter. Animals in the Russian Kuril and Commander Islands eat almost only sea urchins on some of the islands; on others they take fish and several different mollusks, including mussels (Shitikov *et al.* 1973, in Riedman and Estes 1990). In the Aleutians on Amchitka Island, Karl Kenyon in an early study (1969) analysed over 300 stomachs of sea otters, and found that more than 50% of volume was taken up by fish, 37% by molluscs and 11% by sea urchins. On that same island in the 1990s, Jon Watt *et al.* (2000) watched more than 11,000 sea otter dives, which produced sea urchins (*Strongylocentrotus* spp.) in 56%, fish in 8% and various molluscs in 10%; the rest could not be identified, but was probably mostly sea urchin. The fish consisted mostly of Pacific smooth lumpsuckers *Aptocyclus ventricosus*, taken only in winter and spring, when, because of their large size, they made up 62% of the estimated food biomass. Other fish species taken included sand eels *Ammodytes hexapterus* and rock greenling *Hexagrammos lagocephalus*. In summer and autumn sea urchins totally dominated the diet.

In the seas around the Kodiak archipelago sea otters ate almost no fish: clams dominated their food (58–68% of observations), almost all the large butter clam *Saxidomus giganteus*. Mussels made up an additional 22% of observations; the rest consisted of crabs, sea urchins and others (Doroff and DeGange 1994). Sea otters in Prince William Sound also fed mostly on butter clams, and to lesser extent on crabs (Fig. 7.16), especially Dungeness crabs *Cancer magister*, horse crabs *Telmessus cheirogonus* and mussels *Mytilus edulis* (Calkins 1978; Dean *et al.* 2002; Estes *et al.* 1981; Garshelis *et al.* 1986). Along the south-eastern Alaskan coast sea otters did not take fish, but mostly butter clams and some crabs, small sea urchins, mussels and other prey (Kvitek *et al.* 1993).

Along the Californian coast predation on fish by sea otters is 'extremely rare' (Riedman and Estes 1990). Along some parts of the coast the animals eat mostly abalone *Haliotis* spp., rock crabs *Cancer* spp. and red sea urchins *Strongylocentrotus franciscanus* (Fig. 7.17); elsewhere the diet is much more diverse with various species of crabs and clams, snails, mussels, octopus, barnacles, clams, worms, starfish and others (summary and detailed reviews in Riedman and Estes 1988a, 1990).

Figure 7.16 Sea otter eating crab on the surface. © Jane Vargas.

Figure 7.17 Sea otter with sea urchin (*Strongylocentrotus* sp.) in California. In many areas this is its main prey species, with large effects on the ecology of the area. © Richard Bucich.

The causes of the wide variation in diet are several. There are, of course, regional differences in the availability of different food categories, and sea otters respond to these. But there is more to it. First, at least some of the large differences in the fauna of potential food species are caused by the sea otters themselves, as discussed in Chapters 4 and 8. In consequence of the heavy and very significant sea otter's predation pressure, the benthic fauna along coasts where otters have been established for a long time is substantially different from that of coasts where the expanding sea otter populations have just arrived.

The data suggest that, when present in large numbers, sea urchins are the favourite food. Where sea urchins are absent or have been removed by the otters themselves, the animals take to clams or to other prey, such as fish. Where there are large molluscs, such as abalone or butter clams, they may be a principal food item. However, it is not yet clear whether, in the absence of sea urchins, otters switch to a different type of prey such as fish that was always there, or whether the absence of sea urchins causes increased availability of fish (because of changes in the kelp vegetation).

Second, within any one population of sea otters, individuals specialize in different foods—idiosyncrasies that are maintained year after year and passed on from mother to pup. Jim Estes and colleagues (2003b) demonstrated that, over an 8-year period along the Monterey peninsula in California, one female subsisted very largely on sea urchins, two others on mussels, one on both mussels and sea urchins, one mostly on abalone, another almost exclusively on turban snails, *Tegula* spp., and others had different favourites again. There was a striking absence of generalists, and, intriguingly, the diet of weaned pups was more like that of their mother than that of other adults. Male otters also show specializations, and some repeatedly take birds such as grebes, cormorants, gulls or scoters (Riedman and Estes 1988b; VanWagenen *et al.* 1981). Jim Estes *et al.* (2003b) argued that different prey types require specific motor skills from the otters, which have to be learned, and therefore these specializations would bring distinct advantages.

We will see later that the preference for sea urchins, as well as the overall large variation in sea otter diet, has important implications in the role of sea otters in the Pacific kelp ecosystems. This role is especially significant, not only because sea otters often occur in large numbers, but also because they eat large quantities. In captivity, an average 20-kg sea otter will eat a quantity of food of about 20–25% of its own bodyweight, some 4–5 kg (Kenyon 1969), and wild sea otters are estimated to consume an amount of food equivalent to 25–33% of their bodyweight, or 5–7 kg (Costa 1982). These figures do not include the shells of molluscs or sea urchins.

The clams taken by sea otters in Alaska measure, on average, about 6 cm for butter clams and 4–5 cm for other species (Dean *et al.* 2002). This was established from collections of shells cracked and discarded by sea otters, and found by scuba divers at feeding sites. Such clams each supply between 36 and 51 kJ of energy to the otters; crabs were a much better prize, providing between 274 and 505 kJ each (Dean *et al.* 2002). Sea otters need large numbers of such prey to satisfy their energy requirements (Garshelis *et al.* 1986; and see Chapter 9).

The food of giant otters

Because the giant otter is diurnal, its diet has been studied by direct observation of its foraging activities, as well as by faecal analysis. This species is a fish eater, with very little else in its diet, although Nicole Duplaix (1980) found crab remains in some spraints. In the different study areas, most of the food is made up by only a few species of fish, selected from an astonishingly numerous fish fauna. Duplaix mentions that near her Surinam study area 'in 1912, Eigenmann collected 70 to 90 species in one haul of a seine net . . . , and another 60 species in a small creek a few hours later' (Duplaix 1980, p 516). Nevertheless, her giant otters captured only 11 species; most of these were trahiras or wolf fish (the characoid *Hoplias malabaricus*), a fairly large, slow, aggressive predator that lies still in shallow waters, between leaves and branches, and is easy to catch. Other favourites were the similarly slow siluroids or catfish, and perch-like fishes, especially cichlids.

In other parts of its geographical range, for example Guyana (Laidler 1984) and Brazilian Amazonia (Rosas *et al.* 1999; see also the review by Carter and Rosas 1997), the giant otter also specializes in *Hoplias*, siluroids and cichlids. Christof Schenck (1997) analysed 60 faecal samples from 'campsites' in Peru; inevitably, each sample contained contributions from several otters, all mixed together. Identifying fish scales in these samples, Schenck concluded that almost half of the remains came from the small cichlid *Satanoperca jurupari* ('earth eater'), a quarter from the large *Prochilodus caudifasciatus* (a somewhat roach-like fish), and a large number of rarer species. The giant otters did not take (or rarely took) the abundant piranha species. Many of these giant otter prey are either rather slow fishes, such as the cichlids, or they are nocturnal, bottom-living fishes from shallow waters, easy to catch in daytime (like the prey of the Eurasian otters in marine habitats such as Shetland).

Giant otter diet also includes some rather curious items, such as tapir dung and invertebrates extracted from mud. On rare occasions they take a small mammal or amphibian (Duplaix 1980), a bird or a reptile such as anaconda or other snake, and small caimans and turtles. In Brasilia Zoo, giant otters often catch herons that scavenge in their enclosure on fish remains (Carter and Rosas 1997).

The only study in which giant otters were found to eat substantial numbers of crabs was that of Nicole Duplaix in Surinam. Curiously, she never saw crabs being taken, against many fishes, but found quite large quantities of crab remains in 40% of otters' spraints in areas well away from the core areas in the normal, territorial ranges. Possibly, the giant otters' core areas were located where there were more fish; another explanation could be that non-territorial animals, just like the Eurasian otters in Shetland, were more inclined to take substandard prey.

The sizes of fish taken by giant otters are 10–40 cm (review and summary in Carter and Rosas 1997), dependent on species. Duplaix (1980) cited mean lengths for caught *Hoplias* as around 20 cm, for cichlids 10–15 cm, but also incidental captures of tiger catfish of 50–60 cm and peacock bass *Cichla ocellaris* of 35–40 cm. Carter and Rosas (1997) referred to some 'enormous catfish' being taken, with lengths of over 1 m. The main food species in Schenck's (1997) study, *Satanoperca*, is small, on average about 60 g, but the second most common one, *Prochilodus*, had a mean weight of around 450 g.

In captivity adult giant otters consume the equivalent of 7–10% of their bodyweight (i.e. about 3 kg of fish) per day, sub-adults about 13% (Carter and Rosas 1997). In the wild, Nicole Duplaix (1980) estimated that adult otters ate about 3 kg daily. Staib (2002) calculated that the giant otters in her study in the Peruvian Amazon consumed 3.2–4.0 kg of fish per otter per day. It appears evident that, weight for weight, giant otters eat less than, for example, Eurasian otters (see above); this may be a function of their larger body size or of their warmer environment.

Diets of other Latin American otters

The food of the somewhat more secretive **neotropical otter** is known only from faecal analysis. Presumably this carries the same caveats as is the case for Eurasian otters—perhaps even more significantly here as many prey are catfish, which leave little evidence in the scats. There are no quantitative direct observations of the foraging of the neotropical otter, but almost everywhere in its large geographical range fish constitutes the main prey, with the occasional exception where crustaceans dominate the diet. In rivers in Costa Rica, for instance, the otters appear to eat mostly freshwater shrimps such as *Macrobrachium* spp., but even there fish are also important, mostly cichlids and a few other, slow-moving species (Spinola and Vaughan 1995).

Elsewhere, in several areas throughout Brazil and northern Argentina, whether in rivers, lakes or dams, fish totally dominate the diet of neotropical otters, usually cichlids such as *Geophagus* spp., sometimes mostly catfish, or another bottom-living species, the wolf fish *Hoplias* (Gora *et al.* 2003; Helder and Andrade 1997; Muanis and Waldemarin 2004; Pardini 1998; Quadros and Monteiro 2001; see also review in Larivière 1999b). These same studies show that crustaceans, such as the freshwater crab *Trichodactylus fluviatilis*, are important in some areas, and otters even eat the occasional fruit, reptile, large insect, bird or mammal, rarely a few molluscs. In the Brazilian Pantanal I saw neotropical otters take 17 fishes, almost all various species of catfish. In marine habitats neotropical otters similarly eat mostly fish, closely followed in importance by various species of crab (Alarcon and Simoes-Lopes 2004). In general, the fish species taken by neotropical otters are the slow, bottom-living ones; the much faster piranhas, for instance, are left alone. There is no information on sizes of prey.

Further south on the continent, in southern Argentina and Chile, the **southern river otter** has similar feeding preferences. Mostly in streams and lakes, and dependent on the area, either fish or crustaceans dominate the diet, as is the case where it occurs in the sea (see review in Larivière 1999a). In fresh water in southern Chile, Gonzalo Medina (1997, 1998) found crustaceans by far the most important prey remains in spraints, mostly crabs *Aegla* spp., as well as the crayfish *Samastacus spinifrons*. Next in importance were small fish, usually less than 10 cm long, especially cichlid-like species (*Percilia gillisi*) and various introduced salmonids. Medina also found a few remains of birds in the spraints, as well as amphibians and molluscs.

Chéhebar *et al.* (1986, quoted in Larivière 1999a) analysed spraints of southern river otters near lakes in southern Argentina; these also contained mostly crustacean remains. In some marine habitats, however, fish predominate in the diet (Sielfeld 1983, in Larivière 1999a).

The sea cat or **marine otter** occurs on South American Pacific coasts along a large range of latitudes, and its diet varies with the prey assemblages.

However, everywhere its food is dominated by crustaceans, as noted in several studies by direct observation of the animals (which forage by day). In an excellent study at different latitudes in Chile, Rick Ostfeld and collaborators (1989) saw more than 1500 feeding dives; otters came up with crabs in 70%, shrimps in 6%, fish in 20% (Fig. 7.18) and molluscs in 4%. The crabs were mostly *Cancer* spp. and *Homolaspis plana*, the molluscs a large snail *Concholepas concholepas*, and limpets *Fisurella* spp. In these areas sea urchins were abundant, but, as an interesting contrast with the sea otter, marine otters were not interested.

Elsewhere in chile, Medina-Vogel *et al.* (2004) saw crustaceans being taken in 91% of successful dives, fish in 9% and molluscs in fewer than 1%. In spraints from the same areas they found crab remains in 78% (mostly *Taliepus dentatus* and *Cancer edwardsi*) and fish in 20%. Results from earlier studies are similarly crustacean-dominated, as reviewed by Larivière (1998); occasionally birds and small mammals have been noted as prey, and even fruits. Captured fish belonged mostly to families of small, slow species such as blennies *Blennidae*, morwongs *Cheilodactylidae* and clingfishes *Gobiesocidae*.

Where marine otters occur in the same general regions as southern river otters, they tend to take much the same prey, but from different types of coast: marine otters along very rocky and exposed

Figure 7.18 Marine otter having landed a fish, Chile.
© Gonzalo Medina.

locations, southern river otters in more sheltered and forested places (see Chapter 11).

Diets of otters of Asia

In the spectacular forests of western Thailand, the river Huay Kha Khaeng ('River Kwai') and its tributaries provide the habitat for three species of otter. They use the same waters and one another's spraint sites—a setting just begging for a study of the animals' food requirements. With several colleagues I watched the otters and their tracks, and analysed their spraints, and the results showed a striking collection of food preferences (Kruuk *et al.* 1994a) (see Fig. 11.11).

The **Eurasian otter** did much the same thing as in comparable places in Europe: its most common prey was fish (in three-quarters of spraints), closely followed by frogs, and it also ate some small mammals, crabs and large insects. The fish were mostly between 10 and 15 cm long. By comparison, this same species in streams of Sri Lanka (where it is the only otter) eats mostly crabs, far fewer fish and only a few frogs (De Silva 1997).

In the Thai study the Eurasian otters faced potential competition from the larger **smooth otter**. For smooth otters, fish was even more important: we found remains in almost 90% of the spraints, with only a few bits of frogs and crabs, and a few snakes. More than half of the fish remains were of fishes longer than 15 cm. In India the smooth otter takes almost only fish, and very few prey from the other categories. Hussain and Choudhury (1998) established that, in the Chambal River, the large majority of smooth otter prey belonged to two species, the mullet *Rhinomogul corsula* and the catfish *Rita rita*, with mean sizes of 16 cm, although the catfish could be up to 46 cm long. In ponds of Keoladeo National Park, smooth otters also ate mostly fish, with a few molluscs, insects and birds (Haque and Vijayan 1995). In a large Kerala reservoir 96% of the estimated volume of their food was fish, 61% the introduced *Tilapia mossambica* and the European carp *Cyprinus carpio*; a further 21% consisted of the catfish *Heteropneustes fossilis* (Anoop and Hussain 2005).

The big contrast in our Thai study was the third otter species, the **small-clawed otter**. Its diet was dominated totally by the freshwater crab *Potamon smithianus*, with a few frogs and fish thrown in. Almost all of the crabs eaten were small, around 2 cm in carapax width. This otter species was already known as a crab-eater, also in rice paddies (Foster-Turley 1992). Kanchanasaka (2004), too, found that in swamp forest in southern Thailand small-clawed otters ate mostly crabs, followed by snails, fish and snakes.

The one Asian freshwater otter that was missing from our Thai study area was the **hairy-nosed otter**, a rare species usually occurring in forest swamps. What little is known of this otter suggests that, very much like the Eurasian otter, it feeds mostly on fish, and far less on snakes, frogs and crabs, in that order (Kanchanasaka 2001, 2004).

African otter diets

On Rubondo, a beautiful forested, uninhabited island in Lake Victoria, I had the opportunity to study two of the three otter species that occur south of the Sahara: the spotted-necked and the Cape clawless (Kruuk and Goudswaard 1990). **Spotted-necked otters** were everywhere along the coasts, active by day, and it was quite easy to watch them feed, and to collect and analyse their spraints. The murky waters of Lake Victoria are host to a large number of fish species, mostly small cichlids (*Haplochromidae*), locally called furu-furu. Most of the prey of the spotted-necks were furu-furu of 3–6 cm long (size calculated from the size of fish eye lenses in the spraints), but they also took many of the, much larger, introduced tilapia *Oreochromis niloticus*, up to 30 cm long, and the catfishes *Bagrus* sp. and *Clarias* sp., as well as the occasional crab *Potamon niloticus*.

Previous researchers of the spotted-necked otter in East Africa came up with similar results. John Procter (1963) watched them over a 1-year period in Lake Victoria, taking almost only furu-furu; he quoted observations from other observers of this species taking big catfish up to 70 cm long, with occasional crabs. In Rwanda Anne Lejeune (1990) analysed 150 spraints; most of the prey were small *Haplochromis* spp., but when she took the size of the prey into account the actual bulk of the food was mostly the much larger tilapia. Her spotted-necked otters also took a few molluscs and large insects. However, she noticed that the large catfish *Clarias* sp. (Fig. 7.19) was commonly eaten by the otters, caught by the otters themselves or taken from nets of local fishermen, and there were no remains to show for this in the spraints, because otters did not eat the head or vertebrae, and catfish do not have scales.

In South Africa the spotted-necked otter is more nocturnal, and spraint analysis the only means of studying the diet (review by Rowe-Rowe and Somers 1998; Carugati and Perrin 2003a,b). Most of the prey was fish (*Barbus* spp., and introduced trout), but in all studies the spraint contained almost as many crabs, commonly also frogs, as well as the odd insect and bird.

The **Cape clawless otter** shows a different diet profile. In Lake Victoria it ate almost only crab with a few of the introduced *Tilapia* (Kruuk and Goudswaard 1990). Almost everywhere else in Africa the Cape clawless is largely a crab eater, but not exclusively, and in lakes where crabs are absent it lives largely on fish, often introduced tilapia (Watson and Lang 2003). Nevertheless, crabs such as *Potamon* or *Potamonautes* sp. are the mainstay in fresh water throughout their geographical range. Butler and du Toit (1994) confirmed this for rivers in Zimbabwe, where Cape clawless otters also took some catfish (*Amphilius* sp.), frogs and mottled eel *Anguilla bengalensis* (as well as some insects, especially dung beetles, birds and small mammals).

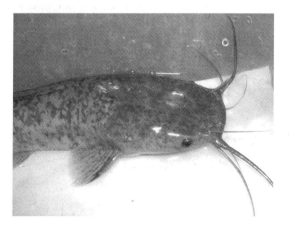

Figure 7.19 Catfish (*Clarias* sp.), Mozambique—a frequent prey of spotted-necked otters. Catfish on all continents are often taken by otters of many species. © Alice Courage.

Similar results have been obtained by Rowe-Rowe (1977) and Carugati and Perrin (2003a,b) in rivers of Natal, and by Ligthart *et al.* (1994) and Somers and Nel (2003) in lowland rivers of the Cape. These last authors identified fish taken by the clawless otters as mostly the introduced tilapia, trout and bass, as well as the endemic Cape rocky *Sandelia capensis*.

Unlike the spotted-necked otter, the Cape clawless frequently occurs in the sea, where its prey is from somewhat different categories. Mike Somers (2000a) watched the animals foraging and analysed spraints in an area east of Cape Town; in his observations crabs *Plagusia chabrus* and *Cyclograpsus punctatus*, as well as the lobster *Jasus lalandii*, were most frequent food items. There were also various, mostly benthic, fishes such as the klipfish *Clinus superciliosus*, octopus and a few molluscs (abalone, *Haliotus* sp.). However, when taking into account the size of prey, it was clear that fish ranked as the most prominent food.

These results largely echoed what had been found earlier by Van Der Zee (1981) and Verwoerd (1987), in areas further east along the same coast. However, Charlie Arden-Clarke (1983), in Tsitsikama along the Eastern Cape coast, found crabs to be the most important foods, and more recently Emmerson and Philip (2004), in the Eastern Cape, observed the spiny lobster *Panulirus homarus* as the most prominent food item (as well as several crab species and fishes). There is no information on differences in availability of these different prey types in these areas.

Apart from the main, dominant food types of crustaceans and fish, Cape clawless otters have gained the reputation for taking the odd bird, such as domestic ducks, coots, geese and even swans (Rowe-Rowe 1978). They also eat many different, large, water insects, reptiles and rarely small mammals (summary in Larivière 2001a). Generally, however, in freshwater habitats crabs dominate the diet, and when feeding in the sea the Cape clawless is dependent on both bottom-living fishes and crustaceans.

Sadly, there are no hard data on the food of the **Congo clawless otter**, or swamp otter, but what little we do know suggests that these animals have some highly unusual feeding habits. Crabs are likely to be their main food, but there is good, anecdotal, evidence that the diet includes especially earthworms, as well as fish, molluscs, frogs, and other small vertebrates and invertebrates (Carpaneto and Germi 1989, in Larivière 2001b; Jacques *et al.* in press). There are detailed observations of these otters catching giant earthworms (see Chapter 9), but research on this species is needed.

Diets: some generalizations and comments

The standard otter, irrespective of species (excepting the sea otter), is a fish eater, with crabs and frogs as secondary foods, and, presumably, in evolution the precursor species conformed to this picture. Thirteen otter species, evolved into habitats all over the world, show themselves remarkably conservative in their diet: usually, given the local aquatic fauna, one can more or less predict, in a qualitative sense, what an otter is going to take there. Otter diet reflects local availability of relevant prey species. There are quantitative differences between otter species, and some show distinct departures from this general feeding pattern—more about these later.

Given this conservatism, it is probably true that other, more detailed, characteristics of the diet occur throughout the *Lutrinae*, and with some care one may be able to extrapolate from one species to another. This is useful, as some of the otters have been studied much more intensively than others.

For instance, several otter species have been shown to take especially those fish species that are slow and bottom living (or resting on the bottom), rather than fast and pelagic ones. This preference prorbably holds true for all otter species. As a similar generalization, in species where males are substantially larger than females, prey taken by males tends to consist of larger fish. There exists a seasonality of diet which reflects variations in availability of different fish, crustaceans and amphibians. The animals provision their young with prey that is different in size from that which the mother takes for herself (in some species larger, in some smaller).

Some populations, and in the very south of South America two species (southern river and marine otters), have a diet with more crabs than fish, probably related to local availability, and fish is still very important to them. Otter species that have evolved with a substantially different diet are, first, the sea otter in the northern Pacific, which eats some fish in some sites, but has become a predominantly invertebrate specialist. We will see that also its foraging behaviour is rather different. Second, there are the three *Aonyx* species—clawless or small-clawed otters in Africa and Asia—which also often take fish, but for which crabs have become much more important. All of the other otter species conform to the classical fish-eating pattern.

Given the aspects of diet that are shared by the different otter species, a relevant question is whether they compete with one another in those regions of the world used by more than otter one species. This is discussed further in Chapter 11; suffice it to say here that the comparisons demonstrate that, despite the similarities in specialization between species, when it comes to cohabitation their interests are substantially different. It has not been established whether, in any of the cohabiting populations, otters have a diet that is actually different from what it would have been had no other otter species been present.

The amounts of food consumed have been studied in some detail only in the sea otter, the Eurasian otter and the giant otter. The sea otter, the only one that spends virtually all of its life in water and therefore needs the most energy, takes a quantity per day equivalent to some 20–30% of its bodyweight. The Eurasian otter needs less than that, probably in the order of 15–20% of its bodyweight, and the giant otter requires even smaller relative quantities. It is likely that the needs of other species are more similar to that of *Lutra*, probably dependent also on climate.

Even such rough approximations show that otters eat very substantial quantities, and this is discussed in more detail in Chapters 9 and 13. In comparison with other mammals, otters have unusually large demands.

CHAPTER 8

Resources: about fish and other prey

Introduction

Recent reports on dramatic changes in some of the otters' important prey species should be of serious concern to otter conservationists. In Europe and North America, populations of some salmonids and eels have been reduced to only a fraction of their former level, in a fairly short stretch of years. On the European continent recently introduced crayfish developed large populations; these may benefit Eurasian otters, but frogs are declining rapidly worldwide, and around Shetland fish populations are crashing.

There have been large changes in coastal ecosystems around the world, and also in our understanding of the causes of these changes. The research on sea otters has been at the start and the centre of much of the relevant recent ecosystem analysis, and the decline in numbers of this charismatic animal, with its consequent effects on habitat and resources, has attracted international attention. However, for most of the other otter species our knowledge of relations with their resources is negligible. Characteristics of prey populations are rarely addressed by ecologists studying mammals such as otters, perhaps because the subjects are less charismatic and more intractable. However, the vital statistics of prey populations are more important to otters than, for instance, the structure of their habitat.

All species of otter need unusually large quantities of food, and shortages have rapid consequences (see Chapters 7, 11 and 12). Questions that need to be addressed, in order to understand the requirements of a foraging otter in any one ecosystem, are:

- Which species and which sizes of fish and other prey are there, and what do otters select?
- What is the prey density, biomass and productivity in different zones and places, and throughout the seasons?
- How do fish and other prey behave; for example, where in the habitat are they at various times of day, and what are they doing when the otters are hunting for them? What is their distribution? What happens after predators remove prey?
- What is the effect of otters themselves on their prey populations?

There are few studies that discuss these problems in the context of otter populations, with sea otter research a notable exception. In the following sections I describe some relevant observations from our Shetland study on Eurasian otters, on the same species elsewhere, and from others such as the sea otter. Some of the conclusions may be extrapolated to interactions between less well studied species, and their role in ecosystems. By providing detail of the studies on fish in Shetland, I hope to inspire similar work on resources of otters in other parts of the world.

Fish numbers and behaviour in Shetland

To understand fully the relationship between otters and their food, one would have to watch actual otter–fish interactions in every detail. The best possibilities for this occur in clear, coastal waters (such as those off Shetland), and with otters active by day. However, try as I might whilst scuba-diving along the Shetland coasts in a dry suit, I was never able to see wild otters find and take even one single fish. Quite simply, I just could not keep up with the otters in the dense forests of huge algae. I did see fish, many of them, and beautiful they were—but even in that I was nowhere near as efficient as the otters. In places where it took a Eurasian otter 10–20 seconds to emerge with a fat rockling, I might be rummaging around for half an hour before I finally found one. The otters' performance was totally superior to mine.

If direct observations of the interaction between predator and prey were impossible, both in the sea and in freshwater areas, the next best thing was various indirect methods of deducing what happened. By trapping fish, and also observing them in an aquarium in Shetland, we were able to obtain data relating to most of the problems mentioned above, except on fish productivity. This we could estimate only in rivers and streams of study areas in north-east Scotland (Kruuk *et al.* 1993a), not in the sea.

It is useful to describe some of the fish observations, just to 'get the feel' of one of the otters' prey communities. Of course, otter populations elsewhere will be dealing with different prey situations, and what we were seeing in our Shetland and Scottish study areas has immediate relevance only for the **Eurasian otters** there. But at least our data give an idea of the kind of relationship between otters as predators and their food species, and it suggested an approach that could be useful elsewhere.

One problem initially was the *embarras de richesse*: there were many species of fish, all fascinating in themselves. Even after more than 3 years of regular fish-trapping along the coasts of our study area in Lunna, Shetland, we kept coming up with new species: we caught more than 30, as well as various crabs and lobsters, and little was known about any of them.

The most common fish around the Shetland shores is the abundant saithe *Pollachius virens*, which used to be staple diet for the Shetlanders. A characteristic member of the cod family, numbers of them swim around the rocks and in open water. We often caught smaller ones in our fish-traps, but only in winter, when they move between the subtidal algae. In summer the saithe are still close to the coast but in more open water, and otters catch them only in winter, when the smaller saithe are 'available'.

The most common benthic fish along the coast Shetland, both in our fish-traps and as a prey for our otters, was the eelpout *Zoarces viviparus* (Fig. 8.1), an interesting northern species with various peculiarities. It is highly camouflaged with its subtle pattern of green, brown and orange; it has a body like an eel and a face like a frog, with bulging eyes and thick lips. Its bones, as found in an otter spraint, are sea-green. A typical specimen is less than 20 cm long, weighing 10–20 g; the largest we ever caught was 28 cm, although it can reach 50 cm (Muus and Dahlstrom 1974; Wheeler 1978). The eelpout produces live young, but very little is known of its behaviour otherwise. In daytime it is found under stones, but in an aquarium, or when diving at night, one notices that eelpout do swim around in mid water like other fish, although almost only in darkness; they are clearly nocturnal (Westin and Aneer 1987). In winter they live in somewhat deeper waters, but later they come closely inshore to mate and reproduce.

Figure 8.1 Bottom-dwelling, nocturnal eelpout *Zoarces vivparus*, the most common prey of Eurasian otters in Shetland.

Figure 8.2 The five-bearded rockling *Ciliata mustela*, nocturnal and bottom dwelling, is the prey with greatest volume in the diet of Eurasian otters in Shetland.

Figure 8.3 Butterfish *Pholis gunnellus* at night—frequent prey of Eurasian otters along all Scottish coasts.

Figure 8.4 Sea scorpion *Taurulus bubalis*, a little active, diurnal and benthic species, often taken by coastal Eurasian otters.

The other main actor on the fish scene was the five-bearded rockling *Ciliata mustela* (Fig. 8.2), also a common prey for the otters elsewhere. This fish shows similarities with the eelpout: somewhat eel shaped, it is also found under rocks and is of comparable size (20–30 g, but up to 200 g in weight). It is common everywhere along the European Atlantic coasts; the fry are pelagic and their huge summer swarms are a frequent food for smaller sea birds. It is strictly a species of the night. There were three other species of rockling, but the five-bearded was by far the most abundant.

The butterfish *Pholis gunnellus* and the sea scorpion *Taurulus bubalis* were very common (Figs 8.3 and 8.4). These are the species seen most often when diving or snorkeling, apart from the very small (and for our purposes irrelevant) gobies. The butterfish are beautiful, with their sharply defined black spots along the back and snake-like movements. They occur intertidally in large numbers, in shallow waters between the rocks, stones and algae, but they also go deeper, well out of reach of the otters (Wheeler 1978). They may be up to 20 cm long, but are light (5–15 g); there is not much flesh on a butterfish, even on a big specimen. They are curiously flattened sideways. They feel slimy, and also spiny because of the hard sharp rays in the dorsal fin—not much of a prize for a predator. Butterfish are mainly nocturnal (Westin and Aneer 1987), but not so exclusively as rocklings.

In contrast, the sea scorpion is active mostly in daytime—if active is the right word, as most of its time is spent lying still on or between the rocks and weeds, waiting for its prey to move. Short and broad, 5–50 g, dark brown and with its head and gill-covers covered with bony knobs and bumps and very sharp spines, the sea scorpion should be well camouflaged and protected against onslaught from an otter. Nevertheless one sees them quite easily when scuba diving, and otters do not seem to be deterred by the spiny protection. They handle them rather carefully, though, often taking even small specimens ashore to eat. There is also a larger relative of the sea scorpion that is taken quite often by otters: the bullrout *Myxocephalus scorpius*. It weighs 60–120 g and is adorned with a strikingly bright red or orange belly, with white polka dots. It is a thrilling sight to see an otter carrying one ashore with the bright flash of red pointing ahead, and the large pectoral fins standing out, sometimes obscuring the otter's forward view. This species is known to be nocturnal in summer and diurnal in winter (Westin and Aneer 1987).

These were the common Shetland fishes that seemed to play the most important roles in the life of an otter, but I could have included several more, such as the ordinary stickleback *Gasterosteus aculeatus* and larger species such as lumpsuckers *Cyclopterus lumpus* and dogfish *Scyliorhinus canicula*; the latter two are quite common in slightly deeper parts of the coastal zone, and popular with otters. Everywhere along those Shetland shores there were thousands of crabs, almost all shore crabs *Carcinus maenas*, in sizes of up to 8 cm in carapax width. Otters had no problems catching and eating them (see Fig. 7.3), but energetically it was hardly worth their while.

Observations of fish behaviour in Shetland

In Shetland I used a glass fish tank, measuring 1 × 0.4 × 0.3 m, installed in front of a window close to the shore, and for many months we kept several of the fish that were of interest to the otters, mostly rocklings, eelpout, sea scorpion and butterfish. Obviously one has to be very careful when extrapolating from captivity observations to what is going on in the wild, but, by using the appropriate combination of several species of fish, the local seaweeds (various wracks) and rocks, we were able to catch at least glimpses of life as it is being lived along the shores. For instance, it was immediately clear that rocklings, eelpout and butterfish were largely active only during the night (Fig. 8.5), especially the first two species. I collected quantitative data by recording whether the fish were swimming or resting, at all times of day or night and without disturbing them, although if necessary I used a small torch.

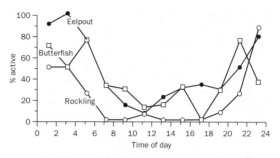

Figure 8.5 Twenty-four-hour activity of prey fish of Eurasian otters in Shetland. Swimming activity of three eelpout, three five-bearded rocklings and four butterfish in an aquarium was determined during spot checks at all hours over 15 days.

When any of these species of fish were not active, they were difficult to find in the aquarium, hidden as they were somewhere beneath stones or in the densest parts of the weeds, beautifully camouflaged and keeping absolutely still. As an experiment I tried several times to disturb them by stirring the weeds in the aquarium with my hand, in the way I thought an otter would have done with its head. The fishes did not budge. I could even touch them without difficulty: there was little doubt that an otter can catch them easily if it is able to distinguish fish from weeds by touch, with its whiskers. However, when the fish were active at night, they reacted strongly to my hand and moved fast: I was quite unable to get anywhere close to them. Clearly the best time to catch these fish is in the daytime, if one can find them by touch, in the weeds or under stones. This observation is of obvious relevance for the otters.

The sea scorpion, which is a frequent prey of the otter, is diurnal, but it never swam about a great deal at any time, it could easily be touched by hand and was always readily visible, perhaps relying on its spiny armour. It seemed to be the proverbial sitting duck for an otter.

From these aquarium observations, and by snorkelling and scuba diving, I had some idea about where and how the relevant fish occurred along the Shetland coast, and about the way in which otters could catch them. This needed to be quantified, to study seasonal and local differences in fish availability along the coast, and to estimate fish densities. To sample the wild populations, we applied small, purpose-built fish-traps, and the results turned out to be quite useful for the understanding of otter behaviour.

Trapping and counting fish

The design of a fish-trap is simple, age-old and used worldwide. It is a cylinder blocked at one end, with a funnel-shaped entrance (Fig. 8.6). I used a heavy-duty, plastic gauze. Each trap was 50 cm long with a diameter of 27 cm. One side of the cylinder consisted of a funnel ending in an opening 7 cm across; at the other end of the cylinder was a small flap for removing fish. Inside we tied a large stone, which kept the trap on the bottom and prevented it from moving about too much. The traps were not baited, attached to a rope, and on days when my students and I were fish sampling we threw ten or twenty of them from the shore into the sea below the low-tide mark, and checked them a day later.

Later, the model was improved by making the cylinder longer (80 cm) and adding a second funnel in line with the first one so that the trap effectively had two chambers: an entrance porch and a holding chamber. This twin-funnel type was more effective in holding fish once they were caught (Kruuk *et al.* 1988). In 1493 'trap-nights', at all times of year, we caught 4141 fish and 1122 crabs, a total of 5263. For further analyses, these figures needed various correction factors, to allow for trap type and for the variation in 'trappability' between species (see below). Table 8.1 gives an overall impression of the species composition of the fish fauna in the Shetland study area, but it does distort.

One of the first things I wanted to know was whether the species that were relevant to otters were, indeed, more active at night, as the aquarium observations suggested. One would expect to catch more overnight in the traps than during daytime and, as shown in Figure 8.7, there was a striking difference, confirmed for several species by a Finnish study (Westin and Aneer 1987).

Table 8.1 Fish caught with funnel traps in otter study area at Lunna Ness, Shetland 1983–1987

Species	% of total ($n = 4141$)	Mean weight (g)
Eelpout *Zoarces viviparus*	44	11
Butterfish *Pholis gunnellus*	8	10
Five-bearded rockling *Ciliata mustela*	13	27
Northern rockling *Ciliata septentrionalis*	1	27
Shore rockling *Gaidropsarus mediterraneus*	2	81
Three-bearded rockling *Gaidropsarus vulgaris*	3	70
Sea scorpion *Taurulus bubalis*	2	16
Bullrout *Myxocephalus scorpius*	1	89
Sea stickleback *Spinachia spinachia*	5	3
Saithe *Pollachius virens*	13	25
Pollack *Pollachius pollachius*	3	19
Eel *Anguilla anguilla*	1	58
Yarrell's blenny *Chirolophis ascanii*	<1	n.d.
Two-spotted goby *Gobiusculus flavescens*	2	n.d.
Montague's seasnail *Liparis montagui*	<1	n.d.
Topknot *Zeugopterus punctatus*	<1	n.d.
Plaice *Pleuronectes platessa*	<1	n.d.
Cod *Gadus morhua*	<1	n.d.
Poor cod *Trisopterus minutus*	<1	n.d.
Three-spined stickleback *Gasterosteus aculeatus*	<1	n.d.
Lumpsucker *Cyclopterus lumpus*	<1	n.d.
Conger eel *Conger conger*	<1	n.d.
Pipe fish *Syngnathus acus*	<1	n.d.
Dogfish *Sycliorhinus canicula*	<1	n.d.

n.d., no data.

Figure 8.6 Standard plastic two-funnel fish-trap *in situ* underwater, for sampling fish populations in coastal habitats.

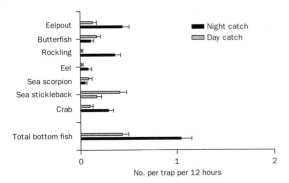

Figure 8.7 Daytime and night-time fish-trap catches of Eurasian otter prey in Shetland over 108 days show that the main prey species are significantly nocturnal ($P < 0.001$, Mann–Whitney U test).

Along the sea shore, animal life appeared to be geared to the state of the tide. It affected the landscape, the birds, the places where we could walk, our ability to see otters and the activity of the otters themselves. The question was whether it affected the activity of fishes. Traps were left out, well below the low-tide line and in the areas where otters habitually foraged, for 3 hours around high tide, 3 hours around low tide and 3 hours for periods in between when the water was rising or falling. Time of day had to be allowed for, so that we had tidal observations both at night and during daytime.

Again, the results were significant: in general, and allowing for the time of day and night, bottom-living fishes and crabs were most active at high tide (0.30 fish caught per 3-hour period per trap) and least active at low tide (0.17 fish per 3 hours), with catches at falling and rising tides intermediate. Shore crabs, too, were least active during the ebb tide (Kruuk et al. 1988). Clearly a predator that specialized in catching bottom-living fish or crabs during their inactive period would find most available prey at low tide, and in daytime.

There were other variations in our fish catches that were more difficult to account for, and probably nothing to do with tide, time or season, and nothing obvious with the weather. Some days all of our traps would catch many fish, of different species, and perhaps the next day with ostensibly the same weather, none would catch anything. One could expect stochastic variation in the numbers, but it was obvious that the observed variation was much greater than that. This is something that every fisherperson knows about, and there must be as many different explanations for these fluctuations in the catch as there are fisherfolk. This variation is just as likely to affect the otters' fishing success as it did our own.

As one would expect, there also were differences in fish numbers along the coast, between bays, shallow sloping coasts, or steep cliffs falling into deep water, between a sandy bottom or large rocks, and many different vegetations of algae. We set fish-traps in various places in the study area that represented the different coastal types, and analysed how the catch in our traps was associated with these differences.

Every species of fish showed its own pattern (Kruuk et al. 1988, 1995). For example, the eelpout was most common along sheltered coasts (i.e. there was a negative correlation with 'exposure'), it avoided *Gigartina stellata*, a typical seaweed of exposed shores, but it was to be found especially amongst the knotted wrack *Ascophyllum nodosum* and bladder wrack *Fucus vesiculosus*, the typical vegetation of the sheltered voes. Shore crabs showed the same distribution as eelpout. However, the other important prey of otters, the five-bearded rockling, was found closely associated with algae of the really exposed coasts in Shetland, *Gigartina*, and the thongweed *Himanthalia elongata*. The other rocklings showed even more avoidance of sheltered areas with *Ascophyllum*, but they were associated with the indicators of exposure, and with large boulders.

Butterfish and bull-rout were found everywhere, with no particular preference for, or avoidance of, any of the factors measured. However, the sea scorpion was a fish of the wilder shores, with almost exactly the same habitat preference as the five-bearded rockling. The two more mid-water fish, saithe and pollack, showed another pattern again; they were found along steep slopes where the water was deep, but with a preference for sheltered areas rather than exposed ones.

Usually the traps were set for 1–2 metres deep at low tide. The otters, however, fished over a much wider zone, sometimes where the water was more than 10 metres deep, or in more shallow places. They would spend different amounts of energy according to the depth at which they were fishing, and their success rates varied with depth (see Chapter 9). It was important, therefore, to discover what the differences in fish numbers were over a range of depths. We compared the catches in a line of six trap sites, from 1.5 to 11 metres deep, in one of the bays, and in summer only (Kruuk *et al*. 1988).

There was considerable variation in the occurrence of different species of fish with depth (Fig. 8.8). Eelpout and butterfish were more common in shallow water, but not significantly so, whereas five-bearded rockling, sea scorpion and shore crabs were found only in places that were less than 4 metres deep at high tide. The larger fish, such as three-bearded and shore rocklings, occurred more often deep down, with some other larger species. Overall, there were many more fish in the shallow strip along the shore, but the mean weight per fish increased with depth, and consequently the total biomass of fish caught per trap did not change significantly with depth.

One of the important implications of these observations for a predator is that, in order to exploit eelpout as well as rocklings, sea scorpions as well as saithe, access is needed to different types of shore, to exposed areas as well as sheltered voes, to steep coasts as well as gently sloping ones. Greater depths should be fished if larger prey are required, for example to provision cubs, and shallow waters, close inshore, are to be exploited for larger numbers of small fish. Ideally, all such areas should be included in an otter's home range.

Seasonal variation in fish numbers, distribution and weights also played a significant role. Otters need to exploit a range of different species of fish, rather than specialize in just one of them. This became evident when we compared the seasonal availability of each species over a 4-year period (Kruuk *et al*. 1988). The data showed that, overall, most species of bottom-living fish (except for the five-bearded rockling) were least common in late winter and spring—a tough time for otters (see Chapter 12). However, every fish species had its own pattern of seasonality, and there was no significant overall correlation between them.

Every year we found a large increase in the total catch of fish in traps during August (see below). This was due largely to one species, the eelpout, which completely dominated the underwater scene during that month. There were no clear seasonal patterns for the five-bearded rockling, but butterfish was most abundant in summer, sea scorpion and shore crabs in autumn, and saithe in winter. Only the sea stickleback, a fish that is a rather small and bony prey for an otter, was clearly more abundant in spring than at any other time.

This seasonality stood in marked contrast to observations of prey for Eurasian otters along the Portuguese coast. Pedro Beja (1995a and b) established that the otters took mostly corkwing wrasse *Symphodus melops*, several shannies, blennies and gobies, the shore rockling *Gaidropsarus mediterraneus* and the conger eel *Conger conger*. Using fish-traps of similar design as those in Shetland, as well as hand-netting and angling, he found that all of these species were present in larger numbers and were also heaviest during winter and early spring, but least available during late summer and autumn.

In Shetland there were also considerable differences in fish populations between subsequent years, although the seasonal pattern generally remained the same. We put out numbers of traps during August, from 1983 until 1988, and there were large, significant differences in the size of the eelpout 'glut' during that month. This variation was highly important to

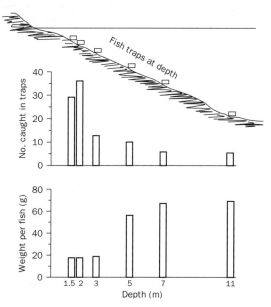

Figure 8.8 Eurasian otter prey species in Shetland, in fish-traps at various depths in catches over 15 days in June 1987. There were more fish in shallow waters, but they were smaller. (After Kruuk *et al*. 1988.)

the otters, especially to the breeding females (see Chapter 11).

To complicate matters, the seasonality of the various fish species was different along various sections of coast, especially for the otters' two main prey species: five-bearded rockling and eelpout. Thus, to exploit a particular prey species efficiently, an otter would have to fish different parts of the coast at different times of year. A two-way analysis of variance showed that this complicated interaction between seasonal and area effects was highly significant for eelpout, five-bearded rockling, three-bearded rockling, saithe and pollack (Kruuk *et al.* 1988). The picture of seasonal food availability was enhanced by fluctuations in the weights of fish, with median weights of eelpout being about twice as high in August at 16 g, compared with 9 g for January. Rocklings showed a steady 25–30 g throughout the year, but in June butterfish more than doubled their weight compared with that in January (12 versus 5 g).

One problem with the interpretation of catch size from fish-traps is that the species differ in the likelihood that an individual will get caught, in their 'catchability'. This means that differences between trap catches (for example, in seasonality, depth, or the effects of tide and time of day) are largely relevant as they stand, uncorrected. However, to interpret trap catches in terms of actual prey density and biomass, corrections need to be applied. To this end, I compared some of the catches with what we found in the area immediately around the traps. This was done by intensive searching, hand-netting, snorkelling and scuba-diving by several people over an area of 20 m^2, where we assumed that we caught every fish present (Kruuk *et al.* 1988).

We discovered that, in an area where we would catch 10 eelpout per trap per night, there would be 6.7 eelpout per 10 m^2; 10 butterfish per trap per night corresponded with 31.1 butterfish per 10 m^2; and 10 rocklings per trap per night with only 0.6 rocklings per 10 m^2. These differences were due to the variation in behaviour of the species: rocklings are more catchable. The comparison enabled us to calculate correction factors, to translate numbers caught in traps into actual fish densities. Unfortunately, we did not get sufficient data on other species. From the numbers of fish caught, their weights and correction factors, I made an approximate estimate of monthly changes in fish biomass in the otters' habitat (Fig. 8.9). In general, fish biomass was higher in summer (over 10 g/m^2) than in winter or early spring (about 5 g/m^2), with a peak in August of more than 60 g/m^2, a veritable glut, due mostly to the annual invasion of eelpout.

We did one small experiment to see what the effects would be when predators, such as Eurasian otters, removed a large number of fish from a small patch. The question was: would new fishes repopulate? This was relevant when studying otters that were repeatedly fishing in the same small patches along the coast, a common strategy (see Chapter 9). During some very low spring tides, we removed and counted all the fish we could find, from under rocks and seaweeds in one area of 20 m^2, at the water's edge. There were 16 fish (butterfish, eelpout and rockling), and we caught and removed 11. The next day we did the same, and again found 16 fish, catching and removing 14 of them. For the count on the third and final day we found 20 fish, a slight increase due to a larger number of butterfish (Kruuk *et al.* 1988). It suggested that fish caught by a predator from a small 'patch' are replaced within 24 hours. Similar results were obtained for butterfish on the Scottish west coast (Koop and Gibson 1991). It is likely that such a patch has a given number of suitable sites for fishes sheltering under rocks, and vacancies are filled up rapidly.

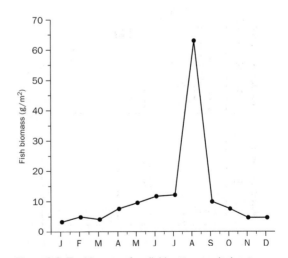

Figure 8.9 The biomass of available otter prey (eelpout, rocklings and butterfish) in Shetland, by month, as calculated from trap catches. There is an annual invasion of prey (mostly eelpout) in summer, with a glut in august.

The results from Shetland on Eurasian otter prey are useful not only to explain behaviour and ecology of the predators on site, but also to suggest similar approaches elsewhere. Based on the use of conventional-type fish-traps, they are subject to problems with interpretation (being dependent on species-specific fish behaviour), but there are ways round this, and these studies enable comparisons of prey availability between areas, seasons, times of day and other variables. Chapter 9 shows that the trapping results explain several aspects of otter behaviour, such as time of day and tide when they forage, and how prey behaviour enables otters to forage repeatedly in the same patches of habitat. They also explain Eurasian otters' seasonality of reproduction (see Chapter 11), mortality (see Chapter 12), and are relevant for understanding spraining behaviour (see Chapter 6).

Selection from prey populations in Shetland

How does variation in fish availability affect the diet of Eurasian otters along sea coasts? Which prey (species, size) do the animals select from what is available by season and site? Within the range of fish available, as in Shetland, they specialize in bottom-living species—they do not aim for open-water species such as mackerel, or saithe and pollack in summer. But does their specialization go further than that? When taking bottom-living marine animals in shallow water, do otters specialize in any particular species, or do they take them opportunistically? Specialization could consist of searching particular sites, or of decisions of whether to carry to the surface the prey items that they encountered on the sea bed.

Clearly, Eurasian otters do select, whatever the mechanism. When we compare the otters' overall diet (see Table 8.1) with what is present along the coast (see Table 7.1), there are differences, also when allowing for biases mentioned for the latter table. For instance, the butterfish, which, after the eelpout, is the most common benthic fish, is not often taken by otters—far less than eelpout. Butterfish are, if anything, easier to find and catch than eelpout in the same habitat; they are smaller, rather sluggish, and when we were scuba-diving we very often saw butterfish, but had to search for eelpout. In contrast, throughout the seasons otters ate five-bearded rocklings about half as often as eelpout, although eelpout were almost thirty times more common overall. Individual rocklings were therefore much more vulnerable to otter predation than eelpout, and eelpout more than butterfish.

The Shetland otter diet cannot be compared directly with overall fish-trapping results, because the traps refused fish above a given size. This was particularly evident in winter when otters frequently took fairly large fish, such as ling, cod, conger eel and lumpsucker, that could not be caught in the funnel traps. It is not possible, therefore, to calculate a 'selectivity index' for prey species. However, there was sufficient evidence to show that otters specialized in rocklings and eelpout when these were available.

Alternative prey species included saithe and pollack in winter. These were always abundant, but only in winter did they enter into the dense beds of algae in shallow inshore water, where they could be caught both by our traps and by otters. Saithe, in particular, were present throughout the year in large shoals in open water, but out of reach of the otters. They vividly demonstrated the point that otters need to catch their prey on the sea bottom or inside dense vegetation. When otters did catch saithe and pollack, they were good-sized, profitable prey items.

Spring was the tough time for otters. During spring they caught many 'other prey', including three-spined and sea sticklebacks, and butterfish—all relatively small, and somewhat spiny, and to all intents and purposes rather miserable food. Of these, only the sea stickleback was actually more widely available at that time than at other seasons; the others were eaten presumably because more profitable species were absent.

An interesting marine prey was the shore crab, especially because most of the time it appeared to be largely ignored by Shetland otters, even in spring. There were masses of crabs, and when scuba-diving they could be seen scuttling away everywhere. In the fish-traps they were caught easily, most abundantly in late autumn and sparsely in spring. I saw a few of the young, newly independent, otters eating crabs, as well as some of the cubs still dependent on their mothers; there was also one very old female, with only stumps for canines, who habitually ate crabs. Apart from these individuals, I noticed that some otters we had provided with a radio harness (which they appeared to feel as an encumbrance) ate crabs for the first week or so. I concluded that crabs must be an inferior type of food.

This result was not at all surprising. Apart from the risk of being nipped by the crabs' ferocious claws, otters had to spend much time in landing them, unlike most fish which are eaten in the water. It was then quite a skill for the otter to remove the carapace, to get at the meat. When they did take a crab, otters did not eat legs or claws, just the soft central contents, which was a relatively small reward for the lengthy handling time. Our study area was quite representative for the whole of Shetland as far as these crab observations were concerned: whenever I walked the Shetland shores elsewhere, it was obvious, from many otter spraints, that crabs were eaten only rarely.

This is in contrast to the many places along the Scottish west coast where crabs were common in the otter diet and often dominated their food, as documented by Jon Watt (1995) on Mull. He found numbers of spraints, strikingly white and pink, consisting of nothing but crab remains. Despite this, Watt showed that, because of the long handling time, it hardly paid energetically for an otter to catch crabs. In fact, unless they were diving in conditions where they had an extremely low hunting success, otters were better off not taking crabs (in terms of quantity of food per time spent hunting) because they were foregoing chances of catching more lucrative fish prey whilst dealing with the crab. Watt's elegant calculations suggested that this would be true for Shetland otters, even more than for animals in his study area on Mull, because Shetland fish as prey are generally larger and more abundant. In Watt's study area it was especially young otters that took crabs (Watt 1993).

Eurasian otters appeared to have more difficulties foraging on the Scottish west coast than in Shetland. By fish-trapping, Watt showed that there were fewer suitable fish as otter prey along the Mull shore; in consequence, otters there ate many more butterfish (a low-quality species) than their Shetland relatives (who had butterfish available, but often ignored them). If the Mull otters caught a dogfish (a shark, often some 60 cm long; Fig. 8.10), they would eat it completely, tough skin and all, whereas in Shetland otters would only whip out the large, oily liver and consume that, abandoning the rest, and I never saw them eat more of a dogfish than just the liver. The Mull otters also ate whole crabs—carapace, claws and everything—whereas in Shetland otters would eat only the contents of the thorax and the abdomen from larger crabs.

Interestingly, if the data from Mull show that all of this potential food can be used, why do Shetland otters waste such a resource, or ignore it, especially as their food is likely to be a limited resource (see Chapter 11)? The answer to this question may possibly lie in the seasonality of the various foods. Many of the fish, and also crabs, are relatively scarce in Shetland just in spring, the only time of year when food is, or might be, at a premium, and when much of the otter mortality occurs (Kruuk and Conroy 1991; see Chapter 12). At other times of year prey is abundant, and Shetland otters can afford to select the best items. Seasonality of food is probably less pronounced on Mull than in Shetland; similarly, along the Norwegian coast there is less seasonality of fish populations (Heggberget 1993).

Apart from the otters' selection of prey by species, also selection by size could be important. The estimated median sizes of fish caught by Eurasian otters in Shetland (see Chapter 7) were considerably larger than those of fish caught in our traps. This was not likely to be caused by the traps being selective for size within this range (Kruuk *et al.* 1988). The results suggested, therefore, that otters took the largest, most

Figure 8.10 Dogfish *Scyliorhinus canicula*, left by Eurasian otter on low-tide algae in Shetland, after eating its liver.

profitable, prey items throughout most of the year, although they were less selective in spring, during low prey abundance. On Mull and along the Norwegian coast, the fish that otters caught were of similar size to those caught in the fish-traps (Heggberget and Moseid 1994; Watt 1995).

Eurasian otters and prey populations in freshwater habitats

Salmonid fishes were an important component of the Eurasian otter's diet in Scotland (see Chapter 7). In most freshwater streams in the north-east, brown trout *Salmo trutta* and Atlantic salmon *S. salar* dominated the fish fauna (Durbin 1997; Kruuk *et al.* 1993a). Their age-class distribution showed many 1-year-olds (3–5 cm in summer), fewer 2-year-olds (6–10 cm) and even fewer that were older than 2 years (Fig. 8.11). These and many other data were collected mostly by my colleague David Carss, by electro-fishing.

During electro-fishing, a section of, for example, 50 or 100 metres of stream is netted off on both sides, and fish are stunned by a pulsing electric current of 400 V between mobile electrodes a few metres apart. The operators fish out the entire section of stream, and all fish that are stunned are netted, measured, recorded and collected. The procedure is repeated three times, with 30 minutes between, with progressively fewer fish being stunned and netted. The total numbers of fish caught, and the differences between the three catches, are used in a formula to calculate the total numbers of fish present (Zippin 1958). The method provides a fairly accurate assessment of fish populations (Bohlin *et al.* 1989).

We found in one of the streams, the Beltie Burn, a tributary of the River Dee, that salmonid biomass varied between 9.2 and 14.4 g/m^2. This is a commonly occurring density, and similar to values found elsewhere in Scotland, England, Denmark, Norway and North America (Bergheim and Hesthagen 1990; Egglishaw 1970; Elliott 1984; Mortensen 1977; Newman and Waters 1989). The biomass of other fish in the Beltie Burn, mostly eel, was only 0.5–1.6 g/m^2.

The total salmonid population in our study areas was not evenly distributed through the rivers and streams, but, interestingly, it was concentrated in the narrower streams and tributaries. These small tributaries had a higher biomass per area than the wider rivers (Fig. 8.12). Durbin (1993, 1997) established that, of the two species studied, brown trout concentrated in narrower streams, whereas salmon tended

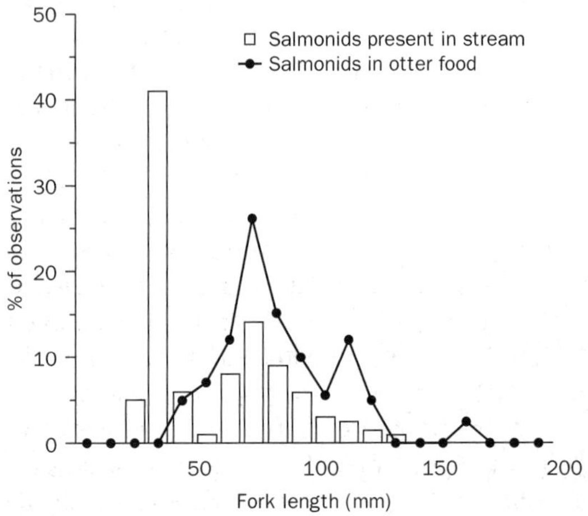

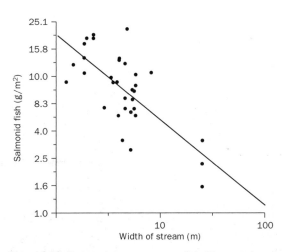

Figure 8.11 Salmonid size selected by Eurasian otters in Beltie River, north-east Scotland. Otters ignore Atlantic salmon and brown trout of size less than 5 cm. Electrofishing results from 2718 fishes; diet calculated from 40 salmonid vertebrae in spraints. (After Kruuk *et al.* 1993a).

Figure 8.12 Narrow streams carry more fish: biomass of Eurasian otter prey in Scottish streams and rivers of different widths (electro-fishing results on salmonid fish) ($r = -0.71$, $P < 0.001$).

to stay in wider waters. The differences in prey biomass in streams of different widths were reflected in the behaviour of the otters (see Chapter 4).

In addition, there were other fish species, such as minnows *Phoxinus phoxinus*, sticklebacks *Gasterosteus aculeatus* and lamprey *Lampetra planeri*, that were all eaten by otters. However, the species that, next to salmonids, was most important as otter prey in many places in the north-east Scotland up to the 1990s was the eel *Anguilla anguilla* (Fig. 8.13). David Carss showed, again with electro-fishing, that in some of the small streams there were as many as 2–16 eels per 100 m^2 (in one stream as many as 80 per 100 m^2). In terms of biomass this was 0.8–5.1 g/m^2, and electro-fishing for eels in shallow Scottish lochs showed 1.4–4.0 g/m^2. These figures are similar to those for eel biomass in streams elsewhere, in England, Ireland, Denmark and Norway (between 1.0 and 7.5 g/m^2; references in Carss *et al.* 1999), and only a little lower than our estimates of benthic fish biomass along the coasts of Shetland (see above). In the Scottish lakes, there were also many perch *Perca fluviatilis* and pike *Esox lucius*. In a Portuguese stream, Beja (1996a) found up to 20 eels per 100 m^2, as well as some 140 cyprinids of various sizes.

Ranking fish biomass by species in a number of river and lake sites in Scotland, Carss found that salmonids dominated in rivers and streams, in some more than in others, with eels coming second. In two lochs eels dominated, in one more than in another, followed by perch, then pike. When then we assessed otter diet from spraints collected in these same places, we found the same ranking of fish species as in the electro-fishing results: prey was taken according to availability (Carss *et al.* 1998b). This result was similar to that of earlier studies, for example by Wise *et al.* (1981) and Chanin (1981) in lowland England, where otters took roach *Rutilus rutilus*, bullhead *Cottis gobio*, salmonids, eel, perch and pike largely according to their relative abundance in the ecosystem (as measured by netting and electro-fishing).

Nevertheless, Eurasian otters obviously select fish sizes, at least to some extent. The otter diet, from measurements of salmonid vertebrae in spraints collected at the same time where places from the salmonid population was assessed, showed that the predators ignored the youngest age-class, and took any fish older than that in relation to availability (Fig. 8.11). In contrast, Libois and Rosoux (1989) measured eel vertebrae from otter spraints along rivers in western France, and compared them with the size of eels caught by electro-fishing. They concluded that otters took eels of all size classes with similar frequencies as their occurrence in the population: there was no evidence of size selection.

In Hungary, Lanszki *et al.* (2001) studied otter predation in a species-rich ecosystem, which also included some fish ponds. Over a 6-year period they sampled the fish fauna of 19 species with small-mesh nets, and compared numbers and sizes with what they found in spraints of Eurasian otters. As expected, the researchers found generally close correlations between otter diet and fish presence, but there were interesting deviations from the general pattern of 'otters take what there is'. The animals had a significant preference for fish between 500 and 1000 g, and they avoided fish heavier than 1 kg, irrespective of species. Fish sizes below 500 g in weight were taken irrespective of size or species. In addition, the otters had a clear preference for fish that lived in dense aquatic vegetation, and in shallows. Species was not important, but habitat and size mattered—in a way similar to what we found in Shetland.

Just as in the sea, there are distinct seasonal fluctuations in the biomass of fish populations in streams and lakes that are different in various parts of the Eurasian otters' geographical range. In northern parts such as Scotland, spring is the lean time for otters; for instance, during May the biomass of salmonid populations is lowest, and highest during October (Egglishaw 1970). This is due largely to

Figure 8.13 Eels (*Anguilla anguilla*) in a Scottish stream—an important, but disappearing, prey.

temperature-related growth of fish, and is likely to apply to most species. However, there are compensations, provided by prey species other than fish.

Frogs and crayfish, which Eurasian otters take throughout freshwater areas in Europe and Asia, are seasonal even more than fish, and some of them are available only for short times of the year. Jean-Marc Weber (1990), for instance, noted that in Scotland frogs *Rana temporaria* were available to otters when they hibernated in mud at the bottom of streams and during mating, when they congregated in dense groups in marshes in early spring. Numbers taken by otters were closely correlated with numbers of frogs available in each marsh, with frogs making up to half of the prey items in the otters' diet. This occurred just at the time of year when fish numbers were lowest. Similar results were obtained by Lanszki *et al.* (2001), with toads *Bufo bufo* and frogs *Rana ridibunda* and *R. dalmatina* (Fig. 8.14). In the Iberian peninsula, too, amphibians are available to otters mostly during their breeding season, toads in winter, and frogs in spring and early summer (Beja 1991, 1995a,b).

The Louisiana crayfish *Procambarus clarkii* was introduced into Spain in 1973, and is now thriving in freshwater habitats throughout most of the Iberian peninsula. The species has become an important resource for many predators (Correia 2001). Amongst them are the otters, which consume large quantities (Delibes and Adrian 1987), especially in summer (Beja 1996b); the crayfish are less accessible at other, colder, times of year in their burrows.

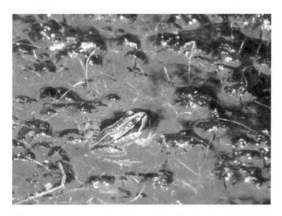

Figure 8.14 A green frog (*Rana perezi*) in Portugal. After fish, frogs are the most important secondary food of Eurasian otters.

Summer is the time in Portugal when the native fish resources are at their lowest, which makes the presence of the exotic crayfish especially useful to otters as an additional resource.

Effects of Eurasian otters on their prey

The presence of otters would be of little consequence to populations of various species of fish if quantities taken by the predator were relatively small, compared with the production of fish biomass—unless, of course, otters were highly selective for particular categories within each species. What proportion of fish productivity is consumed by otters, and what are the potential impacts of this?

Numbers and biomass of fish and their annual productivity were studied in one of our projects in the rivers Don and Dee and their tributaries in north-east Scotland. We concentrated on the otters' main prey species, brown trout and Atlantic salmon, and compared the results with predation by otters (Kruuk *et al.* 1993a). First we estimated the annual production for each fish age group by a standard method using electro-fishing, from changes in the size distribution of the populations over the growing season from May to September (Ivlev 1966; Ricker 1975). Our calculations showed that in one of the tributaries, the 32-km Beltie Burn, the salmonid population produced about 16 g per m^2 per year, very largely as fish smaller than 15 cm (Kruuk *et al.* 1993a). Interestingly, therefore, annual fish productivity was higher than the 'standing crop' at any one time (up to 14.4 g/m^2). The data from other parts of the study area were fairly similar.

For an approximate estimate of otter consumption from these salmonids, we radio-tracked otters that had been injected with the harmless isotope ^{65}Zn as zinc chloride (see Chapter 4). We estimated the total 'otter time' spent in any one tributary or section of river, then calculated the proportion of salmonid fish in the diet from spraint analyses, and total food of otters as described in Chapter 7. Combined, these calculations gave an approximate figure of otter consumption of salmonid fishes in each area.

As an example, the Beltie Burn, with 7.9 ha of water surface, was used full-time by two adult radio-tagged otters, a 7-kg male and a 5-kg female, over a period of 8 months. Over the 6 months for which the isotope was detectable, we collected 273 spraints along the Beltie, of which 74% were from our radio-tagged otters (there were visits from other otters, mostly males). This meant a total of $100/74 \times 2 \times 365 = 986$ 'otter days' per year, or an average otter biomass of 0.22 g/m^2. Spraint analysis suggested that 90% of the otter diet consisted of salmonid fish. Food intake of otters was conservatively estimated as 12–15% of body mass per day. Taking all of these approximations together, we calculated that otters on the Beltie Burn consumed 8.6–10.8 g salmonid per m^2 per year, accounting for 54–68% of the annual production of salmonid fish, or 60–118% of the mean 'standing crop' (Kruuk et al. 1993a).

These figures are rough estimates, but they present an order of magnitude, and the proportions of fish productivity taken by otters in the Beltie appeared to be representative for many streams (Kruuk et al. 1993a). Predation is relatively high, therefore, especially when taking into account that other fish predators are also present, such as herons *Ardea cinerea*, goosanders *Mergus merganser*, mink *Mustela vison* and other fish.

I should stress that our estimates do not automatically imply that otters have a large effect on fish density. For instance, otters may be taking the part of the fish population that is surplus to the stream's carrying capacity. Ultimately, the effects of otter predation on fish density can be assessed properly only by experiment. However, one can say, at least, that these predators play an important role in the interaction between fish populations and their environment. Otters do not just take the odd fish from the river. It has been demonstrated that, through predation on invertebrates such as chironomids and other insect larvae, insectivorous fish may enable the growth of large beds of algae, especially the common *Cladophora* spp. (Power 1990). This means that otter predation on predatory fish, such as salmonids and sticklebacks, could have substantial effects throughout the freshwater ecosystem.

When Eurasian otters take large adult salmon from the spawning grounds in rivers such as the Dee in Scotland, the effects of such predation could be considerable, as I discovered jointly with David Carss. The relatively large fish (up to over 6 kg) are taken mostly by male otters, at an approximate rate of one salmon per otter per day (Carss et al. 1990). Only part of the prey is eaten. On one spawning ground we estimated numbers of salmon 'redds' or spawns, and we assumed equal numbers of male and female salmon (Webb and Hawkins 1989). From the number of carcasses of 'ottered' salmon we calculated that 23% of the fish present were killed by otters. Interestingly, however, most of these were male salmon (81%). Males are more vulnerable as they stay longer near the spawning sites, and they frequently travel upstream and downstream crossing riffles, which are the usual sites where otters catch them.

It is often argued that the effect of vertebrate predators on prey populations is small, or at most instrumental in 'regulating' some other major environmental influence. One usually assumes that the role of predators in a community is not really very dramatic, for instance that they do not actually eliminate prey. Recently, however, some strong 'top-down effects' have been demonstrated, showing convincingly that predatory carnivores may maintain populations of prey species at levels much lower than would be the case if those predators had not been there (see. for example, the review in Estes et al. 2001; Kruuk 2002).

One problem is that predators may have an impact, but without leaving directly visible evidence. For instance, a prey population may be absent or have disappeared from a place where a particular predator species occurs. In our freshwater study area, two Scottish lochs were used by up to ten otters from 1975 until about 1995, in a habitat ideally suited also to coots and grebes. The floating nests of these species are potentially highly vulnerable to otter predation. However, they never bred there, and when coots used the lochs in winter they were frequently preyed upon by the otters. It was only when otter numbers declined to one or two after 1997 that several coots started breeding. Predation may have affected the coot and grebe populations, but actually demonstrating this is difficult (Kruuk 2002).

In Shetland it seemed unlikely that Eurasian otters would have a major influence on the inshore fish communities. They fished only in shallow, relatively narrow, strips of water, usually less than 3 metres deep, and the prey species appeared to go down to greater depths than that. In other words, otters only creamed off the edges of prey populations and their productivity.

The low potential for a limiting role of otter predation in the populations of prey species in Shetland was evident especially for eelpout, the fish that otters took most frequently. Large numbers migrated into the shallows of the study area in summer, producing a glut of prey for otters during a short period. But some favoured, and less abundant, fish species that spawn in shallow water could possibly be hit harder by otter predation, for example the lumpsucker, which spawns near the low-tide mark. Shetland otters may also have influenced numbers of one of their main prey, the five-bearded rockling, which is confined to a zone less than 5 metres deep. We estimated that otters annually removed more than a quarter of the 'standing crop' of the local rockling population (Kruuk and Moorhouse 1990). Crucially, however, it was not known what proportion this was of the productivity of the population, although undoubtedly otter predation took a quite substantial part of it.

Prey availability for American otters

North American fish faunas are richer than those of Europe. Britain has only 42 indigenous species (Maitland 1972), all of Europe fewer than 200, whereas the Mississippi basin alone supports between 230 and 300 (Rieman et al. 2003). In Canada, the Fraser River basin is considered depauperate compared with other North American rivers, and has 53 species of fish (Richardson et al. 2000). However, many of the species are declining in numbers, often as a result of land management practice, including woodland burning (Rieman et al. 2003), and, despite the larger numbers of species, the actual biomass of fish is often relatively low. In the Canadian Fraser River, for instance, fish biomass remained stable over a 21-year period at about 3.4 g/m^2 (Richardson et al. 2000). Elsewhere on the North American continent, fish biomass is higher, at around 10 g/m^2 (Newman and Waters 1989), with annual productivity at levels similar to those in Britain (see above).

As in Europe, the most important North American freshwater fish species as prey for river otters are fairly slow bottom-feeders on invertebrates, such as various suckers (Catastomidae), sculpins (Cottus spp.), bullhead (Ictalurus spp.) and several others (see Chapter 7). The many different salmonids that are taken by river otters may be fast swimmers, but they spend part of the day or night quietly on the bottom or in cover, and are then vulnerable to otter predation. Similarly along the North American coasts, the fish taken by river otters tend to be slow bottom-feeders (Bowyer et al. 1994). However, little, if anything, is known about the effects of river otters on the populations of these species.

Little is known also about the habits and populations of otter prey in South America. There are large numbers of species of fish; as an example, in the Rio Negro in the Brazilian Pantanal, giant and neotropical otters were selecting from 260 species (H. Waldeman, personal communication). The species preferred by otters are again the slow bottom-feeders, whereas the multitude of more pelagic or open-water species are left alone. Typically preferred prey of giant otters, the trahiras or wolf-fish Hoplias spp., are sluggish ambush hunters, lying quietly in wait in cover (Rosas et al. 1999); they, as well as several cichlids, inhabit parts of water bodies with many submerged branches and litter banks, where otters may surprise them. The many different catfish (Siluridae), popular prey for all otters, are also slow-moving bottom fishes, whereas the piranhas, sometimes taken by giant otters, concentrate in the open, deeper parts of ox-bows.

We cannot even guess at the ecological impact of any of the four Latin American otter species, not even in the case of the relatively well studied giant otter. Assessment is made especially difficult because of the vast numbers of fish species involved.

Ecosystems, and effects of predation by sea otters

In sharp contrast to the above, the effects of sea otter predation on their prey, with the subsequent reverberations throughout the ecosystem, have been subject to many highly important and fascinating studies. The main focus of this has been provided especially by Jim Estes and his colleagues, along the Pacific coasts of North America. Summarizing the sea otter's diet from their research, Chapter 7 showed the species as a predator on mainly sea urchins, large molluscs, crabs and fish. The consequences of this predation produce perturbations throughout the coastal ecosystems. The sea otter story has become a paradigm of top-down disruption of an entire ecosystem by a single predator, and, although the initial simplicity of the story is no longer tenable, the central theme stands.

The most important part of the sea otter effect on the environment is their impact on the sea urchins of the rocky bottoms along the fringes of the north-east Pacific, which seriously affected the huge kelp forests. Originally, it had been believed that the ocean's nutrients regulated the miles and miles of kelp, that this constituted the food for sea urchins, which then amply provided the sea otters (Estes 2002). An important paper (Paine and Vadas 1969) demonstrated that sea urchins actually had strong limiting effects on the kelp assemblages. This made otter researchers sit up, urchins being a main item of diet, and when they monitored the return of sea otter populations in Alaska they were able to demonstrate large-scale reductions in urchin *Strongylocentrotus* spp. numbers, and the corresponding appearance of enormous kelp beds (Estes and Palmisano 1974).

When, after an absence of a century (caused by the fur trade) the sea otters returned to rocky subtidal shores of Aleutian islands, coasts of Alaska, California and elsewhere, they entered 'sea urchin barrens'. Large tests were in abundance, and in many places their grazing had completely eliminated the large algae. However, once sea otter populations increased to densities quite unknown for any other otter species, the urchins declined to a level at which only very small ones, or individuals in well protected crevices, survived. This enabled kelp to grow again, and to dominate areas of coast. Underwater forests of *Macrocystis*, *Laminaria* and other large algae emerged, providing shelter for fish and a range of invertebrates. Decomposed algae provided detritus and a foundation for other faunas in parts of the sea bed, well away from the actual kelp forests (Breen *et al*. 1982; Estes and Harrold 1988; Estes and Palmisano 1974; Estes and VanBlaricom 1985).

Where the sea bed was soft and muddy, urchins and large algae were scarce, but even there the arrival of sea otters caused major changes. For instance, around the Kodiak islands sea otters, feeding mostly on the large butter clam *Saxidomus giganteus*, had dramatic effects on clam numbers, removing virtually all of those that were larger than 5 cm (Kvitek and Oliver 1992). Another large and potentially profitable species of mollusc, the horse clam *Tresus capax*, escaped the otters' attention because it burrows deep down, often more than 40 cm into the sediment. Not only did the sea otters seriously affect the butter clam population, but in the process they also dug over the sea bed very efficiently, exposing many old shells (and depositing the shells of clams they took). This provided new attachments for large algae and sea anemones, and a demonstrable proliferation of sea stars *Pycnopodia helianthoides*, which attacked the smaller molluscs. Thus, the interference of the sea otter caused profound perturbations over large areas.

The kelp beds that were created through the intervention of sea otters also affected the sea bottom elsewhere, especially inshore from the kelp beds themselves. The detritus generated by the dying fronds provided nutrients for many detritus feeders, for instance dense mats of mussels and barnacles, as well as many small crustaceans which, in their turn, provided food for fish such as rock greenling *Hexogrammos lagocephalus* that attracted fish-feeding birds, such as cormorants (Duggins *et al*. 1989).

All of this was in Alaska. When researchers started investigating kelp beds, sea urchins and sea otters elsewhere, especially in California, they found a more complicated scenario. Sea urchin populations were preyed upon and affected not only by sea otters, but also by other predators such as a fish called the sheephead *Semicossyphus pulcher*, and they, in turn, were fished by people, as were the sea urchins themselves. Kelp beds came and went also for reasons other than sea urchin grazing. Nevertheless, the role of sea otters was highly important, but because of the 'ecological redundancy' (i.e. the role of one disappearing species being taken up by another) their impact was less immediate and often cushioned (Paddack and Estes 2000; Steneck *et al*. 2002). In the case of red abalones *Haliotis rufescens*, in California as in Alaska, sea otters effectively removed from the population all of the large individuals that previously had been the target of commercial and recreational fisheries (Wendell 1994).

The many detailed recent studies, involving sea otters in the Pacific coastal ecosystems, have shown that predators play a large and vitally important role, much larger than was suspected previously (Estes

2002; Jackson et al. 2001). The 'top-down' importance in population control was underscored even more emphatically when, in the 1990s, sea otter numbers in Alaska declined sharply, they themselves being affected by predation by killer whales (Springer et al. 2003; see also Chapter 12). Sea urchin populations boomed once more, creating large barrens and wiping out the kelp (Estes et al. 1998).

All evidence, therefore, points to a key role of sea otters in these large ecosystems, with their influence reaching much further than their immediate prey species. Compared with any of the other otter species worldwide, their impact is vastly greater. We shall see that, curiously, one of the sea otters' attributes that makes this impact possible is the structure, the insulating capacity, of their fur—the very same property that also almost caused their extinction, through the fur trade.

Prey availability in Asia and Africa

Research on otters in these regions is still sadly deficient, and what has been done has understandably concentrated on the animals themselves rather than their resources. Fish, amphibian and crustacean faunas are almost fantastically rich, with large numbers of species in even one single lake or stream. In the river I studied in western Thailand, three species of otter were feeding and 107 species of fish were recorded, as well as 41 species of frogs and toads (Nakhasathien and Stewart-Cox 1990); there were also large numbers of crabs (Kruuk et al. 1994a). Almost nothing is known about the behaviour or numbers of any of these potential prey, and the impact of predators in such rich ecosystems will be difficult to assess.

In the Chambal river in India, otter researchers were dealing with 37 species of fish, potential prey for **smooth otters** (Hussain and Choudhury 1998). The predators, however, concentrated almost entirely on just seven of these, and especially on two (Hussain and Choudhury 1998, p 245):

Rhinomugil corsula and Rita rita were the preferred prey eaten throughout the year. This may be due to greater availability of these species. R. corsula is mostly found in shoals of 4–200 along the water's edge and congregates near waterfalls and rapids. Otters in groups were often observed to hunt this fish in semicircles moving against the river current. Rita rita is found under and between stones and in cracks and crevices and is a sluggish bottom-dwelling catfish.

In Kerala, India, smooth otters concentrated their predation on introduced, exotic fish species, tilapia T. mossambicus and carp Cyprinus carpio, and the researchers suggested that this could have beneficial effects for the local fish species that were suppressed by the incomers (Anoop and Hussain 2005). This, however, has yet to be demonstrated.

The large East African lakes are renowned for their species richness in fish; for instance, in Lake Victoria about 300 species of cichlids have been reported (Barel et al. 1985) as well as various catfish, lungfish and others. Lake Malawi contains well over 500 different cichlids (Albertson et al. 1999). In these lakes otters such as the **spotted-necked** take many cichlids (Kruuk and Goudswaard 1990), catching them in crevices and between branches of overhanging trees, or they catch catfish on the bottom. Not surprisingly, there are no data as yet on fish biomass and availability. I noted, however, that in Lake Victoria in 1989 these otters took fewer small cichlids (Haplochromis spp.) than in previous studies (Procter 1963), after populations of these cichlids had been sharply reduced by the introduction of predatory Nile perch Lates niloticus, and by overfishing (Barel et al. 1985; Harrison et al. 1989).

For the **Cape clawless otter**, a crab eater, we also have little knowledge of food availability, or production of the freshwater crab populations, except that large numbers of these crabs Potamonautes spp. (Fig. 8.15) live especially in somewhat eutrophic rivers (up to 36 per m^2) (King 1983; Somers and Nel 1998; Turnbull-Kemp 1960). When the clawless otters feed off the sea coast, they may take spiny lobsters Jasus lalandii, as well as many crabs and fish (Somers 2000b). There are shades of the sea otter story along the Pacific coasts, when in South Africa spiny lobsters play an important role in inshore animal communities in the sea, often seriously affecting densities of sea urchins (Mayfield and Branch 2000; Mayfield et al. 2001; Tarr et al. 1996; Van Zyl et al. 1998). However, whether clawless otters, much lower in densities than the sea otter, have any effects on spiny lobster populations (and hence on sea urchins) is unknown, and rather doubtful.

Figure 8.15 Freshwater crabs (*Potamonautes spp.*) are the major component of food of small-clawed and Cape clawless otters.

Conclusions, and global changes in otters' resources

We know reasonably well what most species of otter eat. In contrast, our knowledge of the characteristics and status of the resources themselves is very thin, with a few remarkable exceptions. The gap in our knowledge should be of major concern in the direction of future research, if only because of the importance of resources as determinants of otter numbers (see Chapter 11).

As an example of research into resource characteristics, I have detailed some of our observations on **Eurasian otters** in Shetland and Scotland, suggesting practical ways in which such study might be approached elsewhere. For that species, we now have at least some data on fish activity patterns and behaviour, numbers, biomass, and in some cases productivity; we have ideas about factors that determine when otters eat crabs and when they do not. We know that these otters prefer sluggish, bottom-living prey, and that some, if not all, species of fish are vulnerable at times when they are inactive.

There are leads emerging about the effects that otters exert on their prey, about the role that they play in aquatic ecosystems. There was evidence that in our study sites, in both sea and fresh water, Eurasian otters accounted for a substantial proportion of the biomass and productivity of some of the prey species. In itself this does not imply an effect on prey populations and food levels further down the ecosystem, but the potential is there: it exposes a dire need for further research—all the more as another species of otter has been shown up with a key role in its ecosystems.

The **sea otter** can occur in much higher densities than other otters, for reasons to be discussed Chapters 11 and 13. This animal is part of a chain of profound perturbations in the coastal underwater environment of the Pacific, affecting entire forests of giant algae and enormous densities of urchins and molluscs. The otters themselves are subject to predation, which causes wide changes in densities (see Chapter 12). These two species, the sea otter and the Eurasian otter, have been studied in greater depth than any of the others, so if independently they show either a key role in their ecosystem, or a distinct potential for one, we should expect more of this in the other otters. There is a wealth of fascinating interaction yet to be uncovered. The need for ecosystem-oriented research on otters is apparent, especially at lower latitudes, where the numbers of species in predator–prey systems and competition leave us completely in the dark about the role of otters, and about limiting factors impinging on them.

There is little doubt in my mind that detailed knowledge of interactions between otters and their resources is needed to understand the impact, on otters of all species, of the vast environmental changes that are now being exposed worldwide. The changes in the Pacific Ocean are just one example, where the elimination of large whales by whaling probably caused killer whales to switch to sea-lions and seals as prey. Once those had disappeared, the killer whales started the reduction of sea otter numbers—hence the consequences for sea urchins, and for kelp and its dependant organisms, and for the fauna of large molluscs, and for the human fisheries dependent on clams, etc. (Estes 2002; Springer *et al.* 2003).

In Shetland there has been a precipitous decline in numbers of **Eurasian otters** since the mid-1990s, occurring along almost all of its coastline, with preliminary estimates of fewer than one-third of the earlier numbers remaining, and little reproduction (personal observations, and various local observers). The animals that are left appear to be eating mostly crabs, previously largely eschewed by the Shetland otters (see Chapters 7 and 11). Observations on fish-eating sea birds in Shetland suggest similar catastrophic declines; most likely, there have been big changes in the fish populations around the islands, largely unrecorded.

Throughout western Europe one of the main prey fish of Eurasian otters was the eel. It was a common fish species everywhere, with a large fisheries industry dependent on it in countries such as Holland and Denmark, and with rivers sometimes literally heaving with elvers (young eels). In 2002, the numbers of elvers entering the ecosystems were only 1% of the numbers in 1980, and still declining (Stone 2003). This 99% decline was matched by a 80% decline in Japanese eels, whilst the American eel recruitment has virtually stopped (Castonguay *et al.* 1994). The conclusion seems inescapable that eels are fast heading for extinction. The cause is as yet unknown; it has been proposed that climate-related changes in oceanic currents may be affecting the migration of elvers (Castonguay *et al.* 1994), but could also be due to over-fishing, or something else. Hard data there are none. The recent decline in numbers of Eurasian otters in Scottish freshwater systems (see Chapter 13) is likely to be one of the consequences.

Several other fish species in Europe and America, acutely relevant to otters, are faring badly. In Scottish rivers around the year 2000, the numbers of Atlantic salmon, a main otter prey, were only a fraction of what they had been in the 1970s and earlier. The decline is likely on the one hand to affect otters, but on the other hand to render the smaller numbers of salmon more vulnerable to otter predation, as part of a spiral effect. Salmon in Nordic countries have suffered similarly, although more recently there are signs of numbers increasing again, probably because of restrictions on offshore fisheries (Romakkaniemi *et al.* 2003). In central Europe, the brown trout (an otter favourite) has seen a gradual decline over large areas, and in several Kantons in Switzerland numbers in 2000 were down more than 60% on tallies of the 1970s (Anon 2003a). In many areas of North America, the plight of suckers (*Catostomidae*), often a mainstay of **river otter** diet, is causing serious concern to conservationists, as a consequence of many different aspects of agricultural and technical development, and habitat degradation (Cooke *et al.* 2005).

In Africa there have been major perturbations in the aquatic environment. The introduction of Nile perch *Lates niloticus* in Lake Victoria in 1960 created a large new fishery, caused the decline or extinction of most of the approximately 300 endemic cichlid species, and in 1985 contributed 80% to the total fish biomass of the lake (summary in Kruuk and Goudswaard 1990). This was a massive environmental change. Many of the cichlids were important food to **spotted-necked otters**, whereas Nile perch do not feature in the otters' diet.

Frogs and toads, a secondary yet important part of many otter species' diet, are declining rapidly worldwide, with some 43% of amphibians dwindling in numbers, many critically endangered (Stuart *et al.* 2004). The suspicion is that in North America and Europe the cause is mostly habitat disappearance, in East Asia consumption by people. However, herbicides, fungal disease and stronger ultraviolet light have also been mentioned as culprits. Whatever contributes to the frogs' demise, it is likely to affect the food resource of several otter species.

However, some of the monumental changes have favoured otters, or are likely to do so. For instance, in Spain and Portugal, as well as in Kenya, a North American native has been introduced, the Louisiana crayfish *Procambarus clarkii*. The populations of these immigrants have exploded into very high densities. **Eurasian otters** in Iberia have taken to this new source of food (established in the 1970s) with gusto, and in many places crayfish now dominate their diet (Beja 1996a; Correia 2001; Delibes and Adrian 1987). Similarly in Kenya, spraints of the **Cape clawless otter** are now dominated by remains of Louisiana crayfish, whereas before their introduction in about 1970 (Lowery and Mendes 1977), in the same areas, freshwater crabs were the mainstay of the diet (unpublished observations). The arrival of crayfish may be a simple increase in otters' resources in these areas. However, as yet we know little of the effects of these immigrants on other species in the ecosystem, on freshwater crabs, on native crayfish and on various other potential prey for otters. They certainly cause large changes in the aquatic macrophyte vegetation.

I have mentioned just a few examples of recent large-scale environmental changes that affect various otters. There have been many more, and no doubt the frequency of such upheavals will increase. We still understand far too little about how such changes alter the ecosystem, and about the nature of their effects on resources for enigmatic predators such as otters. A first requisite now is an increased understanding, for almost all species of otter, of the effects on them of resource availability (bottom-up effects), and of the influence of otters themselves on these resources (top-down effects).

CHAPTER 9

Otters fishing: hunting behaviour and strategies

Introduction

To see an otter in pursuit of its prey, anywhere in the world, is an exciting event: people cannot but be deeply impressed by it. The appreciation by naturalists contributes to the conservation value of these animals, and study of their fascinating behaviour will be of general as well as academic interest. One of the values of such research is that it exposes some of the environmental problems that otters are up against, and provides some insights that are quite counterintuitive. In fact, instead of the carefree animal splashing in the waves, we are shown existence on a knife edge.

Artists' impressions of a fishing otter, the pictures one finds on posters and in popular books, often show the animal in fast chase after one or more fish, fast as lightening. In the wild, however, otters usually take fish (as well as other prey) that they do not have to chase. Clearly, it is important to understand such details if we are to assess food availability and the otters' strategies of exploitation, both for purposes of insight in the species' ecology and evolution, and for conservation management. In this chapter I shall discuss observations on otters' feeding strategies and detailed foraging behaviour in order to understand how they exploit their resources, how they select prey and what determines their success, and what are the costs and profitability.

I want to emphasize that these questions are not just an academic exercise. It would be difficult to extrapolate the results from any one study area, to predict what otters do elsewhere (for example, in areas where they are less numerous), if one did not understand the mechanism of the relationships that are found. For instance, one may be able to generalize about prey preferences, or preferences for particular depths of water, or for fishing at certain times of day, in different Eurasian otter populations and even in different otter species, once we understand why and how Eurasian otters select prey in Shetland, and what are the costs and benefits of diving in certain places. Such generalizations include the interpretation of diet and metabolic costs against the background of fish behaviour and availability, which is basic to the understanding of spatial and social organization, and of the 'carrying capacity' of particular waters for otters.

Many of the questions about foraging of otters are also scientifically fascinating *per se*—questions about diving behaviour and capabilities, the exploitation of food patches, energetics and heat loss. Some of the hypotheses I will discuss are that otters maximize their diving success rate, they fish at depths where they get most prey for least effort, they fish at times of day and tide, and in places when and where their diving success is highest. The answer to several of these ideas and questions appeared to be intuitively obvious, at least to me, but often my intuition turned out to be at least partly misleading.

Fishing behaviour of Eurasian otters

The following edited extract from my field notes describes the most common type of **Eurasian otter** foraging in Shetland:

July 1985, 05.00 h. From my observation hide on the shore I watch an otter female walking on the opposite bank of the bay, probably just getting active. She spraints, then enters the totally calm sea, swimming in my direction. Just the head shows, occasionally a bit of back, sometimes completely submerged. A heron flies over, two black guillemots are fishing nearby, a merganser sits along the shore. When the otter is about 30 metres from me, she dives, with hardly any splash or ripple to show for it: the end of the tail is lifted right out of the water and at the same time she goes down, probably at an angle of some 60°. I can see air rising to the surface, with the chain of bubbles trailing her progress deep down. Obviously she covers very little bottom, just staying in the same few square metres. I know that it is about 4 metres deep there, at this stage of the tide, and I can picture her, rooting about in the kelp, in the forest of *Laminaria* where she can slip between the stems but where I, scuba-diving, get completely tangled up if I try to do the same.

Eighteen seconds after I saw her tail-tip slip out of sight, she emerges, almost cork-like, hitting the surface about 5 metres from where she went down. She was successful: an eelpout about twice the width of her head in length. Head up, chewing, she treads water, showing her throat patch for all the world to see, once or twice aiding the processing of the eelpout with one of her paws. Half a minute later the fish has gone, and down she goes again, in the same place. Ten seconds later she comes up, with another eelpout of about the same size. The following dive lasts 25 seconds, and she emerges with empty jaws, quietly floating on the surface for 7 seconds, before going down again.

The fishing session lasts 24 minutes, during which time she catches 15 fish, almost all eelpout, and all in an area no larger than 20 × 50 metres. She lands just below me, shakes, walks about on the seaweed, and lies down. A bit of a roll, some grooming, then she curls up and sleeps.

In our Shetland study area, this type of foraging was the most common; otters swam along the surface, mostly well within 50 metres from the shore, then dived to the bottom, usually less than 8 metres deep, often going down many times in the same small area. The periods they spent underwater varied considerably. The longest dive I watched was for 96 seconds (one of our radio-tagged otters diving in deep water), but it was quite rare to see dives lasting longer than 50 s. Some published values of mean diving times for otters in the sea are 23.1 s (West Scotland; Kruuk and Hewson 1978), 20.1 s (Shetland; Conroy and Jenkins 1986), 23.3 s (Shetland; Nolet *et al.* 1993) and 22.7 s (Mull, West Scotland; Watt 1991). All of these figures are based on large numbers of observations and are strikingly similar.

The above values are only little different from the predicted mean dive time for a 7-kg aquatic carnivore of 19 s. This prediction uses the formula derived from interspecific comparisons:

$$y = 2.60 - 0.83x + 0.05x^2 + 0.02x^3$$

in which y is the log mean dive time in seconds, and x is log body mass in grams (Kruuk 1993). After an unsuccessful dive, Eurasian otters spent a short time

Figure 9.1 Eurasian otter eating fish on sea surface, Shetland. Fish can be identified by telescope, in this case an eelpout.

on the surface before going down again; the time underwater is generally about three times as long as the time on the surface: 3.1 times as long in my own observations (Kruuk 1993) and 2.5–3.6 times according to Jon Watt (1993).

Prey was consumed on the surface (Fig. 9.1), except when it was large, such as a dogfish, or was to be taken to cubs. Some otters, especially young ones, were more likely to take small prey ashore than others, and when otters foraged close inshore they were more likely to land prey than when they were far out. Some small, spiny prey, such as sea scorpion which was difficult to manipulate, were often brought to land, and when a crab was caught it was invariably taken ashore (see Fig. 7.3). Landing prey is time consuming and must add a great deal to the energetic cost of foraging (Nolet *et al.* 1993; Watt 1991).

We assumed that when otters were fishing they always went down to the bottom. We knew that they did not chase fish in mid-water, because (a) whenever we could actually see what was happening (which involved watching from above in clear water) the otters went right down, (b) the fish they came up with were almost always bottom-dwelling species, and (c) the chain of bubbles suggested that fish was caught on the spot where it was found.

To catch their prey, otters are likely to use both sight and touch (Green 1977), but the sense of touch, with the large vibrissae, must be all important. When our captive otters dived in an observation tank where we could watch them from aside, the big set of whiskers was turned down and forwards when the animals were foraging, providing a maximal contact area for prey disturbed by the ever-questing snout. The fact that in fresh water otters catch their fish mostly in the depth of night also suggests that eye sight is of only secondary importance. Once, in one of our box traps along a small river in north-east Scotland, I caught a large adult male Eurasian otter that was completely blind (white opaque eyes) and in excellent physical condition, demonstrating that its disability had little effect on its foraging success: it used tactile stimuli. Little is known about the visual acuity of otters underwater; it appears to be less than in air (review in Estes 1989).

The general, most common, foraging pattern of Eurasian otters, both in the sea and in freshwater lakes, we called 'patch-fishing' (Kruuk and Moorhouse 1990; Kruuk *et al.* 1990), when otters repeatedly dived and searched in a relatively small area of water (see below), but there were variations. For instance, we recognized 'swim-fishing', when otters moved along the surface, then dived to come up again quite far ahead but going in the same direction, and repeated this at intervals. This occurred, for instance, when otters moved between fishing patches, and it was especially striking when otters were feeding along steep cliff coasts (often in white water), only covering a very narrow strip of sea bottom because of the steep incline. Another foraging strategy we termed 'kelping'; this was when otters were feeding in exposed seaweeds at low tide (Fig. 9.2). Presumably they did much the same thing then as during their usual feeding dives in deeper water, but it looked different because the kelp showed, as did the otter, wriggling under and through the dense mat of fronds.

In freshwater lakes otters' fishing is often less easy to observe, because the animals are more nocturnal, but in some sites they are frequently active early in the morning. The behaviour is similar to that in the sea, with diving times of the same length for given depths (see below), and patch-fishing just as in salt water. Underwater times in the relatively shallow waters of rivers and streams are correspondingly short, and the otters were almost always on the move, with very little patch-fishing. As a variation on the routine of surface times being one-third of dive times, there are occasions when, between dives in lakes, otters can be seen surfacing for only a very short interval, just enough to catch a breath, then going down again. This happens especially in winter, and is sometimes repeated many times.

In all freshwater habitats prey is landed more often than in the sea (where it is usually eaten by Eurasian otters whilst floating on the surface), for several reasons. For instance, in lakes otters often take eels, which are difficult to manipulate, and they have to be dealt with ashore. Otters fishing in rivers and streams catch most of their food close to the bank, and they have the current to contend with when trying to remain stationary. Thus it is easier for the animals to eat on land (Ruiz-Olmo 1995). Carss *et al.* (1990) described otter predation on large salmon in Scotland, where fish were taken predominantly in sections of river where the water was very shallow with riffles. Sometimes the salmon was eaten on the riffles where it was caught (see Fig. 7.13), splashing about in only a few centimetres of water.

In Shetland we saw otters in some of their more rarely used and specialized hunting methods, when pursuing prey other than fish. Sometimes they were catching birds or rabbits, behaviour both conspicuous and curious, but not contributing much to the diet. An otter would dive and come up 30 metres further or more, right underneath a swimming merganser, shag or fulmar—which would explode out of reach if it were lucky, or be dragged below the surface if it were not. Once we saw a fulmar taken on the cliff by a large male otter, which more or less accidentally cornered the bird in its nesting site.

Figure 9.2 Eurasian otter foraging at low tide ('kelping'), amidst serrated wrack (*Fucus serratus*).

Eurasian otters caught rabbits inside burrows. One Shetland otter, initially with a radio-collar, was named Miss Rabbit as she was a specialist in this, with rabbits featuring conspicuously in her sprains for at least a year. She entered rabbit holes close to the shore, emerging with a rabbit which she then consumed a short distance from the warren. I assume that rabbits are caught similarly by otters living in freshwater areas, because some of our otters with radio-transmitters occasionally slept in well used rabbit warrens and their sprains contained hair of rabbits (Kruuk *et al*. 1993a).

In freshwater areas frogs are caught in winter, at the bottom of muddy stretches of streams, and in spring in marshes and shallow ponds, away from the otters' main river and lake habitats. The unpalatable spawn jelly of a female frog caught before spawning is tell-tale evidence left on the bank. Toads are skinned before being eaten, and the characteristic inside-out skin and parts of the skeleton are conspicuously discarded (Ruiz-Olmo *et al*. 1998b; Weber 1990).

Observations such as these, where otters catch relatively large or unusual prey, receive much attention, but the general picture of Eurasian otter predation, as demonstrated in sprain analyses and other quantitative observations (see Chapter 7), is that of diving for relatively small fish, in both the sea and fresh water.

I have never seen any evidence of cooperative fishing in the Eurasian otter, such as reported for the river otter, smooth otter and others (see below), nor have I come across evidence from other authors. Cubs less than 1 year old often dive next to their mother, but I have not considered that as cooperative action.

Once prey is caught by an otter, it is eaten and digested quickly. Stickleback-sized fish are consumed within a few seconds, whereas a 20-cm cyprinid fish taken ashore takes about 3 minutes to eat, a 30-cm fish about 10 min (Ruiz-Olmo *et al*. 1998b). Digestion is uncommonly fast; for instance, experimenting with captive otters in a large pond, David Carss found that, when otters are active, the median time for fish remains to appear in the sprains is only 67 min. When the animals are inactive, the median time more than doubles to 170 min, and the food was better digested, leaving fewer intact vertebrae in the sprains (Carss *et al*. 1998a).

Only rarely does a Eurasian otter leave any part of the prey; usually, everything is eaten (personal observations, and Ruiz-Olmo 1995; Ruiz-Olmo *et al*. 1998b), except when a really large prey is caught. Of the salmon that otters took in north-east Scotland (mean weight 2.8 kg), on average 975 g was eaten and the rest was abandoned, usually including head, tail and vertebrae (Carss *et al*. 1990).

Fishing and foraging by North American otters

By and large, a fishing **river otter** behaves very much like the Eurasian otter, and in the field they would be difficult to distinguish on behaviour alone. There is the same swimming behaviour along the surface, usually just the head showing, then the tail-flip initiating the dive, the foraging along the bottom of river, lake or sea, landing a fish when large, consuming it while swimming if small. River otters fish waters of similar depths, and in Prince William Sound I observed them swim-fishing and patch-fishing just like the Eurasian otters in Shetland. Dive times, too, are similar, with mean underwater times of 21 seconds (441 dives) (Ben-David *et al*. 2000).

However, there are some intriguing differences. For one thing, river otters sometimes cooperate with one another. Such joint foraging is quite unknown in Eurasian otters, and it must relate to the observation that river otters are generally more gregarious (see Chapter 6). In some interesting examples, Sheldon and Toll (1964) described watching two adults jointly driving schools of fish into a small cove and catching them. Serfass (1995) observed four otters:

One September morning in Pennsylvania, about 10 a.m., four river otters were seen in the centre of a large river pool, some 30m long, 10m wide and 1m deep. Observing from behind a tree, Serfass saw two of them swim off to one end of the pool, the other two in the opposite direction. At the pool ends, the otters dived simultaneously, and swam in rapid, zig-zagging pattern towards the centre, and when they converged, fish were leaping from the water. Two of the otters each caught one. Serfass disturbed one of the otters when it was eating the fish under a bank, and identified the prey as a white sucker, *Catostomus commersoni*.

Just like the Eurasian otter, river otters target predominantly small fish as prey, and their entire foraging behaviour appears to be focused on that. However, there are well publicised exceptions; for instance, occasionally also birds such as coots are taken (Bergan 1990; Riedman and Estes 1988b). The same method of capture is used as by Eurasian otters, by approaching swimming birds from underneath. In one of those cases the river otter was seen losing its coot to a bobcat, when the otter was eating its prey on the edge of an ice shelf (Bergan 1990).

The foraging of **sea otters** is a different matter altogether. They live on the water surface more or less all the time, even when they are not foraging. Sea otters float or swim on their back when moving between sites, or between dives. Foraging at any time of day or night, they roll over and go down without a tail-flip (their tail also is much shorter than that of other otters) and stay under water for longer than other species. In a radio-tracking study of 38 sea otters off the Californian coast, Ralls *et al.* (1995) found a mean dive time of 74 s, followed by a mean surface time of 65 s. The animals frequently went down for more than 120 s, and the maximum dive time recorded in that study was 246 s, with young males staying down longest. The theoretically predicted mean dive time for a 30-kg sea otter (Kruuk 1993; see Fish behaviour of Eurasian otters above) is only 48 s, so the animals appear to be more dive-efficient than other otters, and seals.

Once at the bottom, the behaviour of sea otters varies with the kind of prey that is being targeted. They are tactile foragers (Ostfeld 1982), using their enormous vibrissae and paws. For collecting their quarry they use mostly only their forepaws, not their mouth as do the other otters. Sea urchins and the more difficult abalone or mussels are pried loose from rock or crevices, butter clams require some rapid digging to get them out of the mud, and sometimes two or more dives are needed to excavate any one prey (Kvitek *et al.* 1993; Riedman and Estes 1990). Usually, however, a sea otter manages to collect several prey in one dive (Watt *et al.* 2000); it is the only otter species that can do this, and the prey are tucked into the loose flaps of skin of the axilla and taken to the surface. These otters have been known to bring up as many as 45 sea urchins in one go (mean 6), or 26 mussels (mean 7) or 12 sand eels (mean 3). Clearly, this enables a large improvement on diving efficiency, compared with the single-load carryied by other otters.

An interesting phenomenon relates to sea otters and some of their prey, especially in south-eastern Alaska. Paralytic shellfish poisoning is caused by toxic algae, dinoflagellates, which are accumulated amongst others by butter clams *Saxidomus*. At times of algal blooms they may cause massive mortality amongst fish, birds and mammals. Clams themselves do not appear to be affected, and butter clams retain the toxin for longer than any other mollusc. If a sea otter ingests a substantial quantity of the potentially lethal toxin, it behaves as if in serious pain, screaming and twitching, and it hauls out on land. However, in experiments in captivity, sea otters detected poisonous clams very quickly, avoiding them, but, when really hungry, detaching the tissues in which the toxin was concentrated (siphon, gills and attached tissue) before eating the clam. Thus, the dinoflagellates provide clams with an important protection against sea otter predation. Consequently, along the south-east Alaska coast, large, patchy areas in which the butter clams contain high levels of the toxin are devoid of sea otters (Kvitek *et al.* 1991).

The sea otter is one of very few mammalian species that use tools to acquire food (Hall and Schaller 1964). Often, a suitably sized rock is selected from the sea bed and carried around for long periods in one of the axillae. Deep under water it is used to dislodge abalones or urchins, and once the otter arrives at the surface with its booty the rock is used as an anvil, to smash shells, with the otter floating on its back and hammering away (Fig. 9.3). Several individual variations on this theme have been noted, such as using two rocks as anvil and hammer, or using various human-made objects instead (e.g. a bottle to dislodge clams underwater), or smashing clams together. It is an intriguing experience when walking along the coast to hear the hammering sounds from far away across the waves. Tool use occurs especially in areas where the otters feed on thick-shelled molluscs (Riedman and Estes 1990).

Individuality appears to be a characteristic of sea otters in other ways. One male discovered that discarded drink cans often harboured a small octopus, and it specialized in biting them open (McCleneghan and Ames 1976). Some individuals learn to wait for hand-outs of bait fish from tourists; others learn to hunt sea birds. Generally, in any one area almost all sea otters specialize in one kind of prey or other (Riedman and Estes 1990), often with large differences in energetic profitability. Because of such individual preferences, daily foraging times required to meet calculated energy requirements

Figure 9.3 Sea otter bashing clam on rock carried on its chest. © Jane Vargas.

varied between less than 4 hours to over 21 hours (Estes *et al.* 1981).

Jim Estes and colleagues (2003b) showed that these idiosyncrasies last a lifetime, and that they are passed on by learning, from mother to daughter. In one area there are sea otters concentrating on mussels, others on crabs, others on sea urchins, or turban snails, or abalone, and there are also many combination specialists of two or three of such prey, but virtually no generalists. All specialize somehow, and this is likely to be advantageous because of the complicated skills required to obtain prey as different as all these—a jack-of-all-trades is a master of none. The authors argue that diversity of specialization is likely to arise in situations where there is little interspecific, but considerable intraspecific, competition; this appears to be the case for sea otters. In populations where numbers were depressed because of predation by killer whales, there was little difference between individuals in their choice of prey (Watt *et al.* 2000), but in really high-density populations dietary diversity was large (Estes *et al.* 1981).

There is no evidence in sea otters of cooperative foraging: they go it alone. Furthermore, just like river otters, they are subject to kleptoparasitism, from their own species as well as from others. Male sea otters often divest females of their rightful property when they pass through the males' territories (Riedman and Estes 1990). Bald eagles have been seen stealing lumpsuckers from sea otters when the latter are eating, floating on their back on the surface. The eagle, from its perch, spots the surfacing otter, then approaches low and fast, and grabs the fish. Otters may avoid this by rolling over, or with a shallow dive, but eagles are often successful (Watt *et al.* 1995).

Fishing and foraging by other otter species

Giant otters live in family groups, but every individual usually fishes for itself, independently. There are occasional reports of cooperative action, such as groups driving schools of fish into shallow backwaters, or a family going eight abreast driving fish into shallows between sandbanks (Duplaix 1980; see also below). The animals are active in daytime only, and their most common method of fishing consists of shallow dives, lasting fewer than 10 seconds, along banks or into overhanging vegetation, or in flooded woodlands. The members of a foraging group communicate continually, with slight growling sounds (Staib 2002), but there is no evidence of any concerted action.

Because the giant otters often operate in close proximity to one another, observation of individuals is confusing and quite difficult, and there are few good data on actual diving times. When they are fishing in deeper waters, often near floating vegetation, dives may take up to 20 seconds. In about one-third of observations, Elke Staib noticed that in deep open water the animals were fishing with a behaviour that she called 'jump-diving'. The otters were diving repeatedly within a small patch, surfacing almost dolphin-like whilst taking a deep breath and immediately going down again, sometimes with the entire body out of the water. This jump-diving had a strikingly high success rate.

In an interesting difference with *Lutra* and *Lontra* species, giant otters are frequently seen fast-chasing fish, over distances of up to 50 metres, especially in shallow water or close to the surface in deeper areas. They cause a great commotion in the water whilst doing so (Duplaix 1980; Staib 2002). This suggests that vision is highly important during fishing, and the large, bulging eyes of this species lend support to this.

Once prey has been caught, it is eaten on the surface, head first, whilst the giant otter is treading water and feeds the fish into its mouth with the forepaws (Fig. 9.4). Only larger prey are landed, and if giant otters happen on a really large catfish or turtle they may pull it out together and share it with other family members (Staib 2002).

Of the other South American otters, details of foraging behaviour are known only for the marine otter. Until more studies are available, we may have to assume that both the **neotropical otter** and the **southern river otter** catch their prey in a manner similar to that of the river otter. In my own casual observations of neotropical otters, their fishing and diving appeared similar to that of river otters and Eurasian otters. The **marine otter**, a largely diurnal species, was studied by Richard Ostfeld and his team

Figure 9.4 Giant otter eating fish © Nicole Duplaix.

(Ostfeld *et al.* 1989) and by Gonzalez Medina (1995). It also fishes by diving from the surface, like the river otter in the sea, and similarly its mean dive time was 28 (±9) seconds, with the longest dive lasting 64 s. Although these dives were of similar length to those of river and Eurasian otters, marine otters are smaller (3.2–5.8 kg) (Larivière 1998), and one would have expected mean dive times of around only 16 s (see Fishing behaviour of Eurasian otters above). Marine otters appear to be more than averagely efficient divers. Catching crabs involved dives that were 4–7 s longer than those for fish. When marine otters emerged with a fish, the prey was eaten on the surface if small, but medium-sized or large fish or crabs were taken ashore; if other marine otters were present, they occasionally stole from one another.

The foraging behaviour of otters in Asia has been little studied. The **smooth otters** that I observed fishing in the Huay Kha Khaeng river in Thailand were in shallow, running water, and in one observation two individuals repeatedly swam alongside each other from one side of the river to the other, rapidly driving fish in front of them into some reeds, where they could easily be captured. Similar behaviour had also been observed by others (references in Kruuk *et al.* 1994a). Foraging behaviour of the **hairy-nosed otter** is quite unknown, and may be similar to that of the Eurasian otter. The **small-clawed otter**, however, does things differently.

With their curiously long fingers and only rudimentary claws, these small otters poke around under boulders and logs, and into nooks and crannies, debris and vegetation both under water and on the banks where their mostly invertebrate prey is hiding. Crabs, insects and molluscs, as well as frogs and fish, are their quarry; perhaps they can dive and fish in deeper water like the other otters, but I have not seen it and it has not been described by other observers (see review by Sivasothi and Nor 1994). An interesting snippet is the observation of small-clawed otters digging up molluscs and leaving them in the hot sun, so they soon opened up (Timmis 1971).

In Africa, foraging has been researched more thoroughly in a much larger *Aonyx*, the **Cape clawless otter**. Its forefeet have similarly long, sensitive fingers, also used for catching crabs, by feeling under stones and in dense vegetation, in the sea or in fresh water. This is often in shallows, where the otter walks or swims, often foraging with its head above water (Fig. 9.5). Alternatively, an animal may dive in deeper water, in the same fashion as described for the Eurasian otter, its tail showing before going down at an angle of some 60° (Somers 2000a).

The mean dive time of Cape clawless otters foraging in the sea was 21 seconds in both Arden-Clarke's and Somers' studies. This is somewhat shorter than the 26 s predicted for an otter with a mean body mass of 13 kg (Somers 2000a), and in freshwater habitats it was shorter still, at 17 s (Rowe-Rowe 1977). These values probably reflect the shallower depths at which the Cape clawless operates, compared with other otters.

Cape clawless eat small prey at the surface, often holding it in the front paws both when transporting and whilst taking bites from it, and, just like other otters, they take large prey ashore. When carrying it on land, prey may be held with one of the front paws (Rowe-Rowe 1977).

Only a few comments have been published about the foraging of the very similar **Congo clawless otter**, specifically about their remarkable habit of feeding on earthworms in swamps of the West African rainforest. The description of this foraging suggests a

Figure 9.5 Cape clawless otter foraging for crabs in river, South Africa.

very un-ottery behaviour (quoted from Jacques *et al.* in press):

> Worms are located by pushing the forepaws deep into soft mud banks, and clawing through the mud with the fingers, gaze averted. After several seconds the forepaws are withdrawn and the worm transferred from the digits to the mouth. An average of 3 worms per minute are obtained in this manner. Parent otters have been observed feeding immatures, with worms passed from mouth to mouth.

The frequency with which worms are caught is rather similar to that of another earthworm specialist, the Eurasian badger (Kruuk 1989).

Somewhat more is known of fishing behaviour of the third otter in southern Africa, the **spotted-necked otter**. Where we watched them on Rubondo island, in Lake Victoria (Kruuk and Goudswaard 1990, p 324),

> ... the groups mostly moved from site to site as a pack with individuals swimming close together, but during actual foraging they spread out, each one fishing on its own. Almost all fishing was done within 10 metres of the shore ... and mostly within 2 metres. The otters hunted between rocks and stones and especially between branches of shrubs ... moving through the water much faster and more silently, without splashing, than does the Eurasian otter, with shorter periods on the surface between dives. Fishes smaller than about 10 cm were eaten in the water, larger ones were landed.

Along the shores of Lake Malawi, spotted-necked otters were similarly fishing between large rocks and branches, not in open water. However, John Procter (1963) saw them fishing for prolonged periods in waters of up 1.5 m deep, with dives lasting about 15 s 'remarkably constant in duration', and an exceptional dive of 21 s. Procter (1963, p 96) also commented on the unusual agility of this species: for their near-vertical dives, their bodies arched almost clear of the water, and the animals

> ... seem able to turn faster than their prey. Once a small fish which was being chased at full speed near the surface flicked out of the water just in front of the otter's nose and landed where the latter's tail had been, but the otter had already turned and caught the fish as it re-entered the water.

In Lake Muhazi, Rwanda, Anne Lejeune frequently saw spotted-necked otters dive in open water of unknown depth, with mean lengths of dive of 22 s (Lejeune 1989). This was rather longer than the calculated expected dive length for an otter of 4 kg, of 16 s.

In general, therefore, all otters show fairly similar behaviour mechanisms to acquire their prey, with relatively small species-specific variations on the central themes. The sea otter is the most aberrant, and the three *Aonyx* species are different, with the use of their long fingers. There is species-specific fine-tuning of the general foraging behaviour, but, to me, the similarities in their behaviour are more remarkable than the differences.

When to feed: time and tide

Are otters night animals, or active in daytime? They may be either (even individuals from the same species), but some species are decidedly more one than the other. This has little to do with disturbance from people, as is often thought. The issue is of some importance, and not only because it determines when or whether one can see the animals for study, or for public interest. It is relevant in the interaction between otters as predators, and their prey species. I will describe activity patterns for some of the better studied otters.

During the short days of winter in Shetland, **Eurasian otters** showed one clear peak of activity just before midday, but in summer swimming and foraging was spread throughout the daylight hours, with one main peak early in the morning and a lesser one in late afternoon (Fig. 9.6). This was not because we watched otters only during the day, when they were easy to see: when we followed animals fitted with radio-transmitters over 24 hours, they were also active only in daytime. I was often out and about in the morning before the otters were, and I could actually see them leaving their holt. The animals were undoubtedly diurnal, and their feeding was a daytime activity in Shetland (Kruuk and Moorhouse 1990), as on the Scottish west coast (Watt 1991; Yoxon 1999).

This daytime activity was in contrast to Eurasian otters' nocturnal nature elsewhere (Chanin 1985; Kruuk 1995; Mason and Macdonald 1986). The most likely explanation for this lies in the habits of the otters' main prey species, the bottom-dwelling fishes (see Chapter 8). Along sea coasts these fish are active at night, swimming fast and often into mid-water, but hiding under stones and weeds during the day.

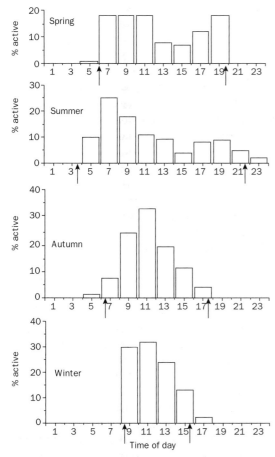

Figure 9.6 Times of day when Eurasian otters are swimming/diving in Shetland, by season. In spring and summer mostly in the morning, and during the short day-length of autumn and winter at midday. Arrows show mid-season times of sunrise and sunset. Number of dives: spring, 4781; summer, 3063; autumn, 3466; winter, 2003.

This nocturnal activity holds true for eelpout, rocklings and crabs, but some other prey are more diurnal. Sea scorpions, bullrouts and butterfish, for instance, appear to be 'active' at any time, but they are slow, sitting still, slow in swimming and slow in reacting to danger—easy for an otter to catch at any time.

In freshwater areas of Scotland otters often caught their prey at night, and it consisted primarily of salmonid fish and eel (see Chapter 7). Where Eurasian otters feed predominantly on eels, in lakes for example, they often do so in daytime (Conroy and Jenkins 1986), just as do otters foraging in the sea. Eels are active at night and not during the day (Moriarty 1978), although even at night they spend long periods still, either on the bottom or partly buried in it. Whilst scuba-diving I saw that in daytime many of the eels just lie quietly on the bottom, some of them easily touched by hand (therefore easily caught by an otter).

Salmonids and many other freshwater fish are often inactive during the middle of the night, motionless on the bottom (Westin and Aneer 1987), with trout under banks or in dense vegetation. Trout also often flee from a disturbance into such cover, and it is there that they are most easily caught by an otter. However, at temperatures below 10°C, salmonids are more active at night, at least in some northern rivers, and their behavioural periodicity is not as simple as was thought at first (Fraser et al. 1993; Heggenes et al. 1993; Valdimarsson et al. 1997). Large male salmon move up and down rivers during the early part of the night, when otters catch them on the shallow riffles (Carss et al. 1990).

In the sea, the state of the tide also has an important effect on the activity of almost all coastal life. Along the Shetland coasts, otters were significantly less likely to be fishing during the 3 hours around high tide (21% of observations), and they were more active during the same periods at falling, low or rising tides (27%, 25% and 27% respectively) (Kruuk and Moorhouse 1990). The preference of otters for feeding at low tide was even stronger than these figures suggest, because it was more difficult for us to find otters foraging in the exposed algae at low water, with all the fronds waving about on the surface. When we did find an otter foraging, the number of dives per observation was greater at low tide than at other times. This showed, again, that otters are active when their prey is not, because almost all the fish on which otters prey are more active at high tide (see Chapter 8).

Less is known about North American **river otters**, although their activity pattern appears similar to that of the Eurasians. In fresh water, they are active mostly at night and during crepuscular hours, and in winter they are more diurnal (Melquist and Hornocker 1983). This is likely to be related to fish activity, as in the Eurasian otter (see Chapter 8). In the sea they are mostly active in daytime (G. Blundell, personal communication, and my own observations).

Sea otters feed during the day as well as at night, with peak activities during crepuscular hours, and in some areas after midnight (see review in Riedman and Estes 1990). Where feeding mostly on crabs, which are active on the open sea bottom at night and more easily catchable then, the sea otters also were active at night. Near Amchitka Island in Alaska, sea otters could be seen active at any time, but it was only at dawn and dusk that they caught fish; at other times it was sea urchins, crabs or molluscs. The observations suggested that it was variation in the activity of fish that determined what the otters caught (Estes *et al.* 1982). Riedman and Estes (1990) emphasized the large differences in activity times between individuals and between social categories, such as males and females with or without cubs. Inclement weather may also affect times of foraging. Generally, therefore, there is no strict single regime of daily activity. Perhaps, with many of the kind of prey in which individuals specialize, there would be little to be gained by operating at any particular time.

Giant otters, which find their prey mostly visually, are strictly diurnal (Carter and Rosas 1997), and at least some of their important prey species, such as catfish and *Hoplias*, are active at night; the same rules appear to apply as for the Eurasian otter. The **neotropical otter** is entirely diurnal in the Brazilian Pantanal, where it feeds mainly on the nocturnal catfishes (unpublished observations), but it may well be more nocturnal elsewhere. The **marine otter** is also diurnal (Ostfeld *et al.* 1989), catching its crabs between the large rocks during any of the daylight hours. There is no quantitative information on the activity patterns of other Latin American or Asian species; I gained the general impression in the field that the Asian otters were largely nocturnal, with some activity in the early morning.

In Africa, both **clawless otters** appear to be largely nocturnal, but both species are also seen foraging in daytime. In the sea, the Cape clawless catches many of its crustacean prey in the early morning (Arden-Clarke 1983; Somers 2000a; and personal observations). The **spotted-necked otter** is mostly diurnal in the open East African lakes, catching small fish especially in the morning, often in narrow underwater nooks and crannies (personal observations in Lakes Victoria and Malawi), but in South African rivers it is almost entirely nocturnal (Skinner and Smithers 1990), just as the Eurasian in similar habitat.

Foraging success

On the face of it, foraging success is easily measured in otters when they are fishing in open water in daytime: one simply counts successful and unsuccessful dives. It is also important to establish a measure of success, because we need to compare the efforts of different animals and populations, in different habitats, and in different conditions. However, there are some serious snags, and these problems have been discussed at some length (Kruuk *et al.* 1990; Ostfeld 1991).

In the Shetland studies on **Eurasian otters** we measured success rates during thousands of observations; overall, 27% of 13,313 dives were successful (Kruuk and Moorhouse 1990). But I became suspicious of what could be concluded from these counts when I noticed how small the variation in success rates was under very different conditions.

For instance, we compared observations of otters that spent most of their time foraging in small sections of coast that we called 'feeding patches'. During this 'patch-fishing', which was clearly the preferred mode (see below), 25% of 2796 observations were successful. In neighbouring sections of coast, though, immediately on either side of these patches, where the otters dived far less often, this rate was 23% (485 observations), and in sections beyond those it was 25% (555 observations). None of these differences in success rate was significant, which was surprising because the otters themselves showed such a clear preference for the small feeding patches. Similarly, when comparing success rates for spring, summer, autumn and winter for all otters over the whole Shetland study, I found rates of 24%, 28%, 25% and 33% respectively, with differences that were statistically significant because of our large sample sizes. Nevertheless this variation was small, where we had expected something much more spectacular (Kruuk and Moorhouse 1990).

It seems likely that interpretation of such success figures is difficult for two reasons. First, dives differ in length of time, frequency, depth, degree of cooling experienced by the animal, and in other ways, so they are not a consistent unit of effort. Second, a dive is not a random event, but the result of a decision taken by the otter, on the surface, which is likely to be based on an assessment by the animal of its chances of success. What we may be measuring, therefore, is the otter's *ability in risk assessment*, rather than the risk itself. For instance, an otter might have a strategy in a certain area (with a given depth and mean prey size), which dictates that it dives only if there is at least a one-in-three chance of catching a fish of a given size there. Such assessment would be based on the animal's previous knowledge of the sites and environmental conditions.

We must be careful, therefore, about how to interpret the simple success rates that I mentioned. Often, they will tell us little about the actual effort on the part of the otter, but in other situations they might be indicative. Ideally, one should collect data on the total energetic cost of catching and handling a prey, and compare this with the award to the animal in terms of calories gained.

Some of the variation in diving success in the Shetland otters, although small, was important when considered together with other information (see Chapter 7). For instance, in springtime success was lower than in other seasons (24% versus 33% in winter), and at the same time the estimated mean weight of prey was lower (only ll g in spring versus 20 g in winter). IN addtion, in spring many of the prey items, (e.g. crabs, sea scorpions and sticklebacks) had low-calorie values (Kruuk and Moorhouse 1990). This variation was statistically highly significant, and suggestive of seasonal difficulties experienced by otters. We did not, however, have data on how much time otters had to spend in the water to catch a prey, or on the seasonal differences in the depths at which prey was caught. Such measurements still need to be made, and are crucial to the understanding of seasonal stress.

There are large differences in the success rates of otters in different areas. Conroy and Jenkins (1986) compared diving success rates of Eurasian otters in a Scottish freshwater loch with those along the Shetland coasts (7% versus 27% respectively). To interpret such figures, one has to take into account that, in the sea, otters took small fish, 20–30 g, from considerable depths (usually 1–8 m), and in fresh water the much larger eels (50–200 g) from waters 1–2 m deep. The larger prey from shallower depths makes a lower dive-success rate acceptable, but more information is needed about the length of time spent by otters in the water.

The variation in success rates of otters foraging along different types of coast in Shetland was considerable (Kruuk and Moorhouse 1990). Along the most exposed shores and the steep cliffs, the success rate per dive was only 12%, but in the larger bays and along more sheltered coasts it was as high as 32%, 25% and 34%. The weights of prey caught in the more exposed areas compensated for this: a fish caught per average dive (including the unsuccessful ones) along the most exposed coast was 19 g, but significantly lower in the two sheltered bays, only 9 and 13 g. Otters along exposed coasts had fewer successful dives, but came up with larger prey. One very small bay did not fit into this pattern; otters there had a low success rate, only 16%, and small prey (mean weight 5 g per dive). However, that was unusually shallow (rarely more than l.5 m deep), so a dive there was short and cheap in energetic terms. All this suggested that there were predetermined chances of success for foraging sessions in different areas, perhaps by otters 'knowing' the particular underwater sites with the stones and holes where fish could be found, with an established expectation of success.

Overall seasonal differences in success rates might be small, but there was substantial variation between subsequent years (Nolet *et al.* 1993). In the main study area in Shetland, success at l, 2 and 3 m deep was 29%, 30% and 38% in 1983, whereas in 1984 it was 15%, 13% and 17%, and in 1985 23%, 20% and 21%, all in the same sites. Success rates can be twice as high in one year compared with the next. The causes of this are as yet unknown, as the annual differences in fish densities (see Chapter 8) did not match the otters' success rates. Because of such problems, in the section on energetics below I argue that, for Eurasian otters and most other species, it is often more useful to estimate success as reward per time spent in water, rather than per dive.

Success rates of other species have been studied less intensively, with the exception of **sea otters**. Ostfeld (1991) argued that their success rates showed important differences, for example between animals specializing in eating invertebrates and those catching fish. He reported that sea otters in the Aleutian

Islands emerged successfully, usually with a fish, in over 90% of dives (Estes *et al.* 1981), but on a Californian coast (taking large invertebrates) in only 35% (Ostfeld 1982). South-east Alaskan sea otters feeding on clams, mussels, crabs and other invertebrates showed an average dive success rate of 81% (Kvitek *et al.* 1993). However, just as for our Shetland data, such variation means little unless one knows more about the nature of the dive (length of time and depth), and the kind and size of prey; sea otters, for instance, often collect several prey items in one dive (see above).

Riedman and Estes (1990, p. 42) noted:

Adult [sea] otters make unsuccessful dives more often than juveniles, although adults also obtain more rewarding but less easily captured prey. Longer dives—and often several dives—are required to capture large prey items, that are less accessible but more rewarding in terms of energetic value (such as abalone and *Cancer* crabs), than are necessary to obtain less valuable prey (such as turban snails . . .).

Beckel (1990) compared the foraging success rates of **river otters** that were either hunting alone, or with two animals together, in a river in Wisconsin. She noted that 76% of dives by a single otter were successful in catching a fish, compared with 36% of dives, in the same places, when there was more than one otter. There was no sign of cooperation between the animals, and the most likely explanation of the difference was that, when there were two otters together, at least one of them was sub-adult and therefore less efficient.

Marine otters showed rather large differences in success rates when diving in separate areas, varying between 37%, 38% and 16% (Ostfeld *et al.* 1989). The otters in the area with the lowest success rate also caught a higher proportion of large prey than the others, especially fish, whereas the other areas produced mostly crabs.

In South Africa, Mike Somers (2000a) mentioned that **Cape clawless otters**, when foraging for quite large fish, crabs and crayfish in the sea, swam and dived like the Eurasian otter. Their prey was large, but their success rate per dive was only 4–13%. **Spotted-necked otters** in Rwanda emerged much more frequently with a fish after a dive, and their success rate was 44%. However, their prey consisted mostly of very small cichlid fishes, with a mean weight of only 3.4 g (Lejeune 1989), and energetically their dives were lest costly in the warm waters. It appears that the overall comparisons of success rates have only limited value, unless many other variables are also taken into account.

Patch-fishing, prey sites and prey replacement

A phenomenon that has attracted a great deal of attention in many studies of foraging behaviour of animals is that they often feed for long periods in relatively small areas. This observation is of importance, amongst others, in hypotheses on feeding efficiency ('optimal foraging'; e.g. Krebs 1978) and in explanations of animal dispersion, territory size and group size (e.g. Bradbury and Vehrencamp 1976; and, for carnivores, Bacon *et al.* 1991; Carr and Macdonald 1986; Kruuk and Macdonald 1985; Macdonald 1983; Mills 1982). **Eurasian otters** in Shetland also showed a strong tendency to fish in certain small 'patches'.

To quantify the 'patchiness' of the otters' foraging effort, I divided the coast of the Shetland study area into 178 sections of about 80 metres. Some sections, when visited, were fished with many more dives than one would have expected if the otters had randomized their movements, and other sections received far fewer dives. This same pattern of usage was seen repeatedly, not just during one trip but over many months, and probably years. When an otter was fishing along the coast, its median number of dives per section was 7.0, but it might dive as often as 61 times in one site (Kruuk *et al.* 1990).

I arbitrarily defined a section as a 'feeding patch' if the mean number of dives per observation there, over the whole study period, was at least 15, that is, more than twice the average number of dives per section. Using this definition, 17 of the 178 sections (10%) were classified as feeding patches, in which otters performed 35% of the 7891 dives.

These feeding patches were not significantly different from elsewhere in their proximity to an otter holt, or to fresh water, and there was nothing specific along the coast that appeared to be different about them. With my students I sampled the fish population in the feeding patches with our fish-traps, and we compared the results with control sites (of the same depth but 100 metres away). There was no significant difference, in either prey species composition or biomass.

There were no substantial differences in rates of successful dives, and there was no correlation between the number of dives in a section and the success rate. Finally, I tested to see whether, perhaps, otters in a 'run' of dives on a patch showed a decreasing success rate. I expected the classical pattern of initial high success, and then subsequent decrease, until the patch became so unprofitable that it would pay to move on again (Krebs 1978). Such a pattern might have been obscured when scoring an overall success rate per patch. In a sample of 20 long runs of dives in one place (total 749 dives), the mean success was 31% for the first ten dives, 30.5% for the last ten, and 31% for the last five dives. There were no differences in success rates, therefore, during each period of exploitation of a patch.

However, during the field observations I noticed that there was something peculiar about the algal vegetation of the sections with feeding patches, mostly kelp (*Laminaria* spp.) These sites had significantly more openings in the otherwise solid canopy of fronds, and more places where we could actually see the bottom from above. There were sandy places with clear boundaries.

We had noticed ourselves when scuba-diving, and attempting to get at the sea bottom, that these openings in the kelp beds could be important. Outside the open sandy places the kelp is difficult, and we had to fight our way through the fronds and stems, but in the open spots we could get straight down, then move horizontally in between kelp stems. Otters probably experienced similar problems.

It was likely, therefore, that the main difference between the highly favoured feeding patches and the rest of the coast was relatively easy access (Kruuk *et al.* 1990): prey availability, rather than prey numbers or biomass. However, it was interesting to note the similarity in success rates between patches and elsewhere, and between the beginning and end of a feeding bout. This suggested that what I, from the shore, called a distinct feeding patch was no more than an assembly of prey sites, in each of which otters could pull fish from under a stone, probably well known to individual exploiting predators. The likely, and simple, explanation for the success rate similarity is that otters visit numbers of known prey sites, which are sometimes far apart and sometimes close together (a patch), determined by access through the kelp.

In Scottish freshwater lochs, similar feeding patches occur (unpublished observations), small areas close to or far from the shore where otters dive scores of times and catch eels, year after year. I have not been able to associate these areas with habitat features, but it seems likely that they are characterized by particular substrates or vegetation. Otters can harvest these prey sites again and again, after a time interval of say, a day (see fish removal experiment in the section Trapping and counting fish in Chapter 8).

I believe that this is a key observation for an understanding of otter foraging and spatial organization. Food resource exploitation is based on patchy availability, which is to a large extent both recurrent and predictable to the exploiting predator. From the otters' point of view, the system would malfunction only if competing individuals utilized the same patches.

Delibes *et al.* (2000) documented Eurasian otters' diet at the beginning and end of short periods of use (several days) of scattered ponds in southern Spain—undoubtedly 'food patches', where eels were the main attraction. The aim of the study was to establish why otters were attracted to, and why they abandoned, such patches. Diet composition (species and size) did not change during the periods of use, but the researchers could not measure the feeding effort involved in this utilization, and the issue was left unresolved.

The home range sizes of **Cape clawless otters** in South African rivers are correlated with the distances between relatively small, high-density food patches, mostly reed beds (Somers and Nel 2004). The animals foraged there, sometimes for hours on end, but there are no data on the quality of the resources in these patches and areas in between.

Depth of dives

Eurasian otters foraged as deep as 15 metres (see Chapter 4), but the vast majority of dives occurred in shallow water. For instance, in Shetland, 54% of 3558 dives were in water that was less than 2 metres deep and 98% were less than 8 metres (Nolet *et al.* 1993). When we compared dives at different depths in the strip of water within 100 metres from the shore, there was a highly significant preference for shallow places (Fig. 9.7).

The reason for the shallow water preference was not immediately obvious. We considered several hypotheses to explain it (Kruuk *et al.* 1985; Nolet *et al.* 1993), such as that otters used shallow waters more because: (1) they could catch more fish there; (2) it might be energetically more efficient, either because more underwater time could be spent on searching rather than on travelling to and from the bottom, or to and from areas further off shore; or (3) perhaps breathing was more efficient during shallow dives.

In fish-trapping observations (see Chapter 8) we did indeed find more fish in shallow water, but they were smaller, and as prey biomass (fish weight per area) there were no significant differences with depth. The otters' fishing success (weight of prey caught per time spent diving) was not significantly greater in shallow water (Kruuk *et al.* 1985). In deeper water prey was larger, but the handling time of prey (time spent in transporting and eating) increased proportionally with the size of the prey (Nolet *et al.* 1993). We rejected the first hypothesis.

Hypothesis 2 suggested that shallow diving migh involve less effort, because of the shorter distance to the surface and/or the shore. There was a clear relationship between dive depths and the duration of dives (Fig. 9.8). Otters' fishing bouts lasted longer if they included fewer deep dives, and the subsequent recovery on land from the hunt was shorter (Nolet and Kruuk 1989; Nolet *et al.* 1993). This indicated that fishing at greater depths is energetically more demanding.

A further variable to be considered is breathing efficiency. For several diving animals, the longer they have been under water, the greater the proportion of total time they have to spend on recovery, on 'catching their breath' (Kramer 1988). Somehow, a diving animal has to establish an optimal balance between underwater time and recovery effort. Houston and McNamara (1994) argued that our otter data supported such an 'optimal breathing hypothesis': the shallower and shorter the dives, the more efficiently an otter can divide its time and effort between searching and recovering. In Eurasian otters the recovery time on the surface, after a dive, is directly proportional to the diving time—it is about half (Nolet *et al.* 1993) (Fig. 9.9)—so this in itself could not explain a preference for short dives, but possibly a greater breathing effort is involved for deep dives. In other words, shallow dives are energetically less

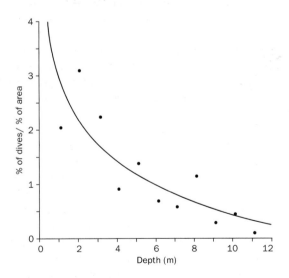

Figure 9.7 Preferring the shallows: numbers of dives by Eurasian otters in Shetland, in sites of different depths. Proportion of 1008 dives, divided by proportion of 100-metre wide strip along coast of the given depth ($r^2 = 0.70$).

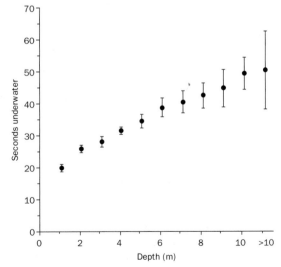

Figure 9.8 Deeper waters involve longer dives: mean times underwater in dives by Eurasian otters in Shetland, at different depths. Values are means and standard errors of 1181 dives, when no prey was caught.

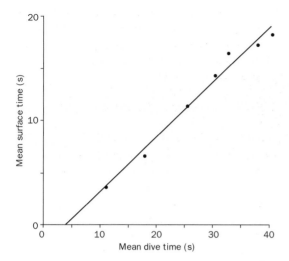

Figure 9.9 After a long dive, a long recovery is needed (Eurasian otters in Shetland). Means of dives at different depths; 1039 observations ($r^2 = 0.98$; surface time = dive time × [0.53 − 2.2]).

demanding than deep ones, because of greater effort involved in deep dives and subsequent recovery, for a similar return of prey.

Incidental observations for various other otter species suggest that their choice of depths for foraging is similiar to that of the Eurasian, with similar energetic consequences. As usual, the exception is the **sea otter**, which behaves differently and goes much deeper. It is also the only otter that takes more than one prey to the surface, as described earlier. Ralls et al. (1995) found by radio-tracking that young male sea otters foraged in deeper waters, at an average of 30 m, whereas, for instance, adult females dived to mean depths of 22 m, with underwater times of well over 1 minute.

In the Aleutian Islands, Kenyon (1969) noticed that, when sufficient food is available, sea otters prefer to forage in (what he called) shallow waters 2–30 m deep, and no deeper than 50 m, with males going deeper than females. Jim Estes (1980), observing sea otters in the same area, commonly found animals feeding at depths of 40 m or more.

More recently, Bodkin et al. (2004) observed more than 180,000 dives of 14 sea otters with radio-transmitters in Alaska. Most of the foraging dives were between 2 and 30 m deep, 85% of dives by females were less than 20 m deep, but half of all dives by males were at 45 m or deeper, with less than 2% of all dives going further down than 55 m. The deepest dive recorded was 100 m.

Energetics of foraging

Almost all otters spend a lot of time asleep. However, when they are not asleep, and we see them in their watery environment, they leave an impression of carefree abandon, of unlimited energy. An ecologist will be wary of such ideas, thinking of the problems that otters have with food acquisition, as mentioned in previous sections, and remembering that prey availability may limit otter numbers (see Chapter 11). One may ask why, therefore, given the amount of time for which otters seem to be doing nothing, they do not expend some more effort and catch more fish? In rivers in north-east Scotland, three otters with radio-transmitters were active and fishing for an average of only 5.2 hours per day, or 20–23% of time (5.6, 5.1 and 4.8 hours, over a total of 56 days spread throughout the year) (Durbin 1993). In Shetland three radio-tagged otters were active for only 2.6–3.4 hours, or 11–14% of the day (Nolet and Kruuk 1989).

The answer to the question of time spent fishing is complicated and important. It amounts to an admission that, in reality, the effort involved in catching fish, or other prey, is unexpectedly large. There is a delicate balance between energy gain and expenditure, which, for otters, are both relatively high. These animals, living in a very special habitat, are in an unusual predicament.

The main problem for a warm-blooded animal living in water is thermo-regulation. Thermo-conductivity is about 23 times higher in water than in air, and the heat capacity of water is 3500 times greater (Schmidt-Nielsen 1983; Williams 1998). Just seeing **river otters** in Canada or **Eurasian otters** here in the north of Scotland, for long periods in ice-cold water (Fig. 9.10) and without the protection of a thick layer of blubber, makes one shiver. **Sea otters** are even more extreme, spending virtually their entire life in cold water. We will see that, although all otters obtain almost all their food from the sea, lakes or streams, the energetic costs associated with their foraging are very largely the costs of being in the

Figure 9.10 Eurasian otter family fishing from ice, in Scottish freshwater loch (Dinnet).

water, of swimming and diving. These costs are high, and in the following sections I will discuss research that has sought to understand this.

Otters are proverbially proficient swimmers, paddling with their hindfeet in synchrony, as well as propelling themselves with undulating body and tail movements (Fish 1994; Williams 1989). One of my students, Addy de Jongh, measured the diving and swimming behaviour of a captive yearling **Eurasian otter** by filming in a large swimming pool and calculating speeds as described by Videler (1981) for fish. When diving, the otter descended at about 70°, at a speed of 0.62 m/s, and when swimming at presumably optimal speed it moved at 1.20 m/s. The animal was not fully grown (total length 76 cm, compared with 1 m for an adult female) and, to extrapolate these figures to adult otters, speeds have to be calculated relative to body length (Videler 1981).

One can estimate, therefore, that an adult female Eurasian otter is able to swim under water at a speed of about 1.5 m/s, or 5.5 km/h (Nolet *et al.* 1993), compared with underwater speeds for sea otters of 0.6–1.4 m/s (Williams 1989). When the animals were searching for food along the bottom of clear water in Shetland, we calculated (from times taken to cover estimated short distances) that they moved much more slowly than that, at about 0.26 m/s, or 0.9 km/h (Nolet *et al.* 1993).

Pfeiffer and Culik (1998) studied oxygen consumption of Eurasian otters swimming in a flume. They calculated mean underwater swimming speeds of 0.89 m/s, or 3.2 km/h. They also calculated that, for a Eurasian otter swimming at a speed of less than 1.3 m/s and in water of 12–15 °C, the energy used for thermo-regulation is higher than the energy cost of moving about. As I will show below, we confirmed this independently in experiments, also with captive Eurasian otters: the underwater behaviour of the animal has less effect on energy expenditure than the temperature of the water (Kruuk *et al.* 1994b).

Even on land, otters have an unusually high basal metabolism. Compared with terrestrial mammals of similar body mass, the resting metabolic rate (RMR) of Eurasian otters is 38–48% higher (Kruuk et al. 1994b; Pfeiffer and Culik 1998; and see review by Estes 1989). In addition, their aquatic lifestyle has an important effect on energy consumption: in water their RMR is 1.7–4.5 times higher than that of similarly sized mammals on land (Kleiber 1975). This, of course, has important implications for the amount of food they have to acquire. With my colleagues in Scotland, therefore, I tried to address the costs of swimming and diving under different conditions; how these costs compare with actual food intake of the animals; and whether they affect the exploitation of prey populations.

First we studied the effects of water temperature, as well as the behaviour of Eurasian otters and the effects of other variables on energy consumption. Energy consumption is normally measured by the amount of oxygen used by an animal. We estimated oxygen intake when swimming and diving; this could then be compared with the energetic content of food actually acquired during foraging periods in different areas. The effects of water temperature on body temperature is discussed in Chapter 10.

The oxygen uptake of **Eurasian otters**, active in water, had to be studied on captive animals (Kruuk et al. 1994b). We kept four otters in enclosures, with a big swimming pool and underwater observation windows. In the pool we made a metabolism chamber, which the otters could enter at will, and in which they could breathe air from under a perspex dome. They could sit on a shelf in the chamber (only their head above water), or they could swim about. The otters were used to the chamber and frequently played in it after entering through the underwater door.

When we wanted to measure the oxygen uptake in one of the otters, we closed the door behind it as soon as it had entered the chamber, and kept it inside for 20 minutes while we recorded the animal's behaviour and measured the oxygen flow under the perspex dome. This experiment was repeated many times on days between August and December, so we had a range of water temperatures (between 2.1 and 16.6°C).

We measured 208 5-minute periods, in which otters were sitting on the shelf for 46% of the time, quietly swimming and diving for 27%, and swimming fast or vigorously scrabbling at the underwater door for 27% of the time. This behaviour was significantly

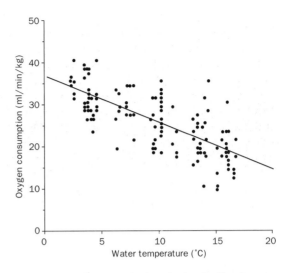

Figure 9.11 Oxygen consumption of swimming Eurasian otters is dependent on water temperature, irrespective of their behaviour (measured in three captive otters). (Adapted from Kruuk et al., 1994b).

dependent on water temperature: the colder the water, the more vigorous the animals' behaviour and attempts to leave the chamber, and during warmer days they were more likely to sit on the bench.

Figure 9.11 shows the relationship between water temperature, T_w, and the otters' oxygen consumption. The animals spent much more energy in cold water: 55% of the variation in oxygen consumption could be explained by differences in water temperature (a correlation of $r = 0.73$), independent of behaviour. Additionally, differences in behaviour (themselves highly dependent on water temperature) of the otters explained a further 14%, so that T_w and behaviour alone accounted for more than 69% of the observed variation in oxygen consumption. We introduced other variables into the equation, such as individual differences between the animals, the otters' bodyweight and the time for which animals had been in the water, but they had little effect.

Interestingly, if one extrapolates the graph in Figure 9.11 to the point where oxygen consumption would be zero, one finds that, as expected, this occurs at a water temperature of 37.6°C, which is approximately body temperature; this gives confidence to our measurements (Kruuk et al. 1994b). We found that, expressed in terms of energy expenditure E (in watts per kilogram of bodyweight; 1 watt = 3.6 kilojoules per hour) and with 'average' swimming behaviour, $E = 12.1-0.40\ T_w$. This approximates the cost to an otter of catching a fish, for a given time in the water, and it can be used to assess foraging profitability.

The effects of water temperature on a swimming semi-aquatic mammal had been studied before, for instance on muskrats *Ondatra zibethicus* (MacArthur 1984). He established that a 1-kg muskrat expended about twice as much energy when diving in water of 3–10°C, compared with diving in water of 20–30°C.

With some assumptions, I constructed a model of the length of time per day that a Eurasian otter of known or estimated weight would have to spend fishing in order to meet its daily energy costs, for different prey availabilities and for different water temperatures (Fig. 9.12). As information for physiologists, I assumed an average calorific content of fish of 4.5 kJ/g (Watt 1991), an assimilation coefficient for an otter eating fish of 0.7 (Costa 1982) and an *RQ* (respiration coefficient) of 20.1 (Bartholomew 1977,

Williams 1986). Prey 'availability' can be expressed as weight of fish caught by an otter per length of time in the water.

In principle, the graph connects 'break-even points': the amount of time a Eurasian otter has to forage each day in order to make up its daily energy deficit, when fishing in waters of different prey 'catchability' and at different water temperatures. The model shows, for instance, that in a time and place where an otter can catch 600 g of fish per hour, it would have to fish for only about 1.5 hours to make up its estimated daily energy deficit. However, when otters can obtain only 150 g per hour they have to fish for almost half a day every day, at least in winter, when in Scotland water temperatures are close to freezing. In summer, 5 hours of fishing per day will suffice, because fishing in warmer water is that much 'cheaper'.

Despite the various assumptions I had to make in this model, it does show that the estimated quantities of fish needed to make up for loss of energy are reasonably close to the estimated quantities that Eurasian otters are actually known to take, in the wild and in captivity, and this is encouraging. The model predicts that, largely dependent on water temperature, Eurasian otters should take between 0.8 and 1.8 kg (representing 3600–8100 kJ) of fish per day for the calorific contents of prey to match energy expenditure. This is, indeed, the range of food quantities that we have estimated previously (see Chapter 7).

The advantage of this energetic approach is that it enables us to measure in the field what we think otters should be achieving. We can observe the time it takes an animal to catch a fish ('fish availability'), and we can estimate the size of the fish either visually and directly, or deduce it from fish vertebrae in the spraints. Water temperature can be measured directly, and with radio-tracking we can now estimate the length of time per day that an otter spends fishing.

The most striking conclusion of the graphs in Figure 9.12 is derived from their curious shape: they show how vulnerable otters are when fish populations are low, especially in colder regions. Eurasian otters can easily cope in an area where they can catch more than 200 g fish (900 kJ) per hour; they need to fish for less than 6 hours per day in winter, and for less than 4 hours in summer. The crucial point is that, if

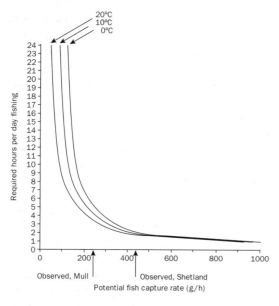

Figure 9.12 Number of hours per day for which a Eurasian otter needs to forage in waters of different prey densities, and at different temperatures. The 'break-even curves' show the length of time for which an otter needs to fish to make up for energy loss at different fish densities. The arrows indicate observed fish capture rates along the Scottish west coast (Mull; Watt 1991) and in Shetland.

that prey availability decreased by half (which is not an especially drastic reduction in fish populations), otter survival would become impossible in winter.

We can now compare this with some observations on fishing success. For instance, in Shetland, where water temperatures range from 6°C in winter to 12°C in summer (Kruuk *et al.* 1987), there was a mean fish capture rate of 390 g/h (Nolet and Kruuk 1989), easily sustained at 2–3 hours fishing per day. However, otters on Mull, along the Scottish west coast, caught only 250 g/h in waters of 14°C in summer (Watt 1991); there, a reduction of 50% in fish availability could make otter existence impossible during the low temperatures in winter. In Spanish rivers, Ruiz-Olmo (1995) estimated that otters caught about 274 g/h.

Fresh water gets much colder in winter than does the sea, at the same latitude. In (unpublished) observations of Eurasian otters in the Dinnet lochs (northeast Scotland) up to 1997, I estimated that, at the end of that period, the animals caught fish (mostly eel) at a rate of 111 g (650 kJ) per hour, which was energetically unsustainable, after a steady year-on-year decline over the previous 7 years from 890 g (4980 kJ) per hour. In the last 3 years of that period, from 1994 onwards, otters did not reproduce in the lochs (and they have not done so since then until the time of writing, in 2005). The animals had been studied there for more than 20 years (e.g. Jenkins 1980; Kruuk and Carss 1996), and where previously up to eight otters resided (producing, without fail, annually one or two litters of cubs), probably only one did presently, part time. Observations became very scarce. The likely cause of this decline was the decline of eels on an international scale (see Chapter 8).

The disadvantage of fishing in cold water is underscored by the observation that otters, only during the cold seasons of winter and early spring, also take other prey such as birds and mammals (see Chapter 7). This, as well as our outline calculations of temperature-related costs and benefits of foraging, suggests that in their (for mammals) unusual habitat, Eurasian otters are often living close to the limits of possible existence. Quite likely, other otters are in the same predicament.

River otters show a high metabolic rate, quite similar to that of the Eurasian otter, and at least in captivity with a comparable diving efficiency (Ben-David *et al.* 2000). It is likely that, for this species, similar energetic constraints apply. Other species also appear to conform to this pattern; for instance, for **small-clawed otters**, Borgwardt and Culik (1999, p 100) concluded that:

Metabolic rates of small-clawed otters in air were similar to those of larger otter species, and about double those of terrestrial mammals of comparable size. In water, metabolic rates during rest and swimming were larger than those extrapolated from larger otter species and submerged swimming homeotherms. This is attributed to high thermo-regulatory costs, and high body drag . . .

This species may be less well insulated by its fur than Eurasian or river otters, as it lives in a usually much warmer environment.

Most observations on the energetics of swimming and fishing by otters have been made on the **sea otter**. This animal is different from the others in that it lives in water full time, and its thermo-insulation is more effective (see Chapter 10). Males with mean 27-kg bodyweight eat 19,680 kJ per day, or 730 kJ/kg daily, and females with mean 20-kg bodyweight eat 14,235 kJ per day, or 712 kJ/kg daily (Garshelis *et al.* 1986). In captivity, Costa (1982) established energy intake as 980 kJ/kg daily. This compares with about 400 kJ per kg per day for Eurasian otters (see Chapter 7), and is therefore considerably higher. These animals spend between 37% and 48% of their time foraging (Garshelis *et al.* 1986), considerably more than the Eurasians, which suggests that for sea otters the relationship between foraging effort and return is of a different shape to that in Figure 9.12—probably a more gradual curve.

Swimming itself is energetically not particularly demanding for a sea otter, especially below the surface. But, even for surface swimming, the metabolic rate of sea otters was only 2.3 times that found during resting in water (Williams 1989).

Generally, the costs of foraging by any otter are, to a large extent, the costs of being surrounded by relatively cold water. The sea otter appears to have gone further in its adaptation to this than the other species. Nevertheless, even for sea otters, as for the others, these costs are unusually high. In return, the benefits are high as well: fish and most other underwater prey have a high energy content. The otters, therefore, invest in a high-risk strategy and, if such energetic foraging happens to be less successful, the consequences will hit the predator hard.

There are some other carnivores with a similarly high-risk foraging strategy, for instance the African wild dog *Lycaon pictus* (Gorman *et al.* 1998b). Its very energetic long-distance running enables the capture of large antelopes, but, also in that species, if the return for effort decreases (in that case through kleptoparasitism by hyenas), the consequences are unusually dire, and foraging becomes unsustainable. For otters, the very high energetic expense of fishing is one of the important factors that makes their existence precarious (see Chapter 13).

The development of foraging behaviour

The increasing success of **Eurasian otter** cubs in catching their own food is described in Chapter 6, following the work of Jon Watt (1993). Foraging improved even after cubs were independent from their mothers, and they continued to become more efficient. One of the problems that young otters had was with their diving technique, or rather their breathing efficiency. The proportion of time they spent under water before recovering on the surface (i.e. the dive-to-recovery ratio) was significantly lower for cubs than for adults, because the cubs' dives (for given depths) were shorter, although their recovery times were the same. This is termed 'diving efficiency', and cubs became just as efficient as adults at the age of about 17 months. However, the success rate (i.e. the number of prey as well as the weight of prey caught per dive) continued to increase until they were 21 months of age, that is, almost 2 years old (see Figs 6.12 and 6.13) (Watt 1991, 1993).

The indications were that Eurasian otters in Shetland matured considerably faster than in Watt's research on Mull. It was rare for otter cubs in Shetland to accompany their mothers until they were more than 10 or 11 months old and, with the odd exception, families split up before the next breeding season (see Chapter 6). Most cubs in Shetland were born around June, and families split up before April. This was perhaps related to better feeding conditions (fish populations) in Shetland than along the coast of Mull, as estimated by fish-trapping (Watt 1995).

Despite such differences between sites, there is no doubt that the development of foraging behaviour in Eurasian otters is a very long affair everywhere, compared with other, similarly sized carnivores. Catching fish is intrinsically difficult for a terrestrial mammal, and in the previous section I have shown that it needs to be done quickly, with as short a time under water as possible. The unusual skill required for this is likely to be the reason for the long dependence of cubs on provisioning by the mother. This, in turn, will probably affect the breeding interval of the females, and overall cub production (see Chapter 11).

In the **river otter**, Shannon (1989) noted that, along the Californian sea coast, three cubs caught their first fish at the age of 3.5 months. At 5 months they took much prey themselves, but were still visibly clumsy, and to a large extent dependent on what their mother provided. At 9 months, much of their food still came from her. The mother deliberately abandoned the cubs when they were about 11 months old, when she was pregnant with the next litter. These observations suggest a development pattern similar to that of the Eurasian otter.

Young **sea otters** generally gain independence more quickly than their Eurasian counterparts, but there are exceptions. As a general pattern, in California pups get their first solid food at around 1 month of age, and by 4 months they eat mostly prey provided by their mother. They were able themselves to collect prey such as molluscs, and break them open with a rock tool, at around the age of 5 months, and they gained complete independence from their mother at about 6 months of age (Payne and Jameson 1984). Other researchers have given the period of pup dependency on its mother as 4.5–9.5 months (Riedman and Estes 1988a; review in Riedman and Estes 1990).

The exceptions to this general pattern in sea otters appear in areas of Alaska, where pups may be dependent on their mothers for a longer, and also more variable, period (review in Riedman and Estes 1990). Many Alaskan sea otters show the same length of dependency as the Californians, but Kenyon (1969) observed that in the Aleutian islands the pups were usually dependent for at least 1 year and probably somewhat longer. Interestingly, this is also the one area where sea otter diet includes a great deal of fish, unlike elsewhere (Kenyon 1969; Watt *et al.* 2000). Kenyon's observations suggest that, just for Eurasian otters, the long period of development is associated with dependence on catching fish.

Giant otters are almost exclusively dependent on fish. They are fast developers, and cubs catch their first fish at the age of about 2.5 months, much earlier than Eurasian or river otters. However, like the cubs of other otter species, most prey is brought to them; they are provisioned not only by their mother, but by several or all members of their group. The age at weaning and independence is not known, but young animals up to 1 year old are still begging for food from the others (Staib 2002), so the period of dependence is a lengthy one.

Probably, therefore, the long development of the cubs' fishing skills, taking many months, is a general otter characteristic. In consequence, the extended involvement of the mother in rearing her cubs is likely to have profound effects on her lifetime reproductive output.

Otter foraging: some conclusions

The lithe and seemingly effortless movements of the fishing otter, and the quick return to the surface with a fish, give an impression of an animal in superior command over its aquatic resources. However, the assessment of an otter's fishing skills is far from easy, and that initial impression is decidedly wrong. Catching a fish is tough—or, at least, catching enough fish is.

I have mentioned that estimation of fishing success is not just a matter of counting the number of times that an otter brings a fish to the surface; time in the water and water temperature also have to be taken into account, and of course the size and kind of prey. Such observations demonstrate that all otter species prefer to take the slowest and inactive prey, usually in shallow waters. The amount of energy spent during fishing is very high—the more so the colder the water—so it is vital that prey is caught as quickly as possible. Once an otter strikes lucky, the reward is relatively high: the average fish, mollusc or crustacean prey is a rich source of energy.

Fishing as a behaviour develops slowly in the young otter, and many of the vital skills involved probably have to be learned. There is the difficulty of quickly recognizing the fishes' hiding places, the skill of grabbing an escapee, even the recognition of likely habitat patches where fishing is profitable. These are difficulties that appear to be associated especially with the exploitation of fish as a resource, where an otter appears to be a relatively clumsy intruder in the underwater environment. Interestingly, sea otters have a long individual development especially in areas where they, too, are dependent on fish, and a shorter one where they eat invertebrate food. Seeing all the problems involved, it is not surprising that otters have an unusually long period of development, often of a year or more of dependence on the mother, with all the consequences for total reproductive output.

The rewards of the otters' aquatic exploitation are high, but so are the costs. The animals have evolved a high-risk strategy, and, if rewards decline, the effects will be dramatic.

CHAPTER 10

Thermo-insulation: a limiting factor

Introduction

The single most important ecological characteristic of all otters—the one that everybody knows about—is the fact that they live in cold water. Some of the energetic problems created by this have been mentioned in the previous chapter, and here I will analyse some of the mechanisms of these constraints, the adaptations in otters to overcome them, as well as their effects on ecology, distribution and survival.

One particular asset of otters plays a vitally important role in their aquatic battle with cold. This is the asset that also made several species of otter the target of violent assault by humankind, which reduced numbers to the brink of extinction: the fur—the main shield that stands between otters and the environment. The unique qualities of otter fur require intensive maintenance, and the consequences of this put severe limitations on the use of some habitats.

The importance of well functioning fur in maintaining otters' body temperature can hardly be overestimated. It is also, therefore, an Achilles heel. Injure it, as after an oil spill, and the animals are doomed.

Body temperature

The normal body temperature of mammals is around 37–39°C. The otters in which body temperature has been studied are no different: theirs is usually around 38°C (39°C in the sea otter; Costa and Kooyman 1982). The challenge to the animal is to maintain that temperature in cold water, and Chapter 9 demonstrated that it spends a great deal of energy in doing this. Not all semi-aquatic mammals are so burdened. The platypus, for instance, has a body temperature of only about 32°C (Grant 1983; Grant and Dawson 1978). Obviously, this will lighten its thermo-regulatory needs (Kruuk 1993). However, otters have not taken to this particular solution, and they have to maintain their high body temperature under testing conditions.

In most parts of Europe, water temperatures are usually well below 20°C, in winter often close to freezing point in fresh water, and between 6°C and 8°C in the sea (Fig. 10.1). Winter water temperatures in Scottish lochs, where my colleagues and I made observations on **Eurasian otter** body temperature (see below), were as low as 0.7°C, in summer going up to about 23.5°C, with an annual mean of 9.4°C (Kruuk et al. 1997b). In neighbouring streams, water temperature goes down to −0.7°C below ice in winter. One may expect similar conditions and colder ones in North America, and in other temperate continental regions. Even in the subtropical parts of their ranges, various otter species will meet water temperatures well below 10°C.

In themselves these low temperatures would not be so much of a problem if the conductivity and heat

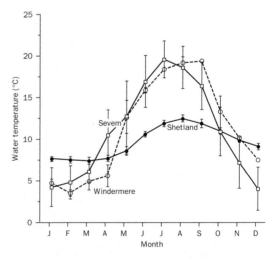

Figure 10.1 Water temperatures in Britain: Shetland 1982-1985, Windermere 1958-1966, Severn 1980-1985. (After Kruuk et al. 1987.)

capacity of water were not so high (see Chapter 9). An otter swimming in 'average' water will lose heat very fast, unless something is done about it, and if the body temperature of any mammal falls below a given value, permanent damage is caused. Insulation helps (see below), but can never be completely effective. Anatomical, physiological and behavioural adaptations are also required.

Once an otter is exposed, or about to be exposed, to cold water (Fig. 10.2) it could, theoretically, follow several possible strategies. For instance, it could increase activity and hence body temperature before entering. Or it could maintain a steady body temperature throughout, generating body heat by increased activity. Or it could tolerate a small decrease in body temperature and exit from water when body temperature becomes critical. In addition, otters could aim to spend as little time in water as possible.

We tested these predictions of body temperature, water temperature and related diving behaviour in the wild on Eurasian otters (Kruuk *et al.* 1997b). When radio-tracking the animals in freshwater lochs in Scotland, we provided four with transmitters sensitive to core body temperature, so we could relate this to behaviour and to water temperature. Such transmitters are inserted intraperitoneally; they are about the same size as ordinary radio-tracking ones, but their working life is about half (4–6 months). The interval between radio-pulses is a measure of body temperature, and this is recorded and analysed with special equipment.

Overall, in 1770 recordings on the four Eurasian otters (on land and in water), the mean body temperature was 38.1°C. There were small variations, between individuals and for different activities. The lowest body temperature measured was 35.9°C, the highest 40.4°C, a difference of 4.5°C. The mean of individual body temperatures for inactive otters was 38.1°C (similar to that of **river otters**; Stoskopf *et al.* 1997)—in one individual as high as 38.7°C, in another as low as 37.6°C. We could not discern any temporal patterns in daily fluctuations.

In 59 observations we were able to measure body temperature at the moment when the **Eurasian otters** entered and left the water, and the length of time for which they stayed in, and we also knew the water temperature. We could then calculate the change in body temperature as well as the rate of

Figure 10.2 Eurasian otter male entering the sea after quick visit on land, Shetland.

cooling, and then relate all of these variables to one another by means of multiple regression. One conclusion was that the body cooling rate in water was on average 2.3°C per hour. For each foraging bout, the mean decrease of body temperature was 1.1°C. These results were similar to those of MacArthur and Dyck (1990) for young beavers *Castor canadensis*.

Second, body temperature was high when a Eurasian otter entered the water, significantly higher than average, as known also for other species such as the muskrat *Ondatra zibethicus* (MacArthur 1979). This is probably due to increased activity on land before diving in. Whilst swimming, body temperature decreased, and at the end of the swimming bout it was not significantly lower than the overall mean body temperature. Figure 10.3 shows a typical run of body temperatures during a foraging trip.

Third, we found no significant effects of water temperature on any of the variables related to body temperature, such as body temperatures at the start or end of the fishing bout, or the difference between the two, or the rate of temperature decrease. This was despite the fact that during our observations water temperature varied between 2°C and 16°C. Neither was the length of the swimming bout affected by water temperature, unlike findings in beavers or muskrats (MacArthur 1979, 1984; MacArthur and Dyck 1990). It was possible that this was caused by otters being more vigorously active when the water was cold, but we could not quantify that.

The measurements indicated that, although there was a substantial drop in body temperature during aquatic activity, at a rate of 2.3°C per hour, otters anticipated this by entering the water with a body temperature that was higher than average. They emerged from the water with a body temperature at approximately average. It was not possible to say, however, whether animals could have stayed in longer, using increased activity in water to boost their body temperature, and still come out with 'average' body temperature. Whatever the case, differences in water temperature had no effect on any of this.

One would expect that cooling effects would restrict the length of time spent by otters in water during any one hunting bout. This may well be the case: generally per foraging trip, Eurasian otters spend an average of only 14.5 ± 10.7 min in the water. Relatively few of these periods are longer than 20 min (Kruuk and Carss 1996) (Fig. 10.4). By these various mechanisms, Eurasian otters maintain their core body temperature within limits that are more or less normal mammalian values. However, they pay a high metabolic price for what the cold aquatic environment inflicts on them.

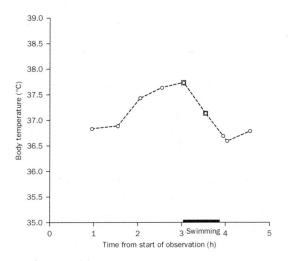

Figure 10.3 Warming up before the plunge: sequence of measurements of core body temperature of a wild Eurasian otter in its natural habitat before, during and after swimming. Observation by radio-telemetry and visual; water temperature 6.7 °C. Dinnet lochs, north-east Scotland. ○ Active on land; and ◘ solid bar, active in water.

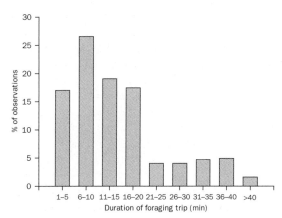

Figure 10.4 Length of 120 fishing trips by Eurasian otters in a Scottish freshwater loch (Dinnet, north-east Scotland). Most foraging trips last for less than 15 minutes. (Adapted from Kruuk and Carss 1996.)

Thermo-insulation

The foundation for the historically massive trade in otter skins (see Chapters 2 and 14) lies in the vastly superior quality of otter fur, in its structural design that provides thermo-insulation and makes the otters' watery exploits possible. The queen of furs is that of the sea otter, but the other species are also better protected by their pelage than almost any other mammal.

Seals, which may live in the same places as otters, are shielded against cooling by a subcutaneous layer of blubber, some 7–10 cm thick, which contains up to 40% of their body mass (Irving and Hart 1957; Tarasoff 1974). Otters, however, have no blubber at all and very little fat; for them, blubber would have some serious disadvantages and fur is all important. A thick layer of blubber would interfere with an otter's agility under water, but, in particular, it would make locomotion on land extremely clumsy (as in seals). It clearly would not be suitable for any semi-aquatic mammal such as otters, mink (*Mustela vison, M. lutreola*), water shrews (*Neomys fodiens*) and others (Estes 1989). In North America the **river otter** has an average of 13% of its body mass as subcutaneous fat, especially around the legs close to the body, and also around the ventral side of the body (Tarasoff 1974).

However, in the 'fattest' **Eurasian otter** dissected in Scotland, only about 3% of its bodyweight consisted of fat as adipose tissue (Pond 1985), and that was mostly mesenteric (inside the body cavity). Most Eurasian otters are as lean as can be, with almost no fat at all—even those living in the colder parts of their geographical range. This compares with, for instance, a seasonal presence of thick layers of subcutaneous fat in another mustelid, the Eurasian badger *Meles meles* (Kruuk 1989), comprising more than 30% of its body mass. There is some fat under the skin of Eurasian otters in the loins, and especially around the base of the tail, but even there small muscles and vascularization are found in the layer between the fat and the skin, and in general this can hardly contribute much to thermo-insulation. **Sea otters** have no blubber or subcutaneous fat, but their skin is about twice as thick as that of the river otter (Tarasoff 1974).

In water, fur is less efficient as an insulator than is blubber, yet all otters depend on fur, and their coat shows special adaptations. Under an outer layer of guard hairs, each 2–4 cm long, there is a layer of under-fur, an extremely dense mat of about 1 cm thick, through which the skin is quite invisible even when one tries to penetrate the fur. When examining an otter skin closely, the under-fur gives the impression almost of being the skin itself, so dense is the hair. The **river otter** has the same arrangement and lengths of hair, but the fur of **sea otters** is about half as long again as that of *Lontra* and *Lutra* (Tarasoff 1974).

One of my students, Addy de Jongh, studied the fur of **Eurasian otters** by electron microscopy. He counted the hairs in bundles of about 20 or 22 under-hairs around each guard hair. There were 20 to 30 of such bundles per quare millimetre, or about 50,000 hairs per cm^2. **River otters** have about 60,000 hairs per cm^2 (Tarasoff 1974). Sea otters go even further, with hair densities of up to 164,000 per cm^2 (Williams *et al.* 1992). Compare this with dogs, which have fewer than 9000 hairs per cm^2, cats with up to a respectable 32,000 hairs per cm^2, and humans with only about 100,000 hairs on our entire head.

During an otter's dive, air is trapped in the under-fur; this is important for thermo-insulation, as in a scuba-diver's dry suit. For instance, it has been shown in polar bears (*Ursus maritimus*) that the thermal conductivity of fur is 20 to 50 times greater when wet than when dry (Scholander *et al.* 1950). It is therefore vital for otters to maintain the air-holding capacity of their fur, even at the cost of considerable effort. It has been demonstrated for **sea otters** that the air-holding capacity of fur is lost very easily, after even moderate fouling of the pelt, or by contamination with oil (Costa and Kooyman 1982), which increases the thermal conductivity of a pelt 2–4-fold (Williams *et al.* 1988).

Marine-living **Eurasian otters** spent quite a lot of time grooming themselves (Fig. 10.5), and especially rolling on grass and seaweed. Bart Nolet, a Dutch student working in our Shetland team, found that three Shetland otters with radio-transmitters spent 6% of their time grooming. The length of each grooming bout was correlated with their previous behaviour in the water (Nolet and Kruuk 1989). The correlation with length of time fishing was not significant, but there was a significant correlation when the depth at which the animal had been

Figure 10.5 Eurasian otter male grooming—fur maintenance is vitally important.

fishing was taken into account (comparing grooming bout length with time in the water multiplied by the depth of dives). This suggested that diving, especially deep diving, stimulated grooming. In later years I have become convinced that the radio-transmitters, which in Shetland we attached to the otters with a small harness or collar, affected our grooming observations, because the attachments stopped otters' access to some small parts of their fur. Anything that interferes with the maintenance of the natural functions of the fur affects the otters' behaviour. Nevertheless, the effect of previous diving on grooming was likely to be real, and not caused solely by the attachment of the transmitter.

We do not yet know whether Eurasian otters living in fresh water spend as much time on fur maintenance as those in the sea, as they are more difficult to observe continuously. However, it is likely that they groom and roll less, in the absence of the effects of salt water on the pelt. The influence of sea water, and the Eurasian otters' reactions to it, opened our eyes to many of the problems of thermo-insulation. In fact, as early as 1938, Richard Elmhirst, Director of the Marine Biological Station near Glasgow, described the otters' use of small freshwater pools along the coast for rinsing off the sea water. Having seen this behaviour in Shetland, I realized that these basic observations held important clues for the otters' distribution along coasts and for their behaviour.

The effect of salt water on Eurasian otter fur was analysed by David Balharry, who did experiments in a project with captive otters (Kruuk and Balharry 1990). Two adult females were kept in a large enclosure with a swimming pool. They could drink under a tap if the pool was dry. For an experiment in which we tried to simulate conditions along the sea coast, we gave the animals an additional fibreglass pool in the enclosure, which I shall call the feeding pool. We sometimes filled the feeding pool with sea water, and sometimes with fresh water. Thus, we could have the otters using the feeding pool with sea water, whilst they did or did not have access to the freshwater swimming pool, or we could make the otters use the feeding pool full of fresh water, again with or without access to the freshwater swimming pool. Most of the experiments were done in October and November, so the water was fairly cold, between 1°C and 6°C.

The two otters were fed five times per day, each feeding period lasting for 25 min. During one feeding period, each animal would get five pieces of haddock, which were thrown into the feeding pool at 5-min intervals. We measured several different aspects of the otters' behaviour during and in between these feeding times, and there were striking differences in what the otters did when they were being fed in sea water or fresh water.

Significantly, the otters paid many more visits to the alternative, freshwater swimming pool when they were being fed in sea water. This happened during 35 of 40 feeding periods in salt water, but only once during the 20 times that they were fed in fresh water.

However, the behaviour of both animals changed quite dramatically when they were being fed in sea water without having access to the swimming pool for a refreshing dip. For instance, between feeds during any one feeding period, the animals used to dive in and out of the feeding pool, perhaps to check whether any food was left, or just out of excitement. They were much more reluctant to do this if they had not had their freshwater dip for one or more days (Fig. 10.6). During the times when they had been fed in sea water for 4 days or longer, without having access to fresh water for swimming, the otters would be clearly miserable. They would sit at the edge of the feeding pool, shivering and reluctant to enter the sea water even during actual feeding, even when the air temperature was not particularly low. We then put a stop to the experiment and turned on the tap for

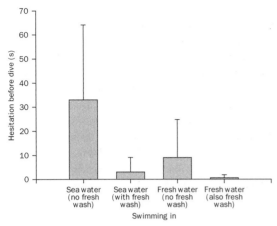

Figure 10.6 The unwashed hesitate before a plunge: experiment with captive Eurasian otters. When being fed in a seawater pool, they are more reluctant to enter the water when no freshwater washing facility is present. There is a similar, but much smaller, effect when being fed in a freshwater pool ($P < 0.001$, Mann–Whitney U test). (After Kruuk and Balharry 1990.)

Figure 10.7 Shaking sea water out of the fur: yearling female Eurasian otter at low tide in Shetland, on bladder-wrack (*Ascophyllum nodosum*).

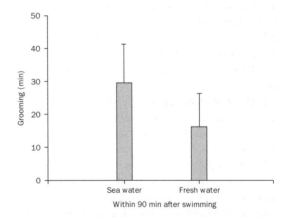

Figure 10.8 Longer grooming after swimming in sea water: experiment with captive Eurasian otters ($P < 0.05$, Mann–Whitney U test). (After Kruuk and Balharry 1990.)

the freshwater swimming pool. Immediately the two would dive in, playing and splashing.

Often an otter gives itself a good shake after it leaves the water (Fig. 10.7). In our experiments we saw that this, too, was much affected by sea water: in the absence of fresh water for washing, the two otters shook themselves significantly more often, and it was clearly the sea water that caused the increase (Fig. 10.8). The same pattern emerged for grooming and licking their fur, which the otters did about twice as much when they had sea water in their feeding pool, compared with fresh water.

These are just some basic quantitative effects of sea water on otters, but to anyone simply watching the two animals when there was only sea water in their enclosure, it would have been obvious that something was amiss even without these statistics. An otter emerging from sea water, after several days without a good rinse in fresh water, appeared to be quite soaked through, with the pelt hanging heavily around it—totally different from the smooth but fluffy coat we had come to accept as normal. We quantified this phenomenon by making the animals enter and leave the feeding pool over weighing scales, so that we could weigh the quantity of water clinging to their fur. This was somewhat rough-and-ready, but the scales did show that (a) the longer the animals stayed in the feeding pool, the more water they absorbed in their pelt and (b) during seawater feeding the amount of water absorbed by the pelt was significantly greater (Fig. 10.9).

The results indicated that sea water interferes with the capacity of the fur to hold air under water. The fur soaks up water instead, thereby jeopardizing the insulating function of the pelt. Microscopic examination showed that when sea water dries in the pelt,

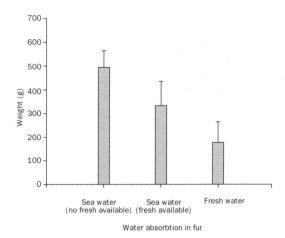

Figure 10.9 Sea water interferes with insulation: experiment with captive Eurasian otters. When fed in sea water in the absence of a freshwater washing facility, more water is retained in the pelt after emerging on land. ($P < 0.001$, Mann–Whitney U test). (After Kruuk and Balharry 1990.)

salt crystals are formed along the length of both guard hairs and under-fur, with many hairs sticking together in small bundles. Perhaps the salt interferes with the lipid secretions from the skin glands on the hairs, instrumental in retaining the air layer (Tarasoff 1974). Possibly, the crystals also interfere with the surface structure of the individual hairs, the scales.

All of this will not come as a great surprise to anyone who has ever swum in sea water without showering afterwards: hair gets sticky and ropy. However, to confirm these results in the laboratory, we did some experiments with pieces of otter skin, measuring temperature at several points inside the fur with small thermo-couples. The pelt was stretched over a copper plate at 35°C (just below body temperature) with cold (6°C) water flowing over the outside. The temperature near the skin, inside the fur, was 29.1 ± 1.1°C, which showed that heat permeated the skin from the inside, and not much was lost through the fur. After the pelt had been rinsed twice in sea water and dried again, there was no significant change in this, but after rinsing five times in sea water the temperature in the same sites inside the fur dropped to 23.0 ± 1.1°C, a significant decrease. There was no significant change in a set of control pieces of otter skin, which had been washed in fresh water. This demonstrated the havoc played by sea water.

Apart from their fur, otters have a few more adaptations against the cold. All but one of the species have a long tail, which functions in propulsion, during diving and with underwater steering, but which is a liability in terms of thermo-regulation, because of its large surface-to-volume ratio. The species that is the most temperature challenged, the **sea otter**, has a tail that is much reduced in length—a compromise between different environmental requirements. At the other extreme, the **giant otter**, in tropical waters, is the species with a long and flattened, broad tail. The **smooth otter**, another warm-water species, also has a tail that is flat at the base. Apparently, these last two can afford a tail that is more effective for locomotion. Most other species are similar in shape to the Eurasian, and some, such as the **river otter**, live in environments that are at least as cold, or more so (Fig. 10.10).

Semi-aquatic mammals of the same size as otters, or smaller, have a problem with heat loss that is more serious than for larger animals, such as seals. Small bodies have a smaller thermal capacity (i.e. they can hold less heat), but the relative surface area is also greater in a small animal. A small animal in cold water cools much more rapidly than a large one with the same thermo-insulation, so size helps.

Keeping warm is therefore slightly easier for **sea otters**, as the most aquatic of all carnivores, as they are larger (two to four times the weight of *Lutra* or *Lontra*). Their metabolic rate is two or three times higher than that of similarly sized mammals (Costa and Kooyman 1982; Morrison *et al.* 1974), higher also than that of other otter species, and this means they have to eat more. Most of their heat flux goes through the large rear flippers, which are sparsely furred and heavily vascularized (Costa and Kooyman 1982; Estes 1989; Iverson and Krog 1973; Morrison *et al.* 1974). These flippers may be used as 'solar panels' to absorb heat, and prevent aquatic cooling, with the animals floating on their back, flippers out of the water (Tarasoff 1974) (Fig. 10.11).

As sea otters leave the water far less frequently than other otters do, if at all, the sea water does not dry out in the pelt and there is less salt encrustation in the fur. Nevertheless, they have to keep their fur scrupulously clean and salt-free by grooming a great deal every day. If the hair gets only slightly oiled, or if the animals get dirty under conditions of captivity, their

Figure 10.10 North American river otter on ice, in Yellowstone Park. © C. Reynolds and J. Stahl.

Figure 10.11 Sea otters keep their flippers above the water to prevent cooling and to catch sunrays. © Richard Bucich.

thermo-insulation is breached and they develop pneumonia very quickly. This has been shown, initially more or less accidentally, when animals died in captivity (Kenyon 1969), and later by other scientists in deliberate experiments to study the effects of oil pollution at sea (Costa and Kooyman 1982; Siniff et al. 1982).

Especially exciting is the grooming behaviour of sea otters. Each day *Enhydra* spends 1–2 hours grooming—longer than any other otter, and whilst afloat on the surface. Generally, after a feeding bout and in a highly stereotyped sequence, the animal first goes through energetic somersaulting and rolling, with vigorous rubbing of the entire body, and actually blowing and whipping the air into the fur (Kenyon 1969). It squeezes and rubs its fur with its forelegs. Then, floating on its back, it rubs its face, neck and paws, and the hind flippers rapidly together. The tail and hind flippers get a good lick, finally the entire body is licked, and the animal wraps itself in kelp, several times (Loughlin 1977, in Riedman and Estes 1990).

In contrast, *Lutra* and *Lontra* never groom in the water but on land, rubbing on grass and seaweeds, drying the fur in the air and relying on this and the use of fresh water to rid the pelt of salt, and to allow air back into it.

Washing in fresh water along sea coasts

Having demonstrated the importance of freshwater baths to Eurasian otters in captivity, I wanted to establish where the animals satisfied this need in the wild. It is not often that one actually sees otters wash and swim in fresh water along coasts, although it happens frequently. We knew this because of the well beaten otter paths to the washing pools, and from following tracks in the snow. Probably, an otter living in the sea washes in fresh water at least once every day. One still has to be fortunate to see it, therefore (especially as it often happens underground, or in darkness). The following are typical observations of the **Eurasian otter**.

April 1986, Shetland. Mid morning, and I am following one of our ear-tagged males. He is a big, territorial animal swimming along the rocky coast, occasionally diving, but mostly along the surface. The otter is surprisingly conspicuous, with his tail floating behind him, swimming slowly about 10 metres out from the rocks. He makes for a protruding rock, lands, spraints—then, instead of diving back into the waves again, hops over a few large boulders, and rapidly climbs the steep grass and gravel bank. He spraints again, then follows an otter path, in his typical clumsy gait, for just about 40 metres, stops, spraints on a large heap of older otter spraints and slips vertically down a narrow hole in the peat. Although I am about 100 metres away, I do not need to be closer because I know the place: the hole is full of black, fresh water, and it goes down vertically for about 1 metre. Another hole in the peat about 5 metres further, also full of water, is connected underground with the first one, and about 12 seconds later the otter pops up there. He shakes vigorously, with a lovely halo of spray around him, then bobs back again to the sea, down the sandy cliff, over the rocks and into the surf, continuing his journey in the same direction.

An early Shetland morning in February 1987. A young female swims along the coast, accompanied by her two cubs, about half a year old. The party lands on a rocky shore at the bottom of a grassy slope, and they run a few metres higher up to a flat piece of ground, settling down to roll and groom, initially ignoring the big spraint site about 2 metres from them.

After 17 minutes the mother gets up, walks to the spraint site, spraints and returns again to her rolling lawn. The party continue to roll, groom, sometimes sleep for a bit. Another 6 minutes and they get up, walk down to the rocks on the shore and all three drink, from a tiny puddle between the stones, fed by a trickle of water running down from the peat above. The puddle is hardly the width of an otter's head, just one of those ordinary seepages of which there are thousands along the shore here. The female dives again, a ripple and a splash, followed by her offspring. Three dives on and the mother lands a big rockling, about 30 cm long; all three eat from it, although most goes to the cubs. One of the youngsters walks up to the spraint site and spraints; then all make their way up the grass bank, past the previous rolling site—and suddenly they are all gone from view, down a narrow hole full of peaty water.

They pop up again from a hole at the other side of a small peat bank, 2 metres away, galloping up the grass slope, soaking wet as they are. I see them for a few more seconds, before they disappear down the entrance of a holt, in another peat bank. When they come out again, more than half an hour later, they are as perfectly dry as if water had never touched them. They gallop again, down and down, straight into the water where the mother dives, in her seemingly endless quest of provisioning the two hungry mouths, with the cubs just behind.

The significant part of this last observation, the watery dip in the peat hole, could so easily have escaped attention, as I am sure happens often.

There were many potential otter bath tubs along the shores of our Shetland area, some no more than a small, deep puddle between the rocks (Fig. 10.12), or a pool in the tiny burns that come trickling down the peaty slopes, others as large as small lochs. The favourite ones were either the rocky pools close to the shore, or otter-sized holes in the peat, vertically down into the groundwater and obviously made by otters themselves in the distant past. Some of these tunnels were neatly U-shaped, so the animals could go in one end and out the other. Or the otters turned upwards again after going down for half a metre or so, into a dead-end underground.

If there is an abundance of the more obvious freshwater washing rock pools along coasts, such as on the Isle of Skye, Eurasian otters show a preference for deep ones, with flat rocks and short grass nearby for rolling. Size of pool was irrelevant to the animals (Lovett *et al.* 1997). Elsewhere along European sea coasts, such as in Portugal, in the absence of pools otters use resting sites along small freshwater

Thermo-insulation: a limiting factor 171

Figure 10.12 Freshwater pool used for washing by coastal Eurasian otters, western Scotland. Note spraint site.

streams; where there are no such streams, there are no otters. The availability of fresh water is the single most important factor determining the distribution of Eurasian otters along the coasts of Scotland (see Chapter 4) and Portugal (Beja 1992, 1996b). The abundance of spraints, near freshwater pools and streams along the coast, in itself testifies to the importance of these places to otters.

Along the Shetland coast there were many other otter baths that were not easily seen in the field. They were located inside holts, sometimes visible from the entrance, but often deep down, needing a major excavation to expose them (Moorhouse 1988). This often explained why some of the otter holts were located in particular, curious positions, for instance on the top of hills, or high up in the hills at the bottom of some small valley, perhaps a kilometre from the sea. Invariably, there was a dip in the rocks or clay underlying the peat, in which fresh water could collect. The extensive tunnel system of a holt made in the peat above has one or more branches that widen into a bathtub, with smooth, well worn sides and rather muddy water in it. Many otter holts actually had running water in them, when they were situated exactly on one of the small drainage lines carrying water from the peaty slopes down to the sea, or when the otters lived in underground, almost cave-like, flow systems (see Chapter 5). Frequently there was nothing to tell from the outside that such washing facilities were present in the holt—we merely found simple entrances, somewhat larger than those of a rabbit warren, with no water visible anywhere near.

During dry times of the year, Eurasian otters abandoned those holts in which the water had dried up. Of seven holts excavated in summer by Andrew Moorhouse (1988), the three dry ones were not in use at the time, but four with full underground bathtubs were well frequented. Associated with the bathtubs, as well as with the more exposed washing places, were often one or more sites at which otters could wring water and air from their fur, usually by wresting themselves through narrow gaps or underwater tunnels—'squeeze holes'. One also finds these squeeze holes in freshwater habitats of otters, probably contributing to fur maintenance in a similar manner, but they are nowhere near as common near freshwater lochs or streams as they are along sea coasts, even where there are just as many otters.

Apart from the Eurasian otter, several others live in the sea as well as in fresh water (see Chapter 2), including the North American **river otter**, and in Africa the **Cape clawless**. It was interesting to see both of these species, from different genera and living in other continents, using the marine habitat with the same freshwater constraints as for the Eurasian otter. Pools are used for washing; dens of the river otter along the Alaskan shores often have running fresh water inside, just like those of the Eurasian, and the freshwater pools are similarly scent marked with faeces. Holts of Cape clawless otters that live in the sea are located next to freshwater streams (personal observation; Van Niekerk *et al.* 1998; Verwoerd 1987), although in brackish mangrove areas a dependence on fresh water was not obvious (in Nigeria, Angelici *et al.* 2005). Several other otter species that occur in the sea as well as in freshwater habitats are likely to show similar, freshwater-dependent behaviour.

Figure 10.13 Even giant otters are exposed to aquatic cooling. © Nicole Duplaix.

There is one exception. In Chile, the **marine otter** occurs only along the ocean coasts, including areas reportedly devoid of any fresh water (Ostfeld *et al.* 1989). It is not yet known how this species copes with the problem of fur maintenance, but it has a quite different type of fur, coarse and rough in appearance, with thick guard hairs (Foster-Turley *et al.* 1990); perhaps it does not have the same maintenance requirements.

Several other small mammals have been able to colonize freshwater habitats, including carnivores, rodents and insectivores. All use a layer of air in the fur as thermal insulation. However, none of them has been able to adapt to the more extreme rigours of a marine environment, with the exception of the American mink *Mustela vison*, which was introduced in Europe from North America. It is common along the sea coasts of Alaska (Ben-David *et al.* 1995), of Norway and in some areas of the Scottish west coast (Dunstone 1993). In Norway I noticed from tracks in the snow that coastal mink also washed in fresh water, often in the same sites as otters. As another, historic example, along the east coast of North America the sea mink *Mustela macrodonta* lived exclusively in marine habitats, until it was exterminated during the nineteenth century. It was much larger than the common mink, larger also than the river otter (Dunstone 1993).

However, for most species of smaller mammals the extra problems of thermo-insulation have been impossible to overcome. Otters are amongst the few exceptions that are successful in colonizing some coasts, but they still remain highly dependent on the vagaries of the distribution of washing water along the sea's rocky shores. It could well be that species such as the spotted-necked otter in Africa cannot access coastal regions because they lack suitable adaptive behaviour in using available washing facilities.

As discussed above, even in freshwater habitats, it is likely that otters' foraging trips are constrained by thermo-insulation in all species (Fig. 10.13). Thus, the ostensibly simple requirement for many species of otter, to be able to maintain body temperature when swimming and foraging, appears to have major consequences. It affects their choice of habitats, their behaviour, and ultimately their distribution and therefore numbers.

CHAPTER 11

Populations, recruitment and competition

Introduction

Whether we are interested in behavioural ecology, the evolution of behaviour patterns or the conservation of a species, a knowledge of population dynamics is vital. This includes especially an understanding of numbers in different habitats, the mechanisms of population regulation, the relative importance of various causes of mortality, and the factors that limit reproduction and recruitment. Such information is scarce or absent for most species of otter, but in this chapter I will discuss some of the data now available, and try to draw tentative conclusions on some of the causes of mortality and of success in reproduction.

Perhaps the main reason for the lack of information on otter populations is the difficulty of assessing population size and changes in species that are highly secretive, nocturnal and sparsely distributed, as well as difficult to catch. Not surprisingly, the diurnal and quite conspicuous sea otter is the only exception, and by far the best studied. However, some progress has also been made with the Eurasian otter, often using indirect sampling methods, involving spraints.

One of the few advantages of studies of mammal populations, compared with birds, is that the age of mammals can be determined quite accurately, from carcasses or extracted incisors, by counting incremental rings in the dentine of teeth. This method has been developed for Eurasian otters by the Norwegian scientist Thrine Heggberget (1984), and it has enabled us to study population composition, patterns of mortality and the cumulative effects of pollutants, amongst other aspects.

In addition, in an exciting development since my previous book *Wild Otters* in 1995, a major advance in our knowledge of otter populations has been made with the use of molecular techniques. They have enabled insights into population structure and history, genetic diversity and gene flow that were previously unheard of. In particular, the extraction of DNA from otter scats, so much easier to come by than body tissue, is opening new doors. Many of the new techniques are still being perfected, including the efficiency of DNA extraction from spraints. New advances in our knowledge of otter population genetics are just round the corner.

Populations: numbers, changes and genetic diversity

One of the most basic questions to be asked by an ecologist is how many animals there are, of a species in any one area. For some mammals or birds this may not be too much of a problem, but otters present difficulties. In Shetland, where **Eurasian otters** are diurnal, it is possible to recognize individuals over stretches of coast of a few kilometres, after intensive observation over a few weeks there. The throat patches are variable and often easy to see (see Fig. 3.2). In South America **giant otters** are similarly distinctive (see Fig. 5.7) and active in daylight, as are **spotted-necked otters** (see Fig. 2.27) in many areas of Africa. These species, therefore, can be counted somewhat more easily. For animals such as the Cape clawless otter this is much more difficult, and often the only evidence of their presence is tracks in the sand (Fig. 11.1), or scats.

Sea otters cannot hide themselves as well as many other species, and populations have been counted routinely in surveys from small boats, from aeroplanes and from the shore. Obviously often some animals will be missed, and it is therefore necessary to calibrate census methods, but such techniques are now well established (Estes and Jameson 1988; Udevitz *et al*. 1995). However, these kinds of field condition are exceptional amongst otter species and populations. For almost all others, including the inland populations of river otters and Eurasian otters, it is impossible to get any idea of numbers without using some indirect way of census, an index.

The most common census method, on which most of our knowledge of **Eurasian otter** numbers in

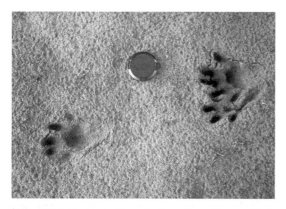

Figure 11.1 Tracks of a Cape clawless otter, evidence of their nocturnal presence. Note long fingers, no claws (size of lens cap 5.5 cm).

Figure 11.2 Sprint site of Eurasian otters in Scotland, often used for population census, but difficult to interpret.

inland waters is based, and from which we deduce most of the pattern of decline and fall in various countries, is the survey of otter scats, the 'spraints' (Mason and Macdonald 1986). Spraints are often the only evidence of the animals' presence; they can be quite conspicuous (Fig. 11.2) and are frequently found in convenient spots, such as under bridges. In one of the standard methods, a carefully chosen 600-metre stretch of bank is surveyed (so not randomly selected). If no spraints are found, it scores a negative, etc. At the end of a survey the percentage of 'positive' sites is taken as a measure of the strength of the otter population. In earlier days such surveys were used to estimate actual numbers of otters; for instance, it was concluded from spraint distribution that in 1975 there were 36 otters in Suffolk (Mason and Macdonald 1986).

However, there are serious problems with this. Clearly, when spraints are present, there are otters, and in areas with many spraints there are likely to be more otters than in regions where spraints are few. But Eurasian otters at least (and probably also the others) also defaecate when swimming, as one can see in both wild and captive animals. The spraints that are deposited on the bank have a specific biological function, apart from elimination (see Chapter 6). Absence of spraints, therefore, does not necessarily mean absence of otters. Along inland waters, where we followed otters with radio-tracking, we found areas where the animals spent a great deal of time (in reed marshes, for instance, far away from open water, and along very small streams), but where despite intensive effort we could not find any spraints. My students and I tried hard, because we wanted to study the otters' food in those sites. Even more worrying was that some of the otters hardly ever sprainted on land at all (which we knew because their spraints were labelled with an isotope).

The implication is that, not only does absence of spraints tell us little about otter activity, but also, when spraints are present, numbers do not necessarily correlate with otters (Kruuk *et al.* 1986). In Shetland we surveyed 21 350-metre stretches of coast for spraints, while also counting the number of directly observed otter visits. There was no significant relationship. In an apparent contradiction later, when studying actual sprainting behaviour, I saw otters deposit more spraints in sections where they also spent more time. The most likely explanation for the discrepancy was that many spraints were deposited on seaweed below the high-tide mark, and these were soon washed off again. Such problems would not arise in freshwater areas, along rivers.

There is also the problem of seasonality. In Britain, otters spraint on land in winter many times more often than in summer (Conroy and French 1987; Kruuk 1992; Mason and Macdonald 1986), and this annual fluctuation is by no means regular, with peaks, troughs and main increases or decreases occurring during different months (Conroy and French 1987). Other regions and other species are likely to show a different seasonality. This makes spraint density, even if controlled for time of year, a tenuous measure for utilization of an area by otters.

Despite such reservations the method of spraint survey to assess otter populations remains popular, because the fieldwork is easy. However, it is controversial, and the merits and objections against it have been argued at length (e.g. Mason and Macdonald 1987 versus Kruuk and Conroy 1987). In the end, one should accept statements on otter density with caution if they are based on spraint surveys. When applied on a large scale, however, such as in the repeated national spraint surveys in Britain, they must be useful, and are likely to indicate trends. Experienced fieldworkers, using a somewhat vague correction for seasonality, found repeatedly that the region in Britain where spraints occur in an area has gradually enlarged, first in Wales and from Wales further into England (Andrews and Crawford 1986; Crawford 2003; Crawford et al. 1979; Mason and Macdonald 2004; Strachan et al. 1990). There is no doubt about the conclusion from the spraint surveys that otters have returned to the country, after being absent for decades. How strong the come-back is, is still open to surmise.

For our own studies we decided that spraint abundance would not be sufficiently reliable as an index for Eurasian otter numbers and distribution, both in Shetland and elsewhere. We resorted to two alternative procedures to arrive at a figure for otter density, one to be used in Shetland and one in freshwater areas. In Shetland the otters use holts on a daily basis, and holts were easy to find and could be used for population assessment. Elsewhere in freshwater areas this method was unsuitable, and we developed a method based on radio-isotope labelling in spraints from otters with radio-transmitters, by calculating the proportion of spraints that was deposited in an area by the known number of radio-otters.

I will describe these methods in some detail, but, more recently, genetic methods have made a new breakthrough. Developments in the extraction of DNA from otter spraints have made it feasible to identify the sprainter from as many as 65% of spraints found along streams on Kinmen island, off the coast of China (Hung et al. 2004). This enables an estimate of the number of individuals in any one area (with some reservations), and it showed one otter per 0.5–0.7 km of stream—a high density, possibly caused by the fact that these animals were feeding in the nearby sea, and only resting along the streams.

In Shetland we estimated the number of used otter holts and the number of adult otters associated with them, in a 100-metre strip along the coast (Kruuk et al. 1989; Moorhouse 1988). We used the number of individually known resident adult females in each female group range (see Chapter 6), and related that to numbers of used holts. Separately, and over larger areas, we estimated how many males and how many vagrants (non-residents) there were for every resident female.

We used a different method in mainland Scotland to assess otter numbers in inland streams and rivers (Kruuk et al. 1993). Otters were fitted out with radio-transmitters and injected with a harmless radio-nuclide (^{65}Zn), which enabled us to recognize their spraints with a scintillation counter. In principle, we established the home range of 'focal animals', the ones with transmitters, and collected all spraints in a given part of that area, for example along a single stream. We assumed that other otters in that stream would be equally likely to spraint along any one bank at any one time. This assumption was based on the observation that there were no significant differences in sprainting rates and seasonality between otters of different sex or social status (Kruuk 1992). Together, the proportion of spraints with the ^{65}Zn, and the proportion of time spent in that stream by the focal otter, allowed us to make an estimate of time spent in that area by all otters present over the study period.

As an example, one male otter (8.0 kg) was a focal animal from November 1989 until June 1990. He spent most time along the River Dee (69 of 125 nights of radio-tracking, or 55%), and foraged along the Sheeoch Burn, a small tributary of the Dee, during 47 nights (38%). He used 11.6 km of the Sheeoch (or 5.1 ha of water), and spent the equivalent of 47/125 × 365 = 137.2 nights per year there. Of 700 spraints collected along the Sheeoch during this period, 78% contained zinc-65 (77.2 ± 5.4% over five collecting periods). This suggests that the Sheeoch Burn was used by otters for a total number of 100/78 × 137.2 = 175.9 'otter nights' per year, with a mean nightly otter biomass of 4.6 kg, or 0.09 g/m^2 water.

The overall, median density of otters in the Rivers Dee and Don and their tributaries was one otter per 15.1 km of stream, but it varied between one per 3 km and one per 80 km of stream, or one otter (8.5 kg) per 2–50 ha water (or 0.02–0.4 g/m^2 water). As shown in Chapter 4, most of this variation in density could be related to the width of streams (Kruuk et al. 1993a).

Comparison of the freshwater data with those on otters along sea coasts in Shetland, at first glance, suggests that Shetland otters live at much higher densities. However, if otter numbers are expressed per area of water rather than length of bank, the discrepancy largely disappears. If we assume the strip of water used by coastal otters to be 80 m wide (enclosing 98% of all otter dives; Kruuk and Moorhouse 1991), the estimate for good otter habitat in Shetland is one animal (6 kg) per 10 ha water (mean 0.06 g/m^2 water), of the same order as densities in streams and lakes.

Elsewhere in Europe, Ruiz-Olmo (1998), using radio-tracking combined with direct observations in northern Spain, estimated Eurasian otter densities in the lower altitudes of the region (300 m above sea level) of one per 2.2 km of stream, declining to almost zero at altitudes over 700 m. Per area of water, these otters occurred at about one per 10 ha at lower altitudes, which translates into about 0.08 g/m^2 water. Eurasian otters are reported in 'densities' of one per 2 km of lake shore and one per 4–5 km of stream in Sweden (Erlinge 1968), and one per 1 km of bank in eastern Germany (Hauer et al. 2002a). However, these last figures were based largely on snow tracking, and it is possible that they did not consider the many very small tributaries to main rivers, which are frequently used by otters in summer but not when covered in snow.

In general, such estimates have to be very approximate, but the results from the various studies are somewhat similar. We do have to keep in mind, however, that animals tend to be studied most intensively in the most profitable places, that is, high-density areas.

Densities of the **river otter** in the mountains of Idaho were estimated by a combination of radio-tracking, direct observation and snow tracking, along streams (Melquist and Hornocker 1983). The researchers found a mean of one otter per 3.7 km of stream, over an area of 158 km, with more than twice as many females as males. Similarly, by radio-tracking in Alberta, Canada, Reid et al. (1994b) found one river otter per 5.7 km of lake shore-line.

Using a different method along the coasts of Kelp Bay in Alaska, Woolington (1984) estimated numbers of river otters from the proportion of individually known animals that he encountered in the field. He found one river otter per 1.2 km of coast. In Prince William Sound, Larsen (1983) radio-tracked river otters and estimated one per 2 km; in that same area Testa et al. (1994) concluded, from a study of radio-tracking and radio-istopes from the radio-otters in spraints, that there was one otter per 1.2–3.6 km. The numbers of river otters per length of coast are comparable to those of Eurasian otters.

The **sea otter**, along those same Pacific shores, occurs (or occurred) in sometimes very large numbers, and there have been reports of aggregations ('rafts') of up to 2000 (Estes 1980). The mean number of sea otters along the shores, from visual counts of 50–340 km in California between 1938 and 1984, was one sea otter per 7.0 ± 1.1 km of coast (Riedman and Estes 1990). Along Amchitka coasts, numbers were as high as about one sea otter per 0.2 km during 1940–1970, but density then declined steadily to one-tenth of that in 2000. Similar densities and declines were observed for the other Aleutian islands, where at present there is about one sea otter per 2 km of coast (Doroff et al. 2003). However, these animals use a wide area of water, so the figures should not be compared with those for other otter species.

There are few data on numbers of any of the other otter species, mostly of the **giant otter**, which is easily observed in daytime. In Nicole Duplaix's (1980) main study river in Surinam there was approximately one otter per 0.5 km of river, or one per hectare of water, at least in the dry season. However, during the rains the forests flooded, and the giant otters moved all over the place. Laidler

(1984) estimated one giant otter per 5.6 km of river in Guyana, and Schenck (1997) in the Rio Manu, in the Peruvian Amazon, one per 5.7 km. Schenck found that in the large oxbow lakes near the river there was about one giant otter for every 14 ha (27 kg, or 0.2 g/m^2 water), a biomass considerably higher than that of otters in most temperate regions.

Along oceanic coasts of Chile, several researchers also based estimates of **marine otter** numbers on direct observations in daytime. Densities were often very high, varying from one per 25 km in the south (the Beagle Channel) to one per 0.15 km and one per 0.23 km in the more central areas of Chile (see references in Ebensperger and Castilla 1992; Medina 1995). In Africa, another diurnal species, the **spotted-necked otter**, was estimated independently by Procter (1963) and Kruuk and Goudswaard (1990) at about one per 1 km along the shores of Lake Victoria, but only where their numbers had not been affected by people.

In conclusion, numbers of otter of all species studied along banks and shores in 'optimal' areas vary between one animal per 0.2–15 km. The variation within species is at least as large as it is between species. Where estimates exist, otter biomass varies between 0.02 and 0.2 g/m^2 water, but not all parts of such areas may have been suitable and used by the animals, for instance where it was too deep.

Population size

With the difficulties of establishing densities and the large variation in results, it is not surprising that, overall, regional population estimates are rare and very approximate. In Shetland we were commissioned by the World Wildlife Fund to estimate the total number of **Eurasian otters** on the islands and their distribution along the coasts. The survey was prompted by an outbreak in 1988 of distemper amongst common seals (*Phoca vitulina*); initially this was diagnosed as canine distemper (Osterhaus and Vedder 1988), to which otters are also known to be susceptible (Duplaix-Hall 1975). As there is frequent contact between seals and otters in Shetland, there was fear that otters could be hard hit by the epidemic, which killed thousands of seals (as it turned out, however, otters were not affected).

For the census we used the otters' holts. As Shetland has some 1500 km of shore, which was more than we wanted to walk, we counted holts in a random sample of 130 5-km sections, grouped them in six different strata of holt density (Jolly 1969; Norton-Griffiths 1973), and then estimated the total number of holts in Shetland, with 95% confidence limits. It was a rather amazing 1185 ± 154 holts.

To relate this to actual otter numbers, we made three assessments in our main, intensive study area. Each month Andrew Moorhouse and I estimated independently from our own observations how many individual otters we had recognized in each home range, and we then pooled our observations of the proportion of otters that we saw with coloured ear-tags. We knew how many ear-tags there were in any one month, so we could make monthly estimates of the population. The three methods gave remarkably similar results (Kruuk et al. 1989), which was not surprising as we knew the population well. The three different estimates were 16.5 ± 0.5 (Andrew), 15.1 ± 0.9 (me) and 16.0 ± 0.9 (ear-tags).

We concluded that along that high-density coast there were about 16 adult otters, or one per 1.2 km. There were 0.33 resident females per holt (or about three holts per resident female), and adult otters numbered 1.83 multiplied by the number of resident females. Of the 16, nine were resident females, four resident males, and the rest were irregulars. Andrew Moorhouse also made some independent estimates on a few of the smaller islands.

From this we calculated the 1988 total population of otters in Shetland: 718 adult otters, of which 392 were resident females. Of course, we had to make numerous assumptions and had various reservations (Kruuk et al. 1989), but, despite these cautions and possible objections, I think we were reasonably on the mark. I am confident that the total number of otters in Shetland in that year was somewhere between 700 and 900 animals, or 0.5–0.7 per km of coast (Kruuk et al. 1989).

In 1993 we repeated the Shetland holt survey, in order to assess the effects on otters of the oil spill caused by the stranding of the tanker 'Braer' (Conroy and Kruuk 1995). There were fewer otter holts used close to the site of the wreck at the south end of Shetland (fortunately a low-density area), but all other parts of the islands showed an increase in numbers of used holts. We calculated the total 1993 Shetland otter population at 836, an increase of some 16%, but within the confidence limits of our previous estimates. The repeat assessment, therefore, did not give cause for concern.

Paul Yoxon (1999) carried out a similar, holt-based, otter survey and population estimate on the Isle of Skye and the small islands immediately around it, about 560 km of coast. He estimated a total of 251 adult otters there, or 0.4 per km of sea shore, with clear concentrations of otters in geologically suitable strata.

At present we only have no more than extremely rough estimates for total otter numbers over larger areas of mainland Scotland, and for the whole of Britain. Harris *et al.* (1995) used some of the above density data, as well as a very approximate estimate of lengths of streams, lake shores and sea coasts. They thought that in the mid-1980s there might be some 6600 Eurasian otters in Scotland and, by using a similar approach, about 750 in England and Wales, with the proviso that at the time otter numbers in England and Wales were rising fast (see also Chapter 14). By now there will be many more in England, but, in any case, these 'estimates' are direct extrapolations from favourable study areas and do not take into account differences in habitat (e.g. altitude, agricultural usage, acidity of streams and suitability of geology along the coasts). They are likely, therefore, to be very approximate.

Similarly, along Norwegian coasts and islands, Thrine Heggberget (1995) attempted to obtain a figure for Eurasian otter numbers for a stretch of approximately two-thirds of the very indented Norwegian coast line, or 34,000 km. This was the part of the country that was most favourable to otters. By snow and radio-tracking eight otters, she concluded that there were 0.4–0.6 otters per km of coast on the islands, and 0.1–0.2 per km along the mainland, producing a total figure for the area of 10,000–15,000 otters. In addition the species also occurs along rivers there, but in unknown numbers.

Although we know little of actual otter numbers in most countries of Europe, and even less about what those numbers were in the past, there is no doubt that, for Eurasian otters, there has been a massive population decline in the twentieth century since the 1950s. There was a total, or almost complete, collapse of numbers in western Europe, and only since the late 1980s has there been a recovery. The recovery is still progressing, now early in the twenty-first century.

The earlier decline was not well documented; rather, it was suddenly discovered that otters had almost disappeared. In several countries (including my own, the Netherlands), otters became extinct where before they had been common. In the 1970s, England was without otters, with a few small peripheral counties excepted (in the south-west, Welsh borders and East Anglia), but even there the decline continued (Jefferies 1989). In the late 1970s, a national survey found only 6% of likely sites in England that showed evidence of otters (Lenton *et al.* 1980). Scotland still had many areas where otters were common (found in 90% of sites in the 1970s; Green and Green 1997), and the species was often seen in the remoter areas of Wales (20% of sites; review in Jones and Jones 2004). But for English otters this was the low point—and one despaired.

However, since then, several national surveys have shown steady and consistent increases; the latest (Crawford 2003) found evidence for otters in England in 34% of sites, in Wales in 71% (Jones and Jones 2004) and in Scotland in 98% (Green and Green 1997). In many countries on the European continent the species has returned or is in the process of doing so. The single, most overwhelming cause for the dramatic decline was pollution, and the withdrawal from use of various chemicals coincided with the return of otters (see Chapter 12).

In North America the collapse of **river otter** populations was not quite as dramatic and uniform as that of the Eurasian otters, but it was equally poorly documented. Probably the same contaminants were involved, as well as intensive trapping (see Chapter 12), but due to our almost complete ignorance of otter numbers, just as for the Eurasian otter, it is not known what happened exactly, and when.

Not surprisingly, given our ignorance in Europe and North America, in none of the other otter species is anything known of regional population sizes, not even of the otherwise well studied giant otter (Schenck 1997). The one exception, as so often, is the **sea otter**, of which numbers have been of concern for a long time, and which is relatively easy to census. However, even for that species, the confidence limits for any estimate are wide, so there is always considerable uncertainty over the assessments. Over the last half century there have been large-scale counts in Alaska and California, using small boats, land-based observers and aircraft. Where the total world population of sea otters in 1911, when persecution officially stopped, was only 1000 to 2000 animals, in the 1980s estimated numbers were between 70,000 and 150,000 (Riedman and Estes 1990).

The increase in numbers varied across the range. Estes (1990) calculated a 17–20% annual increase for populations in the Aleutians, south-east Alaska, British Columbia and Washington State. In contrast, in California, sea otters increased annually by only 4.4–5.5% until 1976, when numbers there declined again by 5% per annum until 1983. Then they started again to increase by 7.8% per year, and after 1994 another Californian decline set in (Estes et al. 2003a) (Fig. 11.3).

The sea otters occupy a coastal belt from California to Russian Asia, but within that there are distinct subpopulations that appear to mix little. For instance, in the Alaskan subspecies of sea otter *Enhydra lutris kenyoni* at least three genetically different stocks are recognized (Gorbics and Bodkin 2001), with discontinuities in distribution. In the years since protection started, however, there have been several transplantations of sea otters from Alaska to further south, in order to encourage resettlement, so some of the distinctions there may have disappeared or become blurred.

The separation between populations in different areas is important when issues of population change are to be considered. Since the region-wide sea otter population highpoint, in the late 1980s, there have been dramatic declines in numbers, at least in some

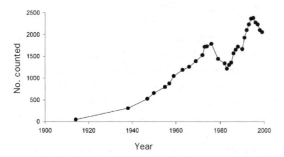

Figure 11.3 Sea otter numbers in California: increases were slow, with periodic declines. (After Estes et al. 2003.)

areas. Around the islands of the Aleutian archipelago, for instance, sea otters were counted by air in 1965, 1992 and 2000, with results of 9700, 8048 and 2442 respectively. The rate of decline in the 1990s was 17.5% per year. These figures do not represent real population sizes, because a proportion of sea otters are missed in the aerial census, but the rate of decline is probably realistic. Allowing for a correction factor to address the numbers that are overlooked from the air, the actual number of sea otters present in the Aleutians in 2000 was 8742 (95% confidence limits 3924–13,580) (Doroff et al. 2003). There was uncertainty at that time, but no doubt that a very substantial decline was in progress, referred to as the 'megafauna collapse' (Springer et al. 2003).

Since then Jim Estes and co-workers have established that this disastrous trend, towards total obliteration of the sea otter in the Aleutians, has continued at an average rate of 29% per year (Estes et al. 2005). The number of sea otters in the Aleutians in 2003 was estimated at 1810, a decline of 63% since 2000, and of 95.5% since 1965. These results are discussed further in Chapters 13 and 14.

In one of the other important sea otter areas of Alaska, the Prince William Sound, the wreck of the oil tanker *Exxon Valdez* in 1989 caused substantial mortality; the number of sea otters that died as an immediate consequence was estimated at 600 to 1000. However, counts of sea otters from boats in the affected area, carried out several years before (1984–1985) and after (1996) the oil spill, detected a sizable increase in numbers, from 651 to 925. The researchers suggested that sea otter numbers had been increasing during the years immediately before the oil spill (Garshelis and Johnson 2001), because the otters' main food, clams, were recovering at that time from the devastation caused by a large earthquake in 1964.

Californian sea otters also declined, first between 1976 and 1984, when every year there was an increase by 1.6% in the number of dead sea otters recovered. After a few years of slow increase, numbers declined again after 1994, by more than 3% per year (Estes et al. 2003a). Even in the periods when numbers were going up, the increase was sluggish compared with that of sea otters in Alaska, and some important mortality factors appeared to be holding Californian animals back (see Chapter 12).

Populations of sea otters everywhere, and several of the other otter species were and are subject to large fluctuations, and some of the causes will be discussed below. Fluctuations pose a risk, and their existence suggests questions regarding connectivity between populations, genetic composition and impoverishment, and gene flow. Genetic impoverishment could reduce the ability of a population to respond to environmental change. If there is sufficient gene flow between populations, there is no problem, but one would expect that severe population fragmentation might pose a danger. Recent studies have made some interesting progress with this.

Gene flow in otter populations

For **Eurasian otters**, John Dallas and co-workers (1999, 2002) analysed DNA from carcasses collected throughout most of Britain. Within the almost 500 tissue samples from the more or less continuous metapopulation of otters in Scotland and on the Scottish islands, they established that the further apart, the more dissimilar the genetic composition of populations. There was little gene flow over distances greater than 100–150 km. This was somewhat unexpected, because of the large home ranges and movements of individual otters. There was no evidence that local or regional fluctuations in the populations had had any effects on the levels of genetic polymorphism. The strength of the distance effect on population composition suggested that effective population sizes, that is, the numbers of otters that are interacting reproductively, are relatively small. This finding is of considerable importance to conservation management.

The greatest discontinuity was in the otters from Shetland (and to a lesser degree Orkney, which is closer to the mainland), which were genetically different and showed a reduced level of diversity. Previously, I had noticed that the Shetland otters looked different, with their striking yellow–white throat patches, which are absent from Eurasian otters almost anywhere else (except in Ireland, where they are the norm). Of course, the distance between Shetland and other areas is such that virtually no otter could be expected to swim across—it is out of sight. However, at present there is no evidence that the lack of genetic diversity in Shetland otters has led to a reduced fitness: they may be well adapted to local conditions. It could well be that a loss of genetic diversity is detrimental to a population only if it occurs suddenly (Saccheri *et al.* 1998; Westemeier *et al.* 1998).

Further south, the Eurasian otter populations in Wales and south-west England were almost the only ones to survive the 1960–1970s English pollution disaster. John Dallas found that the south-west English otters were genetically distinct from the Scottish ones, and both these and the Welsh otters showed less genetic diversity. This was due, he suggested, not so much to recent population declines, but to the fact that there had always been smaller numbers of otters there than in Scotland. There appeared to be little contact between otters from Wales and the English south-west.

Because of the recent history of severe decline and recovery of Eurasian otters on the European continent, several recent projects have used DNA analysis to establish whether there was evidence of a bottleneck in the genetic variability. In Denmark, Pertoldi *et al.* (2001) analysed DNA from roots of teeth, comparing genetic variability in historical (pre-1960) and recent otters. They found no evidence of a recent bottleneck, but their data suggested a decline in genetic variability over a long period, in the order of 2000 years, due to high mortality. The causes of mortality could be any, of course; the authors suggested persecution by people, but an alternative explanation is shortage of prey (see also Chapter 12).

Earlier, Effenberger and Suchentrunk (1999) showed a relatively low genetic variability in DNA from more than 80 otters, mostly from eastern Germany, with a few from Austria, Hungary and eastern Scotland. They also concluded a long history of high mortality, perhaps from persecution. One otter population, within an area west of Berlin, showed increased genetic variability, and it was suggested that this should be subject to special conservation measures. Randi *et al.* (2003) took DNA from 100 tissue samples of Eurasian otters from various areas throughout western Europe. They reported a genetic diversity that was moderately high within and between populations, but no evidence of a genetic bottleneck. Their results suggested a general population decline that started several thousand years ago, probably post-glacial. It is not easy to combine this latter finding with another analysis of DNA variation in continental European otters (Tiedeman *et al.* 2000). These authors suggested that, judging from the distribution of genotypes, European regions had been re-invaded from one central population.

The effects of the enormous fluctuations in numbers on the genetic composition of **sea otter** populations are more pronounced than in the case of the Eurasian otter. Larson *et al.* (2002b) analysed the genetic variation from before the 99% loss of the pre-fur trade sea otter population in Washington State. They took DNA from sea otter bones found at middens of historical Indian village sites. The fossils showed significantly more genetic variation than in five modern sea otter populations: more than 62% of alleles and 43% of heterozygosity has been lost. The genetic diversity before the decline was similar to that found in most other mammals, and the genetic impoverishment was similar to that in another over-exploited mammal along the same coast, the elephant seal *Mizounga angustirostris*. The authors are concerned that the reduced genetic variation could cause inbreeding depression.

The genetic diversity within present-day individual sea otter populations, introduced ones as well as originals, is dependent on the severity of any bottle-necks and on the length of time for which they lasted (Bodkin *et al.* 1999; Larson *et al.* 2002a). One obvious practical implication of this is that, for reintroductions of otters, relatively large numbers should be used (see Chapter 14). Gene flow between sea otter populations is apparently less than one would expect from animals inhabiting a continuous strip of coast. This is suggested also by the fact that there are three regional subspecies: *Enhydra lutris lutris* from the Asiatic part, *E. l. kenyoni* from the Aleutian islands east and south to Oregon, and *E. l. nereis* from California (Cronin *et al.* 1996). But, even within these subspecies, gene flow is restricted, as indicated by differences in genotype frequencies and the presence of unique genotypes amongst areas. There is little or no movement across several suggested stock boundaries within the distribution area of *E. l. kenyoni* (Gorbics and Bodkin 2001).

Otter numbers rejuvenated: recruitment

In Chapter 12 it will be shown that otters face low survival rates and a fairly short life expectancy. Consequently, one might expect high birth rates to make up for this. However, none of the species obliges: cubs or pups are few and far between. There are some interesting observations that may explain reasons for this.

In the **Eurasian otter**, both sexes can reach sexual maturity in their second year of life (Chanin 1991; Hauer *et al.* 2002b). However, it is likely that there is much variation in the proportion of pregnancy amongst young females, and in eastern Germany only one 2-year-old in a large sample was pregnant, as were 20% of females between the ages of 3 and 5 years (Hauer *et al.* 2002b). In Shetland at least some, and possibly most, females started reproducing in their second year.

Litter sizes of Eurasian otters are small, reaching five (Hauer *et al.* 2002b), but usually they remain well below that number. In Shetland from 1980 to 1985, the mean number of cubs per litter was 1.86 ± 0.61 (102 litters), and in our main study area it was 1.64 ± 0.73 (28 litters; Kruuk *et al.* 1991). Considering sizes of litters when they were first observed in Shetland (with cubs about 2 months old), 13 (46%) of 28 observations consisted of only one cub, 13 of two cubs, and in the remaining two there were three and four cubs each. Elsewhere, too, litters were small in coastal areas, with recorded means of 2.0 cubs in Norway (Heggberget and Christensen 1994) and 1.55 along the Scottish west coast (Kruuk *et al.* 1987). In inland areas family sizes are somewhat larger: 2.8 in the Netherlands (when otters still occurred there; van Wijngaarden and van de Peppel 1970), 2.4 in east Germany (Stubbe 1980), 2.4 in Poland (Wlodek 1980) and 2.5 in inland areas of the UK (Mason and Macdonald 1986).

The numbers of cubs that are born per adult female per year are more difficult to determine. In Shetland a total of nine simultaneously resident females produced over 5 years a mean of 9.2 cubs per annum (1.0 cub per adult female per year). Some individual examples: female F1 had one cub in 1984 which she lost, then one cub in 1985, and she disappeared in 1986. F7 reared one cub in 1984, none in 1985, two in 1986, and she disappeared in 1987. Her 1984 cub (F16) stayed in the same range, reared her first cub in 1986, and probably none in 1987. F33 was quite old when we first caught her (judging from tooth wear), and she had no cubs in 1985, 1986 or 1987, then disappeared. F11 had litters of two cubs in both 1985 and 1986; F44 had one cub in 1986, and three in 1987, of which she abandoned one (Fig. 11.4).

Figure 11.4 Eurasian otter cub, 2 months old, deliberately abandoned by its mother.

These observations do not reflect pregnancy rates of females, because litters may have been lost before or even after birth (inside the natal holt) without being registered. Otter females may abort, for instance as a consequence of bacterial infection (Weber and Roberts 1990). In Norway, Heggberget and Christensen (1994) found litter sizes of 2.5 at the time of birth, and of 2.0 once the cubs were mobile and outside the natal den (more than 2 months later). In east Germany, Hauer *et al.* (2002b) calculated a mean loss of fetuses between birth and first emergence of 29%; they observed 2.0 ± 0.8 mobile cubs per female, but 2.9 ± 0.6 embryos per female.

As far as we know at present, the Eurasian otter does not have a delayed implantation (Chanin 1991), so a zygote starts development immediately after fertilization, unlike some other otter species (see below). In some post-mortem examinations, however, unimplanted blastocytes have been found, and it is possible that a short delayed implantation has been overlooked in this species (S. Broekhuizen, 2005, personal comunication). Unusually for a mammal of this size, reproduction of otters in Britain is remarkably unseasonal. Cubs may be born in any month—a fact that has been noted by many authors (Harris 1968; Stephens 1957). On the European continent births are more common in spring and summer (Danilov and Tumanov 1975; Erlinge 1967; Hauer *et al.* 2002b; Reuther 1980; van Wijngaarden and van de Peppel 1970), but, in general, in many places otters are far less synchronized in their breeding season than any other carnivore (Corbet and Harris 1991).

Surprisingly, though, we found otters in Shetland with a distinct breeding season (Kruuk *et al.* 1987). The copulations that we saw all took place in spring, and the length of gestation is 61–74 days (Wayre 1979). When calculating 2–2.5 months back from dates of our first observations of otter cubs during their first weeks out of the natal holt, there appeared to be a clear birth peak around June (Fig. 11.5). From similar observations, a slight synchronization of births was found also for the coasts of north-west Scotland, but far less pronounced (see Fig. 11.5). When in Shetland we recorded mere presence or absence of cubs (irrespective of size or age) accompanying adult otters, at different times of year, a clear seasonality was evident (Fig. 11.6). The data demonstrate a more or less steady proportion of family parties moving around in the population from October onwards, with families splitting up around April.

The difference in the timing of Eurasian otter breeding between Shetland and elsewhere in Britain was striking. Probably, this is not due to a more pronounced seasonality in climate in Shetland: if anything, the differences between winter and summer are smaller there (Berry and Johnston 1980; Kruuk *et al.* 1987), tempered by the seas around the islands. Having noted the relationship between otter mortality and food availability, we therefore postulated a similar hypothesis to explain the seasonality in breeding. This suggests that the breeding season is timed in such a way that the greatest energy

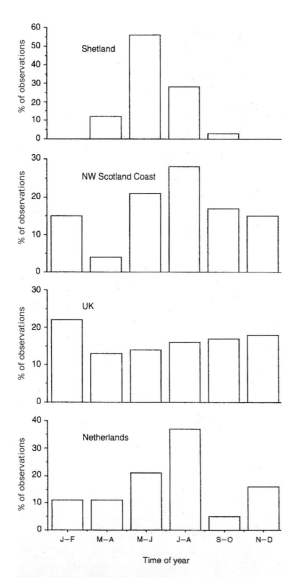

Figure 11.5 Eurasian otters breed seasonally in some areas, elsewhere at all times. Numbers of litters observed in four areas of north-western Europe: Shetland, 34; Scottish west coast, 47; all of UK, 168; The Netherlands, 19. (After Kruuk et al. 1987.)

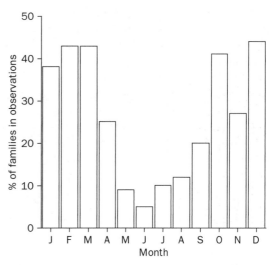

Figure 11.6 Eurasian otter families are rarely seen in summer: 1505 observations in the study area at Lunna Ness, Shetland, 1983–1987.

requirement of the female coincides with the greatest availability of fish.

The energetic needs of female mammals are highest at the time of lactation (Loudon and Racey 1987; Randolph et al. 1977; Widdowson 1981). In fact, Oftedal and Gittleman (1989) estimated that the energetic requirements of a female during just one day at peak lactation exceed the total energetic contents of a litter at birth. In the case of Eurasian otters, the cubs are provisioned with fish starting at the age of about 2 months, and they are weaned slowly over a period of several months. We may expect, therefore, peak lactation between 1 and 3 months after birth, in Shetland between July and September.

We did, indeed, register an up to 10-fold increase in August in the number of eelpout in our study area (see Chapter 8). Eelpout was one of the most important prey species year-round, by far the most important during summer, especially in the areas where most females have their cubs, near the shallows. This supported our hypothesis (Kruuk et al. 1991). In fact, we found that the relationship between seasonalities of otter breeding and fish availability may go even further.

We had also established that, during autumn and winter, otter diet includes more relatively large fish, also related to availability (see Chapters 7 and 8). Female otters need these relatively large prey to carry to their offspring: they take only one fish at a time, and often cubs are waiting ashore whilst the mothers fish (Fig. 11.7). Clearly it would not be economical to carry small prey over a long distance to the cubs, so large fish are needed (see Chapter 6). This means that, in Shetland, cubs were growing up and being provisioned by their mother just at the time when more large fish were available, making for increased efficiency of the female's provisioning behaviour.

If seasonal increase in food availability is an immediate causal factor underlying the timing of otter breeding, it is likely that the actual amount of prey in the territory, during the peak of lactation, has an effect on recruitment, on the number of cubs raised per female. This suggestion is interesting, particularly as the number of fish caught in our traps during July and August over several years varied a great deal. To test the prediction we compared the numbers of otter families and cubs in our Shetland study area, observed during winter months between 1984 and 1993, with the numbers of potential otter prey caught in our fish-traps during the preceding month of August (Kruuk *et al.* 1991).

There were strong, significant correlations between both numbers of cubs and numbers of families in the study area, and the numbers of prey in August (Fig. 11.8). However, there were no correlations with fish numbers at other months, so cub numbers were not affected by fish at some other time. Along the coasts of Norway, Heggberget and Christensen (1994) noticed that cubs could be born any time of year, but there was a clear peak in summer and autumn. Just as in Shetland, prey availability (determined by fish-trapping), was highest at a time when cubs were about 2 months old (Heggberget 1993). All this strongly suggests that food supply, during a critical period, sets a limit to the number of otter offspring that joins the population.

The relation between otter breeding success and prey numbers happens not just along sea coasts. In one 40-hectare north-east Scotland freshwater loch, Davan, the diet of Eurasian otters and their reproductive success has been monitored, with several studies in a period spanning more than 20 years (e.g. Jenkins 1980; Kruuk *et al.* 1998). Eels constituted almost 90% of prey throughout this time until about 1990, when they began to decrease in importance, and in 2003–2004 eels were less than 50% of items caught (unpublished observations). At the same time, the size of eels caught by otters declined by about half, and I noticed that the otters' capture rate of eels declined from over 6 per hour in the 1980s, to about half that in the 1990s. The otters' foraging trips became longer. The result of these and other changes was that the overall energy intake per hour foraging declined substantially (see Chapter 9), and we estimated that the observed return for effort would be insufficient to sustain reproduction. No

Figure 11.7 Eurasian otter families in Shetland stay together for the first year. Female and two full-sized cubs, 10 months old.

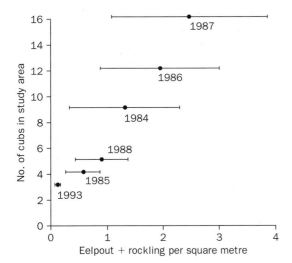

Figure 11.8 Number of cubs versus density of main prey: when fish is abundant, Eurasian otter cubs are many. Low prey biomass during July and August in Shetland is followed by few Eurasian otter cubs in the following season. Eelpout and rockling density was estimated from fish trapping. Values are means and standard deviations ($r = 0.99$, $P < 0.001$). (From Kruuk 1995.)

July 1986. Just by chance I see one of our females carry a small cub down the hill (see Fig. 6.9). She comes from the natal holt which she has used over the last couple of months, and I had been expecting her to emerge with her offspring for some time now. The natal holt is about 300 metres inland, quite high up on the hill, and she carries and drags the little one all the way down, to a large otter holt on the shoreline which has been out of use for several months. On arrival there is another cub sitting in one of the entrances, so what I am watching is not her first trip. She and the two cubs disappear underground, and for 2 hours nothing more happens. By then it is afternoon, and I see her emerge again, going down to the shore, alone. She splashes about in the partly exposed algae, knotted wrack, and comes up with a large bunch of the weeds in her mouth, which she takes into the nearby holt. Ten minutes later she does the same, obviously nest building. Another hour passes before she emerges again, this time going uphill, back to the natal holt. Soon she is back at the coastal holt, dragging, carrying a third cub. Then all settles down, with the family inside the holt, out of sight; from a distance, I can quietly keep an eye on it.

Later, the evening of the same day, the otters are on the move again. The mother emerges, three cubs follow, all going down the bank, scurrying around on the seaweed. The female dives in, comes up, watches the cubs, then swims back again and grabs one of the cubs in the scruff, dives: the little one's first exposure to water. Ten seconds later she surfaces, still holding the cub, heading for a small rocky island, about 200 metres away. She leaves the cub there, returning almost immediately to her two remaining offspring, which are sitting close together, piping anxiously. Again she grabs one, taking it across to the island, submerging with the cub for long periods. But this time she does not come back after landing on the island, and the third cub stays behind, whistling loudly, no doubt clearly audible to its mother.

After 20 minutes the cub launches itself into the water, obviously distressed. It swims out, a fluffy ball quite high out of the water, on the smooth surface of the Voe, whistling, and about 50 metres out it begins to swim around in circles, apparently totally disoriented. The mother takes no notice; there is no doubt that she can hear the calls of the cub. Slowly, the cub sinks deeper into the water, clearly struggling, still calling, occasionally submerging. Half an hour after it began its swim, it finally disappears under the surface.

We made three such observations of cub abandonment in Shetland, all very soon after emergence from the natal holt. It could be, therefore, that previous to this, a 'deliberate' reduction of litter size also took place inside the natal holt, unrecorded.

other relevant changes in or around the loch were noticeable, it was just the eels that were disappearing. The once commonly sighted otters (up to eight residents) became rare occurrences (probably only one resident), and where from the 1970s until 1994 every year one to two litters of otters were born, since then until now, 2005, no more have been seen.

It is not clear exactly what brings about the striking correlation in Shetland, between summer prey population and numbers of Eurasian otter cubs in winter. Heggberget and Christensen (1994), analysing Norwegian otter carcasses, calculated that females ovulate most often during spring, and in that way cause a summer peak in births. One additional explanation could be that cubs die during their first 2 months of life in the holt, before they emerge and can be counted. It was not possible to verify this. However, on several occasions we have observed deliberate abandonment of individual cubs by mothers, immediately after the family leaves the natal holt, when the cubs are about 2 months old (see Fig. 11.4). The following describes one of those incidents:

Mortality of Shetland cubs could be assessed from the time they were first seen following the mother. Possibly, some deaths of cubs in the first few days after emergence went unrecorded, because they may have been out for several days before we spotted a family and could start monitoring them. Of a sample of 33 cubs (18 litters) that we watched from emergence or a few days afterwards, a total of six (18%) died within the following 4 months—mostly from causes unknown, though in one case a cub was killed by a farm dog. A further three (10%) died in the next 2 months; after that, when the cubs were about 8 months old, some could have been large enough to start independent life, and we could not be sure whether disappearance meant death or departure. From such observations (in an admittedly small sample), we estimated total mortality of cubs, outside the natal holt, as 18% for the first half-year of life, and from carcass returns (which are not very reliable for that age) as 12% for the second half-year (Kruuk and Conroy 1991).

There is an interesting difference between the reproductive systems of the American **river otter** and the Eurasian one. The river otter has a clearly delayed implantation as a normal occurrence (references in Melquist and Dronkert 1987), with the embryo 'dormant' for up to 10 months. Only then follows a 60–63 days gestation before birth in spring. Delayed implantation occurs in a number of mustelid species and other mammals (see review in Mead 1989). The effect of it is that the female river otter can mate immediately after losing its cubs or after weaning, or later at any other time of year, and always produce cubs early in spring. In the Eurasian species, even if a delayed implantation would be physiologically possible (as yet an uncertainty), it is uncommon.

Why should the two species, living at the same latitudes, have different physiological mechanisms to cope with seemingly similar environmental requirements? There is no satisfactory functional explanation, as is true for many other cases where delayed implantation has been observed. The **sea otter**, also in North America, is a delayed implanter as well (Sinha *et al.* 1966). Other otters probably do not delay, but evidence is only anecdotal.

The **river otter** reaches sexual maturity in its second year of life, like the Eurasian species, and there is a large amount of data from analyses of carcasses of fur-trapped otters. There is considerable variation in the pregnancy rates of 2-year-old females (e.g. 20% in Minnesota, 43% in Alabama, 55% in British Columbia; Melquist and Dronkert 1987), whereas females of 3 years and older were pregnant in 61% and 91% of cases. In Oregon, Toweill and Tabor (1982) found that 99% of all adult females were pregnant.

Litter sizes of river otters are also similar to those of the Eurasian, varying between one and five, but usually there are one to three cubs. Mean fetal litter sizes were 2.3 (Hamilton and Eadie 1964), 2.7 (references in Melquist and Dronkert 1987) and 2.7 (Tabor and Wight 1977). The last authors noted that there were 2.3 cubs per female at the beginning of independence. Older females almost all become pregnant annually (Melquist and Dronkert 1987; Toweill and Tabor 1982) and they give birth in early spring, between December and April. The observations show that, although the American river otter and the Eurasian one belong to different genera, and show differences in the pattern of implantation, as a general picture the two are remarkably similar in their reproductive ecology.

Not so the **sea otter**, with its single pup and its mother in permanent attention. Its reproduction is clearly seasonal in Alaska, although births can happen at any time of year. The peak of pupping is around May (Schneider 1973, in Jameson and Johnson 1993). In California, however, there is almost no seasonality in pupping (Riedman *et al.* 1994), with perhaps a slight preference for February to April (Sinniff and Ralls 1991). The period between mating and birth is also extremely variable, as copulation often takes place in late summer and autumn in the entire geographical range (Riedman and Estes 1990). There is a delayed implantation after mating, but the exact length of gestation proper (after implantation) is not known. The total time between known copulations and 25 births was estimated as 5–7 months (Jameson and Johnson 1993).

Sea otters do not start breeding until quite an advanced age. In Alaska, one female gave birth for the first time when 5 years old; in a sample of nine females in California their ages at first pupping were 3 years (two cases), 4 years (one otter), 5 years (four), 6 years (one) and 7 years (one), so quite variable, but about 5 years old on average (Jameson and Johnson 1993).

Populations, recruitment and competition   187

Figure 11.9 Giant otters playing. Fully grown cubs may stay in the family group for several years. © Nicole Duplaix.

As soon as her pup is weaned or has died, a sea otter comes into oestrus again within days. However, there is disagreement over the length of birth intervals: according to Jameson and Johnson (1993), once they start reproducing 'female sea otters are essentially always pregnant or caring for a pup', and after their first pup some 85–90% of females produce another one annually. Kenyon (1969) and Garshelis et al. (1984), also in Alaska, reported that most females give birth only once every 2 years, and Riedman et al. (1994) calculated a 407-day interval in California.

The young pup is dependent on its mother for at least 6 months, with various studies in Alaska and California showing an average of 5–6.5 months (see review in Monnett and Rotterman 2000). However, it can be much longer, especially in Alaska: in the Aleutians, Kenyon (1969) found mothers looking after their pups for at least 1 year. One might speculate that this is linked to the Aleutian sea otters' preying on fish, which may require a longer learning period for the cubs (see Chapter 6).

What limits pup production more than anything else is death in the first month or so of the pups' existence. In Kodiak, Alaska, 66% of pups of 2-year-old mothers died before weaning age, compared with 0% of pups of 5-year-old mothers; three-quarters of the losses occurred within the first month after birth (Monson and DeGange 1995). In Amchitka, more than half of pups died before weaning (assumed to be at 120 days; Monson et al. 2000), in Prince William Sound 33% (Monnett and Rotterman 2000), and amongst Californian sea otters 35–40% (Riedman et al. 1994). The suggestion was that many pups died because their mothers were in bad condition, needed more time to forage and consequently had less time on the surface, to nurse and keep the pup warm (Monson et al. 2000). There is much mortality amongst pups during bad weather conditions (Kenyon 1969), possibly because the mothers' foraging is less productive, or mothers may lose pups between dives, in the choppy seas.

In the sea otters, therefore, even more than in the Eurasian otters of Shetland, we find a pattern of high loss of very young offspring. Monson et al. (2000) have referred to this as a strategy of 'bet hedging', enabling the animals to make use of episodic favourable conditions (such as a population explosion of lumpsuckers in Alaska; Watt et al. 2000), whilst cutting losses early if need be. It enables females to concentrate efforts on their own survival, when conditions are suboptimal.

The restriction of sea otters to producing only a single pup (twins are exceedingly rare and do not survive) is unique amongst otters. It is an adaptation to a habitat very different from that of any of the other species, and an adaptation that is also the norm amongst other sea mammals, such as pinnipeds—one of the many evolutionary convergences between sea otters and seals.

From other species of otter we generally know little about reproduction. **Giant otters** are rather better studied than most, and were shown to be quite seasonal. Almost all cubs (96% of 27 litters) are born in the 5-month dry season of the Amazon, between May and September (Staib 2002) (Fig. 11.9). The mean litter size, as observed after the young started emerging from the den, was 2.3, with the largest family of 4 cubs.

Interestingly, the researchers found a strong correlation between litter size and the number of adults in the group in which the litter was born, which varied between two and six. One interpretation of this could be that the related, and helping, group members enabled a female to produce more offspring. Alternatively, and more likely, it could be that a high-quality group territory allowed larger group size (sub-adults staying on for longer before dispersing), as well as facilitating a larger litter (Fig. 11.10). Whichever of these two mechanisms applies, it does suggest that resources are an important factor in reproductive output, just as in the species discussed earlier.

The age of two female giant otters was known when they first gave birth—they were 3 and 5 years old—and three others were likely to have been 4, 5 and 5 years old (Staib 2002). They were therefore of similar age to Eurasian otters in Germany with their first litter (Hauer *et al.* 2002b), although much older than the Eurasian otters in Shetland. The gestation period has been established as 65–71 days (Wünnemann 1992, in Staib 2002), but whether there is a delayed implantation needs verification. Staib (2002) observed several copulations that occurred probably just after parturition. This, together with the clear seasonality of births, suggests that a delayed implantation is a possibility.

Many cubs die early, more than half during their first year after they were first sighted outside the den,

Figure 11.10 Small giant otter cubs are provisioned by all members of the group. © Nicole Duplaix.

at the age of 2 months. When they are still inside the den, and very small, there must be substantial mortality caused by cannibalistic male intruders, given the various published references to this (Duplaix 1980; Mourão and Carvalho 2001), but it is not possible to assess its ecological importance.

In central India the **smooth otter** also proved to be a seasonal breeder (Hussain 1996). There is a gestation of 62 days (in captivity; Desai 1974). With mean litter sizes of three (seven litters of two to four cubs), they were born in winter, October to November, which is also the dry season. At the age of about 8 months cubs were fishing for themselves, though it was not known to what extent they were independent of their mother (Hussain 1996). **Small-clawed otters** are the only species in which larger litters have been recorded, up to seven cubs (in captivity; Foster-Turley and Engfer 1988).

The **Cape clawless otter** along South African coasts breeds in summer, between December and February. Once the cubs come outside the holt, litter sizes vary between two and three (mean 2.3; Verwoerd 1987).

Although information is lacking for many of the otter species, what we do have suggests a fairly uniform picture (with the striking exception of the sea otter). With the odd exception, numbers of offspring are few, with high early mortality rates. There may be a clear birth season or not; this is variable (within species as much as between), and for different species and areas there is evidence that resources are critical for breeding success. The period of dependence is unusually long. Altogether, one cannot help but think that the reproductive output and performance of otters falls short of what one would expect of animals that need to compensate for difficult conditions of adult survival.

Some conclusions on recruitment

In our Shetland study area the average adult **Eurasian otter** female produced 1.0 2-month-old cub per year, translating to 102 cubs per 100 females, of which an estimated 74 would survive to the age of 1 year. With an (assumed) even sex ratio, there would be 37 females amongst these recruits, which will produce their first litter at 2 years of age. With an estimated female mortality rate of 8.3% during the second year of life (Kruuk and Conroy 1991; see Chapter 12), 34 2-year-old females would be joining the population annually, for every 100 adult females present. This compares with a calculated mean annual mortality rate amongst females of 31% (see Chapter 12). Despite the uncomfortably small sample sizes, these data suggest that recruitment and mortality are fairly similar, leaving little leeway if things go awry.

One interesting aspect of the juxtaposition of recruitment and mortality in the Shetland otters is the suggestion, which comes back again and again, that both are strongly affected by prey availability. Recruitment is correlated with a summer influx of fish into inshore waters, and mortality coincides with low fish numbers in spring (see Chapter 12). There is no evidence that these two food variables are in any way linked, but, of course, they could be under some conditions. Which of the two mechanisms is most important in linking otter populations to numbers of prey is as yet undetermined.

Otters in ecosystems, and competition with other species

As I am focusing on just some single species of predator, there is a danger in seeing these animals out of context. Otters are an integral part of large, complicated ecosystems, and whatever affects them, affects an entire community. Conversely, whatever is changing in the aquatic community is bound to have some effect on otters, sometimes very dramatically, as in the marine coastal communities where sea otters live (see Chapter 12).

Many such ecosystem effects are on a very large scale. For instance, fish communities, affecting otters as discussed above, are being altered by climate change throughout regions, for instance in the North Sea (Genner *et al*. 2004), and eel species are declining on a continental scale (Wirth and Bernatchez 2003), as are Atlantic salmon. The causes of these changes may be thousands of kilometres from the actual effects. In general, freshwater fish are amongst the categories of animals with the highest rates of extinction (Jenkins 2003), and the changes in the ecosystems in which sea otters occur have been related to whaling (Springer *et al*. 2003; Williams *et al*. 2004). Large-scale aquatic ecosystem changes may also be due to aerial pollution, or even to aquatic (mercury) pollution a long distance away (Gutleb *et al*. 1997).

On a smaller scale, fish predators such as otters are likely to be affected by the occurrence in their ecosystems of other fish-eaters, as competitors. Where prey availability sets the 'carrying capacity' of an area for otters, one may expect that the presence of other animals with similar diets will affect otter numbers. It is a scenario of interspecific competition, and one set of competitors is other species of otter.

Competition between otter species is not common, as huge regions in the world are inhabited by only one species, and often when there is more than one they live in separate habitats. **River otters** and **sea otters** both use the Pacific coast of North America, but river otters stay closely inshore and feed on small fish (Bowyer *et al.* 1995); sea otters go out much further and feed on large invertebrates, much less often on quite large fish in deeper waters (Riedman and Estes 1990). Consequently, there is no effective competition between these two. Similarly, there are coasts in Chile where the **marine** and the **southern river otter** both occur, the marine otter along very exposed, rough and rocky shores, the southern river otter in much more sheltered parts (as well as in rivers and freshwater lakes), which keeps them well apart. These two species have quite similar diets. In this case it is not known whether this habitat separation is due to one otter actively excluding the other, or to an inherited habitat preference (Ebensperger and Botto-Mahan 1997).

In contrast, the **neotropical otter** appears to occur wherever the **giant otter** is found, in the same rivers as well as elsewhere, and in the Brazilian Pantanal I have seen them forage within sight of each other. There is a difference in micro-habitat between the two, with neotropicals spending almost all of their time within short distance of the bank, and giant otters foraging all over the river, taking larger fish (Muanis 2004; Muanis and Waldemarin 2004). The solitary, slower and smaller neotropical otters are more vulnerable to attack by piranhas, which live in deeper waters of rivers and oxbows, and almost half of the neotropicals in areas of the Pantanal show evidence of previous injury from piranha (especially tail mutilation; see Chapter 12).

In south-east Asia the **Eurasian otter** shares rivers and streams with the **smooth** and **small-clawed otters**, and the potential for competition is there. The diets of the three are rather different (Fig. 11.11), as is their use of different parts of the rivers and of micro-habitats (see Chapter 4). However, there is much overlap between them on all scores, and they even use and scent-mark each other's latrines (Kruuk *et al.* 1994a). I found some evidence from tracks in the sand that *Lutra* actively avoids *Lutrogale*. The two most similar species in that general region are the Eurasian and the **hairy-nosed otter**, but as far as is known (there is much uncertainty; Sivasothi and Nor 1994) the only area where the two actually overlap is a small part of southern Sumatra: the geographical distributions of the two are almost entirely exclusive.

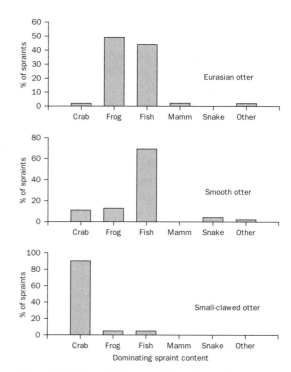

Figure 11.11 There is little competition for prey between sympatric Eurasian, smooth and small-clawed otters in Huay Kha Khaeng, Thailand. (Data from Kruuk *et al.* 1994a.)

In southern Africa the **Cape clawless** and the **Congo clawless otter** are very similar, and there are relatively small regions where the geographical ranges of the two overlap. However, nothing is known about interactions between the two, and where they occur in the same area they are potential competitors, despite dietary differences (see Chapter 7). More importantly, both species also share much of their geographical ranges with the **spotted-necked otter**, and there are interesting observations on overlaps of interest, or lack thereof (Carugati and Perrin 2003a and b; Kruuk and Goudswaard 1990; Somers and Purves 1996). The diets of Cape clawless and spotted-necked are quite different, with fish for the spotted-necked, and crabs, fishes and others for the clawless. Although they live in the same waters, and their habitat selections have much in common, the spotted-necked never occurs in the sea, whereas Cape clawless often do. Where they occur in neighbouring areas, such as in Nigeria, spotted-necked otters stay in freshwater rivers, Cape clawless in the brackish mangroves (Angelici *et al.* 2005).

The African otters share their habitat almost everywhere with the water mongoose *Atilax paludinosus*, which also feeds in water and consumes crabs, insects and small mammals (no fish). Its interests potentially overlap especially with that of the clawless otters (Purves *et al.* 1994; Somers and Purves 1996). However, here again there are distinct differences in foraging methods and choice of micro-habitat, so competition, if present at all, is minimal. The mongoose takes larger crabs than the otter, probably because adult crabs often forage on land at night, but small ones do not. A similar relationship exists between the various otter species in south-east Asia and the crab-eating mongoose *Herpestes urva* (Kruuk *et al.* 1994).

The species that is most often thought to compete with the **Eurasian otter**, for food and space, is the introduced American mink *Mustela vison* (e.g. Dunstone 1993; Lever 1978), despite the fact that otters are some seven times heavier than mink. There is some overlap in diet between them (Chanin 1981; Clode and Macdonald 1995), but there are large size differences in their prey, and there is no evidence that any overlap includes prey species that may be limited by predation. The two often occur in the same habitats in fresh water or along sea coasts, even using the same dens (personal observation).

In North America one finds a similar relationship between the mink and the **river otter**, the two even sleeping simultaneously in the same cover (Melquist *et al.* 1981). However, along sea coasts in Alaska the habitat selection of river otters and mink is slightly different, mink using more sheltered sites for foraging. However, there is much overlap, and the diet of the two species shows many similarities (Ben-David *et al.* 1995).

When mink encounters otter, mink avoids, whereas otters do not appear to take much notice (personal observation). It has been suggested more recently that, in Britain, mink alter their food habits to a more terrestrial diet when Eurasian otters are present (Bonesi and Macdonald 2004a; Bonesi *et al.* 2004), and also that mink numbers decrease when otters move into an area (Bonesi and Macdonald 2004b). Generally, however, the two are so different in size and feeding habits that competition is not likely to be fierce. A similar conclusion appears to be valid in Chile for the introduced American mink and the southern river otter (Medina 1997).

Not surprisingly, there are many animals whose competition with otters is far less subtle than mere interest in the same prey species. Straight kleptoparasites are out there, such as bald eagles *Haliaeetus leucocephalus*, which dive on to **sea otters** that emerge with a fish, and are often successful in stealing it (Watt *et al.* 1995). However, despite the fact that bald eagles are numerous along Aleutian and Alaskan coasts, overall, the amount of fish they take from sea otters is insignificant. From elsewhere there is the record of a bobcat *Felis rufus* stealing a coot from a river otter (Bergan 1990), and one may expect foxes or coyotes to indulge similarly.

Caiman (*Melanosuchus niger*; Fig. 11.12) frequently steal fish from **neotropical otters**, even on land, and in the Pantanal, where caiman are very common, it is not rare to see an eating otter surrounded by several of them (personal observation). Many other species are likely to compete with otters of all kinds, including predatory birds (including herons, cormorants), fish such as pike, and other predators. However, one should not equate a dietary interest in similar prey species, with competition—the latter occurs only if one species reduces the numbers of the prey that can be taken by another predator. For instance, the various vertebrates that feed on the recently introduced crayfish *Procambarus clarkii* in Portugal (including Eurasian otters) do not necessarily compete with one another (Correia 2001).

Figure 11.12 Competition for neotropical and giant otters: caiman (*Melanosuchus niger*) are abundant, eat large quantities of fish and steal from otters.

People may suffer when in competition with otters, as their fishing interests can be similar. In this particular interaction, the aggressive competition from otters, of whichever species, causes fishing people to lose out when otters destroy nets to get at the fish (for example, personal observation of Eurasian and smooth otters in Thailand), or when otters take large numbers of a species important to fisheries. **Spotted-necked otters** in Rwanda take about 15% of the catch from the nets of local fisherfolk (Lejeune (1989), and along Californian coasts fishing of red abalone collapsed after the **sea otters** moved into areas (Fanshawe *et al.* 2003; Wendell 1994). Otters frequently take fish from fish-farms (see Chapter 14). There is no evidence that otters have less prey available because of people's fishing activities, but, of course, otters do suffer from persecution because of their predation—a clear consequence of competition.

Most contact between otters and other fish-eating animals is far less damaging, and should be classed as commensalism rather than competition. These are cases such as the fascinating association between **giant otters** and freshwater dolphins *Inia geoffrensis* in large rivers in Colombia (Defler 1983), where the dolphins appear to swim along offshore from groups of giant otters foraging along the edges. The dolphins take fish that escape the otters. My own observations were somewhat less spectacular, of several kingfishers and pond heron, catching fish that were disturbed and fled from **smooth otters** in Thailand (Kruuk *et al.* 1993b).

As a generalization, competition between other otter species and different fish eaters has little effect on numbers of any of the otter species where they occur at present. The only major exception is competition between humans and several of the otters, resulting in persecution of the animals. One has to accept, however, that the *status quo* between otters and other piscivores has possibly evolved after conflict in the distant past, conflict that is no longer evident. It shows when different otter species have largely exclusive geographical ranges, and when these geographical ranges do overlap we find variation in diet and/or habitat selection.

In general, when present-day, closely related carnivorous species are sympatric, competition between them has little effect, because of clear differences in size (hence differences in prey selection), for example the mustelids in Britain and elsewhere (Dayan and Simberloff 1994). Such regularities are easily upset by introduced species, such as mink (Sidorovich *et al.* 1999). Nevertheless, competition is of little import, in clear contrast to another interspecific relationship: predation (see Chapter 12).

CHAPTER 12

Survival and mortality

Age structure, life expectancy and rates of mortality

To assess the age of individual animals within populations of **Eurasian otters**, several methods have been used, including measurements of eye lenses, baculum and various other bones, tooth wear and incremental rings in teeth—methods usefully reviewed by Morris (1972). Of these, the counting of annual, incremental rings in teeth is the most accurate for animals such as carnivores, and for otters was first developed by Thrine Heggberget (1984). Heggberget used the incremental rings in cementum at the tip of the root of one of the small incisors, or of the much larger canines. A tooth is decalcified, sectioned with a freeze microtome, sections are stained, and rings are then counted with ×64 magnification (Fig. 12.1).

Previous to our observations from Shetland, Stubbe (1969), recognizing adults and sub-adults from size, and Heggberget (1984) collected data on the age of otters that had mostly been shot by hunters, in Germany and in Norway. They found high proportions of juveniles, but such samples are likely to be biased towards younger animals, as they are more easily approached and shot. Similarly, most information on North American **river otters** comes from harvested populations, which showed a relatively young age structure (Stephenson 1977, Tabor and Wight 1977). We therefore had little idea about what to expect from the observations on Eurasian otter ages in Shetland, apart from an intuitive picture of rather long-lived animals—which appeared to be quite wrong.

Eurasian otters can live to respectable old age, up to 15 years in captivity (Chanin 1985). In the wild, we found the oldest animal, a female, killed on the road in Scotland at the age of 16 years. It was all the more surprising, therefore, that in general the age structure of Eurasian otter populations was very young, and in the wild this species has a remarkably

Figure 12.1 Determining an otter's age from layering in the dentine—section through the root of a canine tooth of a 3-year-old Eurasian otter.

low life expectancy (Kruuk and Conroy 1991). The following figures on mortality of otters in Shetland include data published earlier from a smaller sample (Kruuk and Conroy 1991), although the conclusions are similar.

The ages of otters found dead in Shetland (on the roads, or along the shores; Fig. 12.2), are shown in Figure 12.3. Most of the animals had died young,

Figure 12.2 Natural, non-violent death of a 5-year-old female Eurasian otter, Shetland.

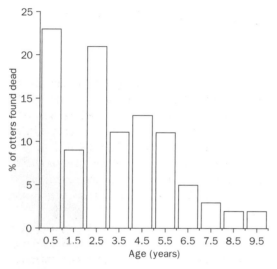

Figure 12.3 Most die young: ages of 113 Eurasian otters found dead in Shetland, 1984–1988.

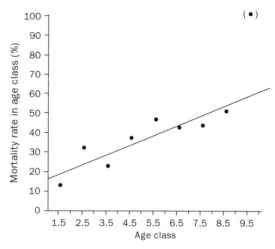

Figure 12.4 Gradual increase of mortality rate with age of 113 Eurasian otters in Shetland ($r = 0.93$, $P < 0.001$). Each point shows the percentage of otters of that age dying in the subsequent year. (After Kruuk and Conroy 1991.)

before they were 3 years old. This phenomenon was probably even stronger in reality than is shown in Figure 12.3, because young cubs, which spent almost all their time underground, were probably never found if they died (see Chapter 11).

The mortality picture changed somewhat when annual mortality was expressed as a percentage of each year class. I assumed that the sample of dead animals was representative of the whole population, that is, I treated it as a population of 113 otters, of which 21 died before the age of 1 year, etc. If annual mortality is plotted against age, a picture emerges of a linear increase in probability of death with age (Fig. 12.4). This is unusual amongst mammals: most species show a pattern of mortality that is high when young, then lower in adult age, and sometimes high again at ripe old age (Caughley 1977). As an example, in a British population of Eurasian badgers there was very high mortality rate amongst cubs and 1-year-olds, then a moderately high death rate amongst adults (mean 22%), but no significant change with age (Cheeseman *et al.* 1987).

As explained above, we should be wary of mortality figures for animals less than 1 year old. To look at the pattern of otter survival, therefore, it is useful to start with animals of 1 year, at the beginning of independence of individuals, and express the probability of death before a given age (Fig. 12.5). In Shetland more than half of the recruits to the adult population died before they were 4.5 years old, and mean life expectancy at the age of 1 year was 3.1 years (calculated as in Southwood [1978], with a sample size of 94 adults).

There were small differences in the life expectancy of otters depending on whether calculations were based on animals that died violently (almost all on roads) or non-violently (i.e. 'naturally'). From the road-kill sample, I calculated a life expectancy of 2.9 years ($n = 50$); from natural deaths this was 3.6 years ($n = 44$). Obviously the overall figure lies somewhere in between, and I will argue below that it is closer to 3.6 years, the expectancy based on

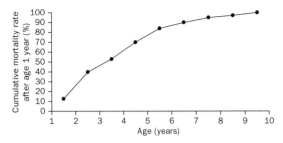

Figure 12.5 Cumulative mortality of 113 Eurasian otters in Shetland: percentage of otters (all categories) that were alive at 1 year of age and had died by the given age. More than half were dead by the age of 3.5 years. (After Kruuk and Conroy 1991.)

Figure 12.6 North American river otter, with fish in winter. © Angie Berchielli.

non-violent deaths. Females can expect to live slightly longer, on average 3.2 years ($n = 54$), compared with 3.0 years for males ($n = 40$).

The low life expectancy for Eurasian otters appears to be a common feature in other areas. In a sample collected between 1982 and 1990, of 146 dead otters from Shetland, 148 from the Scottish mainland and 97 from England and Wales, the median age at death was 4 years for both Shetland and the Scottish mainland, and 3 years for England and Wales (Gorman et al. 1998a). Everywhere the mortality rate increased with age. It was also shown that, over about the same period (1982–1994), the average age at death of otters on mainland Scotland decreased, especially for females, from about 5 years of age at death in the late 1980s, to about 3 years in 1994 (Kruuk et al. 1997c). Most likely this was due to an increase in numbers of young animals killed on roads.

A Spanish female Eurasian otter of 1 year has a life expectancy similar to that of an otter in Britain aged 3.0–3.6 years, and the oldest animals found dead there were 10 (female) and 12 (male) years old (Ruiz-Olmo et al. 1998a). The oldest German otters found dead were aged 16 years (Ansorge et al. 1997) and 15 years (Hauer et al. 2000), with a mean age at death of 4.2 years; populations there had high proportions of young animals, similar to elsewhere in Europe. The available information suggests, therefore, that the survival pattern of Eurasian otters is fairly uniform throughout the region, with a remarkably short life expectancy.

North American **river otters** (Fig. 12.6) have a reported longevity in captivity of 25 years, and the oldest reported river otter caught by fur-trappers was 14 years old (Melquist and Dronkert 1987). The average annual mortality rate for yearling river otters was 54%, and that for adults 27% (calculated from trapped animals; Tabor and Wight 1977). This adult mortality rate is close to the 33% mean annual adult mortality rate of Eurasian otters in Shetland, and to the rather similar figures for Eurasian otters from elsewhere.

For coastal river otters (adults, with radio-transmitters) in a protected population in Prince William Sound, Alaska, the mortality rate was lower: about 10% per year, measured over 20 months. The median age of these animals was 4 years, compared with 3 years for a heavily trapped population in Maine (Bowyer et al. 2003).

There is more information on rates of mortality in **sea otters**. For young pups this is quite high: in Prince William Sound 33% die before the date set as the age of weaning (120 days), and in California 36% (Monnett and Rotterman 2000). Along the coast of Kodiak the pup mortality rate is lower, at 15% (Monson and DeGange 1995). However, the annual adult mortality of sea otters is strikingly lower than that of *Lutra* and *Lontra*: adult females in Kodiak die at a rate of 4–11% per year, males at 9–14% (Monson and DeGange 1995), whereas in California 9% of adult females die each year, compared with 29–33% of adult males (Siniff and Ralls 1991).

Udevitz and Ballachey (1998) calculated survival rates from the age distribution within the sample of 508 dead sea otters collected in Prince William Sound after the *Exxon Valdez* oil spill in 1989. Unlike the Eurasian otter, the sea otter shows the classical mammalian survival pattern, with annual surival rates of 92% in the age range of 2–4 years, 100% in the middle age range of 5–8 years, 81% for ages 9–15 years, and virtually zero for ages 16–20 years. Longevity in captivity is 20 years (Riedman and Estes 1990).

There is virtually no information on the population dynamics of any other species of otter. It seems likely that they follow the same short-lived pattern as Eurasian and river otters, with the sea otter as a longer-lived exception. A **giant otter** in captivity lived to the age of 17 years (Brandstätter 2005).

Causes of mortality: introduction

Many different diseases, accidents, predation incidents, persecution and other misadventures have been mentioned in the literature on various otters (see, for example, summaries in Harris 1968; Melquist and Dronkert 1987). However, we still know very little, either about their relative importance as causes of death, or about the extent to which they compensate for one another. For instance, many otters of different species are killed on roads, where they are much more likely to be reported than if they died in a den or were taken by a predator. In any case, if the 'carrying capacity' of any one habitat (often determined by prey populations) does not allow more than a given number of otters, it matters little whether the otter 'surplus' is killed on roads, by starvation or in some other way. However, if the otter population is below 'carrying capacity', causes of death can be highly important, for example for purposes of conservation management (see below).

Proximate causes of death, and body condition

We obtained some data on causes of death in our study of **Eurasian otters** in Shetland (Kruuk and Conroy 1991). As mentioned above, one major problem in quantifying such information is the fact that various causes of death have different probabilities of being detected, and this affects many studies on the subject. Several authors have emphasized the slaughter of Eurasian otters on roads (Fig. 12.7), in fyke nets set for eels, in lobster creels, by hunts and various other means (Chanin 1985; Chanin and Jefferies 1978; Jefferies *et al.* 1984; Mason and Macdonald 1986; Philcox *et al.* 1999; Stubbe 1980; Twelves 1983; van Wijngaarden and van de Peppel 1970; and others), but no data have been presented enabling the evaluation of these observations against other causes of mortality.

In Shetland the same problem applied, but to a somewhat lesser extent: the habitat is more open and accessible than in most other otter areas, and otters dying in remote places are more likely to be found. However, it was still impossible to ascertain the magnitude of the detection bias, which must have been large.

Of a sample of 113 Eurasian otter carcasses collected in Shetland, 49% were killed by vehicles (Kruuk and Conroy 1991). Nevertheless, we concluded that, with the large bias favouring road kills, traffic mortality was a relatively small factor of unknown magnitude. Other kinds of violent death in the sample included one otter drowned in a lobster creel, and four that had been bitten to death, by either other otters or dogs. Some 46% died from

Figure 12.7 Eurasian otters braving traffic and industry: traffic sign near oil terminal, Shetland.

non-violent causes (see Fig. 12.2), with symptoms that included haemorrhaging in the stomach or intestines (8% of all carcasses), liver abnormalities (lesions, hepatic neoplasm; 2%), pneumonia (1%), and poisoning by oil or paint (2%). However, for many of the non-violent deaths no immediate cause could be found, either because there was no significant abnormality (9%), or because the carcass was too decomposed (25%). One clue to the cause of death came from the seasonality of mortality (see below).

Considering only the 24 bodies of non-violent deaths that could be analysed, gastrointestinal haemorrhaging occurred in 37%, a symptom that may be associated with starvation. One other characteristic of the non-violent deaths was the poor overall body condition: the animals looked worn out and thin. To quantify this, we calculated an index of condition, K (Box 12.1), now used as standard in otter post-mortem analyses.

Figure 12.8 shows the differences in condition index for otters that died in Shetland, with a mean $K = 0.77 \pm 0.15$ (standard deviation) for non-violent deaths and $K = 1.08 \pm 0.15$ for animals killed violently (Kruuk and Conroy 1991), a highly significant difference.

We also collected data on carcasses of Eurasian otters that died in mainland Scotland. Of 76 bodies analysed in 1991–1992 to check for pollutants, 65 (86%) had died violently, almost all on roads. The mean condition index of the road-kills was 1.04, that for the non-violent deaths 0.78. Values for the mainland and Shetland are very similar, except that on Shetland we found relatively more non-violent 'natural' deaths. A similar pattern for cause of

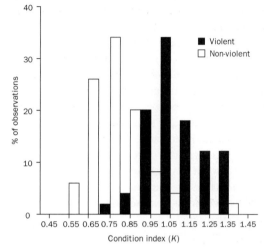

Figure 12.8 Body condition (K) of Eurasian otters in Shetland, at death from violent (48) or non-violent causes (49). $K = 1$ is 'normal'; $K < 1$ is underweight, etc. Otters dying from violent causes (mostly traffic) were in significantly better condition ($\chi^2 = 68.2$, 5 d.f., $P < 0.001$). (After Kruuk and Conroy 1991.)

Box 12.1 Index of condition

To obtain a quantitative expression of body condition, otter carcasses are weighed to calculate the **condition index, K**, which takes into account the substantial differences in overall body length between otters. A similar condition index is used for fish (Le Cren 1951), and can be calculated for any species of otter. It uses the relationship between average weight (W) and average body length (L) of animals, to be expressed as:

$$W = aL^n$$

in which a and n are constants, determined by the shape of the body, amongst other things. In a sample of 25 road-kills (therefore, presumably 'normal' Eurasian otters) we found that, on average, for females a = 5.02, n = 2.33, and for males a = 5.87, n = 2.39, with weight in kilograms and length (straight line from nose to tip of the tail) in metres. So, as weight-for-length, males are relatively heavier than females, on average.

The condition index for each individual is calculated as:

$$K = W/5.02L^{2.33} \text{ for females, and}$$

$$K = W/5.87L^{2.39} \text{ for males.}$$

If, in the above sample of road-kills, we use the calculated constants, the mean condition index $K = 1.0$. This, by definition, is for the average, normal, healthy otter. An animal that is underweight will have a value of K as low as 0.5, and a really heavy Eurasian otter may have $K = 1.4$ or more.

death related to body condition was found in Spain (Ruiz-Olmo et al.1998).

Shooting or trapping of Eurasian otters is illegal in almost all western European countries, but is still common in eastern Europe (Kruuk 2002). In Britain the animals were traditionally hunted with packs of specially bred otter hounds, until the 1960s. However, one still finds some otters killed by unscrupulous game-keepers (Fig. 12.9).

In Germany, Silke Hauer and colleagues (2002a) detailed the causes of death of 1067 Eurasian otters, and found 69% killed by traffic, 6% in fyke nets and 4% shot. Just as in our Scottish samples, this produces a biased view of mortality towards human causes. The study also showed a massive increase in deaths on German roads after the two Germanies joined together—the price of prosperity. Amongst other causes of death, Hauer et al. (2002a) mention pneumonia, starvation, infections and unspecified injuries, peritonitis, tumours, bites, drowning under ice and poisoning.

Otter mortality on UK roads was studied by Philcox et al. (1999), from the details of 673 dead animals. Some 56% were males, and there was a clear seasonality: both March and November saw about five times as many casualties as June. Most of the accidents happened on major roads, close to water (but mostly where roads run alongside, not places where they cross rivers). There was a good correlation between rainfall or river flow, and numbers of otter casualties. This may be explained by otters either avoiding crossing under a road when rivers are in spate, or making more overland trips during wet

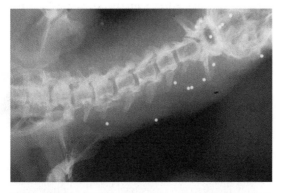

Figure 12.9 Radiograph of a Eurasian otter killed by shotgun along River Dee, Scotland, 1991.

nights (when frogs are active, which would also explain the seasonality of this kind of death). Having followed otters with radio-transmitters walking overland over considerable distances at night, I have little doubt that both phenomena are important.

The most thorough post-mortem analyses of Eurasian otters were carried out on 77 otters from south-west England, of which 83% were killed by road traffic (Simpson 1997). A fairly large proportion, 12 animals (9 males, 3 females) showed bite wounds to face, legs or scrotum, probably caused by other otters, and this was the cause of death in 5 cases.

In summary, the immediate causes of death of Eurasian otters are many. There is no evidence that the ones we see most often (i.e. death caused by road traffic) are also the most important ones affecting the populations. However, with the observed seasonality of death in otters and conclusions drawn from dissections, it is likely that at least one other, ultimate, factor is important. This is discussed further below (population limitation).

For **river otters**, Melquist and Dronkert (1987) and Larivière and Walton (1998) reviewed observations of a number of mortalities. They mention human activities as a major cause, although the same reservation applies as for the Eurasian otter: human-caused deaths are much more visible. Traps are a much more common end for a river otter than for a Eurasian one. Trapping is widespread in Canada and the USA, with many river otters caught legally each year for their pelts. In addition, road traffic and fish nets take their toll, as well as predation (see below) and diseases. In Idaho, five of six deaths of river otters with radio-transmitters were caused by humans (Melquist and Hornocker 1983). Even road-kills often go unrecorded, and this kind of death may be substantial. As an example, on just 16 km of one interstate road in Florida, 15 river otters were killed over a 7-month period in 2000 (Kinlaw 2005). This is a road through prime otter habitat, used by some 60,000 vehicles per day.

A documented example of the extent of mortality caused by trappers every year is the tally for 1978, which was 47,000 river otters in the USA and Canada (Kruuk 2002). The Nebraska Game and Parks Commission website (www.ngpc.state.ne.us/wildlife/otters.asp, accessed April 2005) mentions that about half of the American states, and all of the

Canadian provinces, have otter-trapping seasons. In some recent years, more than 50,000 otters have been taken in North America, and the annual otter harvest in Louisiana alone sometimes exceeds 10,000 animals, usually surpassing that in any other state.

Because of the large number of available carcasses from trapped river otters, there has been a considerable amount of research on ectoparasites and endoparasites, as well as bacterial, fungal and viral diseases. There have also been many relocation and reintroduction schemes for this species when otters were kept in captive conditions for some time (see Chapter 14), and in which there was a great deal of veterinary interest. The results have been reviewed extensively by Kimber and Kollias (2000). Most of the parasites and diseases appear to have little effect on individual otters, but some may be fatal. They include trematodes, salmonella, canine distemper, rabies, feline pan-leukopenia, hepatitis, pneumonia, tuberculosis and many others, but even for the potentially fatal ones the importance in otter populations is not known. Otters in captivity appear to be more vulnerable, as they lose part of their resistance against many different afflictions.

Sea otter mortality is often highly seasonal. At northern latitudes, for instance in the Aleutian islands, many were found to die especially in late winter and early spring, with carcasses washing up on the beaches (Bodkin *et al.* 2000; Kenyon 1969). The actual cause of death in these cases was not identified, but was likely to be related to starvation. Around Kodiak Island, Alaska, most deaths in sea otters with radio-transmitters was caused by (legal) shooting, by Alaskan natives (Monson and DeGange 1995). In later years, the role of predation has become much more prominent, and is discussed below. In contrast, the mortality rate of sea otters along Californian coasts is higher from spring to late summer (Estes *et al.* 2003a), and appears to have quite different causes.

One striking population decline occurred amongst Californian sea otters from 1976 until 1984, when numbers dropped from about 1800 to 1200. This coincided with a large increase in in-shore fish-netting (carried out mostly in summer), and many otters were caught and drowned. Once the practice was legally stopped, and nets were allowed only much further offshore, the sea otter decline ended (Estes *et al.* 2003a). However, another decline started in 1994 (see Chapter 11), and its cause proved more difficult to pin down.

Several researchers have noted that many Californian sea otters succumb to disease, and hundreds of beached carcasses have been analysed. Estes *et al.* (2003a) found that a large proportion of animals died during their prime middle age. They also noticed a considerable number killed by sharks (see below), and large numbers of carcasses in which the cause of death could not be identified (therefore quite likely to be from disease). At the same time, Kreuder *et al.* (2003), in a highly detailed analysis, pointed out that in 64% of the otter carcasses found along the Californian coast the animal had died from disease as a primary cause. Amongst the pathogens, the most prominent was *Toxoplasma gondii*, which turned out to be a highly significant affliction. Not only does this protozoon cause encephalites in sea otters and in many other terrestrial mammals, including humans, and is thereby an important direct cause of death, but interestingly it is also closely associated with other causes of death, such as heart disease, acanthocephalan infection (thorny-headed worm) and shark attack (see below). None of the Alaskan otters appears to have been in contact with this protozoon.

Toxoplasma is best known from domestic cats, and infections in people are usually a result of contact with cats. In California, many domestic cats may have been exposed to the pathogen, or actually carry the disease. The route of infection goes through faeces, and this is the way in which other mammals become involved. It seems quite probable that, given the high human (and domestic cat) population density along areas of the Californian coast, the sea otters there are prominently exposed to *Toxoplasma*. Coasts where sea otters have especially high exposure to the pathogen are those with a high run-off of fresh water.

In summary, evidence suggests that in California a sluggish speed of increase in numbers, then a decline in the population of sea otters, is largely due to increased mortality amongst adult animals. It is likely to be caused by a disease organism, *Toxoplasma gondii*, which may well ride on the coat-tails of our civilization. In other parts of their range, sea otters are disappearing even faster, a consequence of predation that may also be driven by the activities of humans (see below).

Effects of pollution

There is little doubt in anyone's mind that chemical pollution has been the cause of the dramatic, sharp, and in some places ultimate decline of **Eurasian otter** populations in England and several other countries in Europe. It started in the 1950s, and continued into the 1970s and 1980s (summaries by Jefferies 1989; Mason and Macdonald 1986). Only late in the 1980s and during the 1990s did the trend begin to be reversed, and at the turn of the millennium otters in Europe had recovered in many areas, or at least they were on their way. Similar events took place in North America.

The otters' rapid disappearance from many waters of Europe and North America coincided with a massive increase in the use of organochlorines as insecticides in agriculture, and for various purposes in industry. Several of these substances, such as DDT derivatives, dieldrin, lindane and several toxic heavy metals, were found in damaging concentrations in the tissues of dead otters. Many other carnivores, fish predators, birds of prey and others suffered similar fates, demonstrably due to pesticides and other chemicals (Newton et al. 1993). Most of the compounds that were held responsible for the wildlife deaths have now been taken out of use, and many of the species that were affected by them responded with recovery, or are in the process of returning. However, we still know relatively little about exactly which compounds were responsible for the twentieth century demise of the otter: the possibilities are many (Mason 1989; Mason and Macdonald 1986; Roos et al. 2001).

Moreover, although a great deal is known of the possible effects of various relevant chemicals on individual otters, there is ignorance about their role in populations, which is a very different proposition. Newton (1988), in a seminal paper on the effects of pollution in birds of prey, pointed out that the (often well known) concentration of a chemical that causes death in half of exposed individuals (LD_{50}) bears no relationship to concentrations required to cause a population decline. The reason for this is compensatory mortality or reproduction: some populations may suffer high mortality through pollution, but still maintain their density (possibly limited by availability of resources). Other populations may decline at even very low levels of contamination, because of the loss of just a few individuals that received a high dose. In most of the research on chemical contaminants in otters, of whichever species, there is no information on populations.

The data needed to assess the identity of the pollutants that caused the large decline of Eurasian otter populations in the twentieth century were never collected, and we will never know for sure what happened. However, the otter decline took place at about the same time as that of many raptors and fish-eating birds, which were similarly badly hit, and which were better researched. For their disappearance, at least in Britain, there was good evidence that the use of dieldrin was responsible (Newton et al. 1993). In all probability, therefore, dieldrin was also the culprit in the case of the Eurasian otter, as has been argued strongly by Jefferies (1989).

Dieldrin has now largely disappeared from the environment, but some of the other, potentially seriously nasty, compounds have not. Possibly, they still play a role in depressing otter numbers. For instance, at the time of writing in 2005, Eurasian otters have still not returned naturally to some parts of western Europe (Netherlands, parts of Germany and France, Switzerland). In Shetland and parts of north-east Scotland there is evidence of recent declines in otter numbers. One cannot exclude a potential involvement of contaminants, although other factors may be responsible. A study of such compounds, in their present-day role in mortality of Eurasian otters, may therefore provide further insights in their effects on otter populations (Kruuk and Conroy 1991, 1996; Kruuk et al. 1997a).

There are several good overviews of relevant contaminants, and concentrations in which they are found, for example in Britain by Mason (1989), in Sweden by Olsson et al. (1981) and Roos et al. (2001), in the Netherlands by Broekhuizen (1989) and Traas et al. (2001). In studies with my colleagues we analysed liver samples for organochlorines, first from 113 otters from Shetland, and then from 116 otters from different parts of Scotland. The results were complicated, as expected, but we could draw some tentative conclusions (Kruuk and Conroy 1991, 1996). The samples were analysed for various polychlorinated biphenyls (PCBs), for the organochlorines DDE (the breakdown product of DDT), dieldrin (HEOD), lindane (BHC) and mercury (Hg). Some carcasses from Shetland were also analysed for cadmium (Cd), lead (Pb) and selenium (Se).

Almost all of these, with the exception of PCBs and mercury, occurred in concentrations well below the levels known to have any significant lethal or sublethal effects in individual mammals or birds (e.g. Bunyan et al. 1975; Jefferies 1969; Robinson 1969; Wren et al. 1988). With the exception of mercury, none was correlated with age of the otter, so they were not likely to accumulate and produce complications later, at least not at the rate at which they were consumed at the time of collection. Further, none of the concentrations of other compounds was correlated with the otters' body condition, and of the substances analysed, therefore, only PCBs and mercury were candidates for causing possible deleterious effects in present-day populations.

PCBs alter the metabolism of vitamin A in the body (as does dieldrin), causing developmental irregularities (including fetal resorption and abortion) and increasing the risk of infections and cancer. Some authors have suspected PCBs of playing a major role in the demise of otters in Britain and elsewhere in Europe (Mason 1989; Mason and Macdonald 1986; Mason and Madsen 1993; Roos et al. 2001). They may be distributed as aerial or aquatic (industrial) pollutants, so they could have an impact almost anywhere. PCBs occur as a complex of different congeners with different values of chlorination; in particular, those with high chlorination values (for example, those numbered 118, 128, 138, 153, 170, 180 and higher, which were also common in our studies) tend to affect the physiology of animals. We undertook analyses for all the compounds separately, but for the present purpose I use the total value of PCB (following Roos et al. 2001).

It was shown in the laboratory that PCBs cause failure of reproduction in female mink (at concentrations in liver lipids of 50 parts per million (ppm); Jensen et al. 1977), and later this value was accepted uncritically as a yardstick for damaging concentrations in otters. Mason (1989) suggested that otter populations with PCB values of 50 ppm (lipid in liver) or more should be declining. This prediction was not borne out in our work: the apparently dense and healthy population of otters in Shetland in the 1990s showed PCB values with a mean greater than 210 ppm lipid (5.5 ppm wet weight) in the liver (Fig. 12.10). This included lactating (i.e. successfully reproducing) females, with values up to 20 times higher than the concentration causing reproductive failure in mink. On mainland Scotland we found a lactating female otter, shot by a keeper in 1987, with 1097 ppm PCB (lipid).

There was no correlation between PCBs and age, so otters appeared to be able to metabolize or excrete this pollutant, at least at the level of concentrations found in Shetland (in fact, they appear to mostly metabolize it; Smit et al. 1994). There was a significant negative correlation between the body condition of otters and the PCB concentration in the liver; the most likely explanation for this was that, in animals in bad condition, body fats are mobilized and consequently the contaminants of those lipids are moved into the liver.

There was no evidence that the Shetland otter population was suffering from the burden of PCBs, and our results, from the apparently healthy, dense and in the 1990s increasing Shetland otter numbers, do not support the Mason hypothesis. Similarly, otters in north-east Scotland showed no obvious decline in density over some 20 years, although their PCB levels were such that they would be called 'of concern' when compared with those in the mink studies (Smit et al. 1994). PCBs at the above

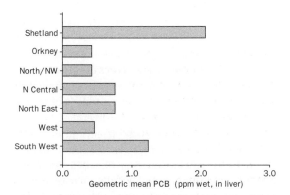

Figure 12.10 Concentration of PCBs in the liver of Eurasian otters from different parts of Scotland: Shetland otters carried the highest PCB burden. Number of carcasses analysed: Shetland, 14; Orkney, 14; north/north-west, 10; north central, 10; north-east, 33; west, 18; south-west, 13. (After Kruuk and Conroy 1996.)

concentrations appeared to be relatively harmless in otter populations, but it is possible that they have deleterious effects on individual otters. Whatever the case, the results show that one should not extrapolate results obtained from individual mink in captivity to populations of otters in the wild.

In the south-west of England, strong negative correlations were found between levels of vitamin A in Eurasian otter livers and the presence of several PCBs as well as dieldrin, between 1988 and 1996 (Simpson *et al.* 2000). This was a period of major recovery of the otter population in Britain, and of significant declines in pollution levels in the animals. Especially remarkable was that otters with highest levels of dieldrin and lowest levels of vitamin A also showed serious eye malfunction, with detachment of the retina in 26 of 131 otters (Williams *et al.* 2004b); this ailment could easily have been overlooked in many previous post-mortem examinations.

Eurasian otters from other areas in Scotland, including Orkney, the west and south-west of Scotland, and the north-east, showed lower PCB values than those from Shetland. The explanation is that the waters and sediments of the northern North Sea and the north-east Atlantic Ocean are known to contain high concentrations of PCBs, brought by the North Atlantic currents and through atmospheric deposition (references in Wang-Andersen *et al.* 1993). Animals such as pilot whales around Faroe are now deemed dangerous for human consumption, because of their high levels of PCBs (Simmonds *et al.* 1994). Pregnant women in the Faroes had dangerously high concentrations of PCBs (Fangstrom *et al.* 2002), and arctic foxes in Svalbard carry a similarly high burden (Wang-Andersen *et al.* 1993). Yet, in none of these species is there evidence that numbers are declining. However, there is also no indication in the area that PCB concentrations in the animals have decreased over the last few decades.

In Spain, as in Scotland, Eurasian otter populations have returned or increased over the past few decades, in the face of sometimes still considerable burdens of PCBs (Ruiz-Olmo *et al.* 1998c). Populations of North American **river otters** in Ontario, as measured by annual numbers trapped, did not show evidence of an effect of the high PCB levels in the nearby Great Lakes (Wren 1991). All of this does not, of course, exclude the possibility that in some countries (such as the Netherlands) PCB contamination is such that it would affect otter reproduction to the extent of preventing a self-sustaining population. So far, however, there is no firm evidence for this.

Apart from PCBs, the presence of mercury in **Eurasian otters** in Scotland and Shetland could also be of concern (Kruuk *et al.* 1997a). Mercury, in its methylated form, has potentially lethal effects on many wild carnivores, causing damage to the central nervous system with symptoms of lassitude, loss of coordination and paralysis (Wren 1986). It occurs naturally in the environment; in the waters near Shetland, for instance, there are remarkably high concentrations of methyl mercury in fish, due to submarine volcanic activity (Carr *et al.* 1974; Davies 1981). Because of mercury, many freshwater eels in these islands are unfit for human consumption, by World Health Organization standards (Kruuk *et al.* 1997a), so their effect on a fish-eating carnivore could be even more serious. Elsewhere, naturally occurring mercury is far outweighed by that produced by agriculture and by industrial sources such as mining, smelting, use of fossil fuels, waste incineration and others, with much of the produced inorganic mercury converted into the lethal methyl mercury (Lindquist and Rohde 1985). Its occurrence is declining (Newton *et al.* 1993).

Methyl mercury affects individual otters. In **river otters** in the USA some unpleasant, but useful, experiments by O'Connor and Nielsen (1981) showed that river otters fed with mercury-laced fish died after 6 months, with a mercury level of 110 ppm (dry weight) in the liver. By comparing mercury levels in road-killed **Eurasian otters** of different ages in Shetland, we found that the animals accumulate it over the years, with a mean concentration of 10 ppp and a maximum of 65 ppm in the liver (Fig. 12.11).

Many sea mammals absorb mercury in relatively high quantities, but are able to use selenium to counteract its toxicity (Koeman *et al.* 1975; Reijnders 1980; Smith and Armstrong 1978). Thase species accumulate and store selenium at the same rate as mercury, and on analysis the above authors found a near-perfect correlation between the two elements in species such as seals and dolphins. Wren (1984) suggested a similar relationship in semi-aquatic freshwater mammals in Canada. However, in our

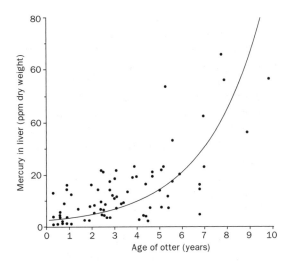

Figure 12.11 Older Eurasian otters in Shetland have significantly more methyl mercury in their liver ($r_s = 0.63$, $P < 0.001$). (After Kruuk and Conroy 1991.)

Shetland otters there was no correlation between mercury and selenium levels, suggesting that in these animals no such selenium mechanism has evolved to cope with the mercury problem (Kruuk and Conroy 1991).

On the Scottish mainland, we found a close correlation between annual rainfall and mercury concentrations in Eurasian otters, consistent with an industrial and then atmospheric origin of the mercury. Here, the mean mercury concentrations in otter livers was 9 ppm, similar to that found by Wren (1984) in **river otters** in Canada. Just as with PCBs, there was a negative correlation between mercury concentration and physical condition (K) of the otter, probably with the same explanation.

Summarizing the occurrence of contaminants in **Eurasian otters**: the use of laboratory tests on mink to set critical danger levels for contaminants in populations of wild otters is suspect, and should be avoided. At present there is no evidence to suggest that any of the organochlorine compounds or heavy metals is exercising a deleterious effect on otter populations. Even if individuals are affected, this does not necessarily translate into a population impact. We have no good data on the identity of pollutants in past, dramatic, population declines of otters; most evidence points to a significant role for dieldrin, although others, such as PCBs and mercury, cannot be exonerated completely.

For the North American **river otter** there have been similar concerns about the effects of chemical pollution, although the role of pollution in twentieth century population declines was probably not as dramatic as for the Eurasian otter. Dieldrin was widely used in the USA; high levels were found in river otters around 1980, for example in Georgia (Clark 1981), and are still present in some areas, in fish and water birds in Florida (Rumbold *et al.* 1996; Marburger *et al.* 2002). Kinlaw (2005) suggested that physical abnormalities, associated with dieldrin and its subsequent vitamin A deficiency, perhaps cause river otters to be more prone to be killed on roads.

Where the species did disappear in the 1960s to 1980s, it has been quite slow in returning, despite generally low levels of contaminant organochlorines and heavy metals (Elliott *et al.* 1999; Halbrook *et al.* 1996). As for the Eurasian otter, there is no good evidence to indicate which compounds were responsible for previous population crashes. In Tennessee, of all the bird and mammal fish predators studied, only the otter was deemed to be at some risk because of the presence of PCBs and mercury. However, this assumed the same sensitivity as found in mink (Sample and Suter 1999), which is just as unlikely as for the Eurasian otter.

In recent years several marine populations of **sea otter**, **river otter** and **Eurasian otter** have been hit by large oil spills, in well publicised incidents. By far the most substantial was caused by the stranding of the *Exxon Valdez*, in March 1989, in the Prince William Sound of Alaska, which killed both sea and river otters. Oil affects otters in three different ways: it destroys the insulation of the fur, poisons the animals when they ingest it, and kills potential prey. Once oil has entered the ecosystem, effects may be felt for many years because of the presence of various breakdown products, and some of the oil itself may be buried and released later.

The effects of the *Exxon Valdez* spill on coastal **river otters** were studied over several years in detail, by Terry Bowyer and his team, who compared the oiled area with a clean one nearby (Bowyer *et al.* 2003). In the oiled area 12 river otters were found dead immediately after the spill, but this could have been only a fraction of the total. Over the following years 27 radio-otters died, and of those only four would have been found if they had not been equipped with radio-transmitters. Consequently the number of river otters killed by the oil could well have been 80 or more, mostly perishing inside dens or under cover somewhere away from the coast.

River otters that came in contact with the oil, but did not die immediately, were likely to have a more difficult struggle to survive than clean ones. Their blood was shown to have lower haemoglobin levels, their oxygen consumption increased, they dived less often and their prey capture rate decreased (Ben-David *et al.* 2000).

The effects of the oil on river otters wore off quite quickly. In the first year after the accident, their body condition was lower than that of otters in a control area, but 3 years later this difference had disappeared. The presence of oil derivatives in the otters themselves was assessed from the presence of characteristic components, 'biomarkers' such as haptoglobin (which do not necessarily translate into deleterious effects on otters). These had largely disappeared after 3 years, and there were no further measurable differences in otter diet, density or home range size.

Sea otters were probably much harder hit by this same oil disaster. There were a number of different estimates, with the latest taking into account several possible biases: a total of some 750 sea otters died in the Prince William Sound (confidence interval 600–1000; Garshelis 1997). This was considerably lower than some of the earlier assessments (e.g. 2650; Garrott *et al.* 1993). It proved difficult to determine what proportion this constituted of the population in the sound.

Fortunately, until now, no oil spills of similar magnitude have occurred in the coastal habitats of **Eurasian otters**. The only recorded otter deaths from such a source were in Shetland in 1978, when the tanker *Esso Bernicia* spilled oil whilst docking. At least 13 otters died, with symptoms of gastrointestinal bleeding, probably after ingesting the oil during grooming. There were no noticeable and significant effects on populations (Baker *et al.* 1981).

Apart from the North American river otter and the Eurasian otter, the only other otter species for which chemical contaminants have been cause for concern is the **giant otter**. In that case, mercury pollution of rivers is caused by gold mining, and in Amazonian giant otter habitats many fish carry concentrations of methyl mercury that far exceed recommended maximum levels for human consumption (Gutleb *et al.* 1997; Uryu *et al.* 2001). However, at least in the Peruvian Amazon, the scats of the giant otters contained relatively low levels of mercury (Gutleb *et al.* 1997), which is open to various interpretations. It is not known whether populations of the animals are affected. Any potential effect of this pollution may well be hidden, for instance by the devastations caused by poaching (Schenck 1997).

In general, therefore, we do not have good evidence that present-day populations of otters, anywhere, are diminished by any specific contaminant. If individual animals are affected and die early (which is quite possible), compensatory mechanisms in populations may mask the effects of this, as explained above. Nevertheless we should be extremely cautious, because, as the past has shown, we may notice the pernicious influence of pollutants only well after the otters have disappeared.

Recently (since the late 1990s), I and other observers have noted a marked decline in otter numbers in Shetland, and, similarly, numbers are well down in some freshwater areas on the Scottish mainland. In such cases it is possible that pollution is rearing its ugly head again, and this should be checked. It is more likely, however, that declining fish stocks are the culprit (see below).

Food availability, starvation and population limitation

The effects of lack of food as a cause of mortality are complicated because starvation acts together with other environmental stresses. It is rare to find any animal dying just from starvation, as usually some other, more immediate, cause can also be identified and labelled as the culprit. In England and Germany, starvation was deemed the sole cause of death in none (Simpson 1997) and only 1.3% (Hauer *et al.* 2002a) of inspected otter carcasses. However, the role of starvation may show in different ways. For instance, if food shortage were involved as a major factor causing mortality, one might expect more otters to die (from whatever proximate cause) at times of low food availability, and with associated low body condition.

In some high-density areas, mortality of **Eurasian otters**, other than by violent death, is indeed highly seasonal (Kruuk and Conroy 1991; Kruuk et al. 1987). In Shetland we found most 'natural' deaths in spring, with 60% of all natural deaths occurring between March and June. This was a significant seasonality, in sharp contrast with the even distribution of violent deaths throughout the year (Fig. 12.12). Similarly, we reported that, on the Scottish mainland, 42% of non-violent deaths occurred in April alone—more than in any other month. Violent deaths, such as 73 cases of road death, were evenly distributed over the months (12% in April), a significant difference (Kruuk et al. 1993a). The seasonality is paralleled by the availability of important prey species in both sea and fresh water: fish populations are at their lowest in early spring (see Chapter 8; Kruuk et al. 1987, 1988, 1993a). Furthermore, low prey availability occurs at a time of year when the otters' metabolic requirements are high, owing to low water temperatures (see Chapters 9 and 10).

Despite such evidence of seasonal starvation, we found no seasonality in the body condition of otters: animals killed on roads or by other violent means showed the same relative bodyweight throughout the year (Kruuk and Conroy 1991). However, when an animal dies 'naturally' in spring, its bodyweight is significantly lower. The likely explanation for this is that otters have only small fat reserves at any time of year, and that when an animal starts losing condition (i.e. starts using those reserves) it is likely to die quickly.

If starvation were an important mechanism of population limitation in otters, one would expect the animals to take a large proportion of the available prey population. In addition, one would predict that during the time of low prey availability otters would switch to suboptimal types of prey such as carrion, birds and mammals. Both predictions have been confirmed in rivers and streams of mainland Scotland (see Chapters 7 and 8).

The presence (numbers and time spent) of Eurasian otters in Scottish inland rivers and streams (see Chapter 5) was significantly correlated with fish biomass, and hence productivity, although the sample was small (Kruuk et al. 1993a) (Fig. 12.13). Similarly, absence of otters in southern Norway was explained by the decline of fish populations due to acid rain (Heggberget and Myrberget 1979), and in Spain a positive correlation was found between otter density and fish numbers (Ruiz-Olmo et al. 2001). With all the various strands of evidence, it seems likely, therefore, that at least in some areas prey shortage was an important ultimate factor determining otter

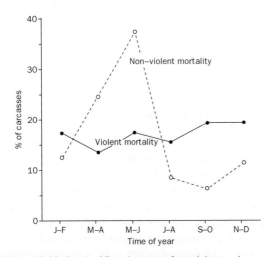

Figure 12.12 Death of Eurasian otters from violent and non-violent causes in Shetland. Mortality from non-violent causes ($n = 49$) was seasonal, peaking in late spring ($\chi 2 = 20.1$, 5 d.f., $P < 0.01$). There was no significant seasonal variation in violent deaths ($n = 53$). (After Kruuk and Conroy 1991.)

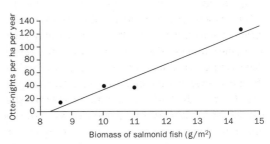

Figure 12.13 Eurasian otters spend more time where fish biomass is highest. Data from radio-tracking otters and Zn-65 in spraints; fish biomass from electro-fishing in a small sample of streams in north-east Scotland ($r = 0.97$, $P < 0.05$). (After Kruuk et al. 1993a.)

numbers. It may affect mortality, as suggested above, and possibly also reproduction (see Chapter 11).

There is evidence that the upper limits of **sea otter** numbers may be determined by the 'carrying capacity' of the habitat, and the interaction makes for a fascinating story. In the Aleutian islands Attu and Amchitka, Estes (1990) demonstrated that in areas where sea otters are newly arriving and prey (especially sea urchins, shellfish and crabs) is plentiful, populations increase through reproduction at their maximum rate, 20–24% per year. Once populations are at 'equilibrium density' (the habitat's 'carrying capacity'), mortality from starvation increases (Kenyon 1969), especially amongst young otters. Numbers do not increase any further. After some years at this density, the sea otters have to spend up to three times as long every day to obtain sufficient food, as prey numbers decrease owing to high predation by otters.

Next, because of the otter-induced reduction in sea urchin numbers, sea urchin grazing of kelp is reduced and, where sea urchin barrens existed previously, kelp forests are able to establish. This sets the scene for a quite different ecosystem, with substantial populations of fish in the dense stands of kelp. Sea otter diet changes from invertebrates to fish. The kelp forests have a significantly higher carrying capacity for sea otters, and another, higher, 'equilibrium density' is established.

This intriguing scenario was confirmed and amplified by Monson et al. (2000), comparing populations between Amchitka and Kodiak islands before the recent decline set in (see below). There were equal birth rates in the two areas, but in the equilibrium population at Amchitka the percentage of pups being weaned was about half of that in the growing population at Kodiak (52% versus 94%). The rest died mostly within the first month of life. Adult mortality rates were similar in the two areas. A similar comparison was made by Monnett and Rotterman (2000), between sea otters in Kodiak and Prince William Sound. Until the early 1990s, early pup survival was probably the primary factor regulating sea otter populations.

Predation on otters

At a first glance, it would appear that predators could hardly make a significant impact on otters. One would expect the odd cub to be taken and an occasional adult casualty, but nothing significant in terms of populations. Tony Sinclair divided ungulates into two categories, one that was regulated by 'top-down' processes such as predation, the other controlled by 'bottom-up' mechanisms such as shortages of resources (Sinclair et al. 2003); otters might be expected to fall firmly into the second category. For at least one of the otter species, this now appears to be widely off the mark, and research on a fascinating series of events has demonstrated again that the killing of one carnivore by another can be part of an important process (Palomares and Caro 1999).

There are several published observations of otters killed by predators. For instance, there is the odd record of a pack of wolves killing a radio-marked **river otter** (Route and Peterson 1991), and the chances are that this occurs more often, especially as also domestic dogs are partial to killing **Eurasian otters**; see, for example, one observation in Shetland (see Chapter 11) and three cases in Hauer et al. (2002a). Toweill and Tabor (1982) mentioned several North American carnivores as potential predators of **river otters**, including wolves. However, **Eurasian otters** found with bite wounds in south-western England (16% of 77 otters found dead) were thought to have been attacked by other otters, not dogs, and 4% had actually died from such injuries (Simpson 1997). Such injury from intraspecific attack was almost certainly an underestimate (many victims would have died undetected), yet it appeared to be the second most important source of mortality, after road-kill (Simpson and Coxon 2000).

Cubs may also be snatched by raptors, for example those of Eurasian otters (Hauer et al. 2002a), and in Alaska young **sea otters** are taken quite frequently by bald eagles *Haliaeetus leucocephalus* (Sherrod et al. 1975). Anthony et al. (1999), who found 48 sea otter pups killed by bald eagles on Amchitka, noted that the otter population was probably at equilibrium density with its food supply, hence the effects of bald eagle predation were likely to be minimal.

Sea otters hauling out on beaches in Kamchatka, in reportedly weak condition after winter storms, are a ready prey for brown bears *Ursus horribilis* (A. Zorin, personal communication, in Riedman and Estes

1990). Newly weaned sea otter pups that hauled out along Prince William Sound were killed and eaten by coyotes *Canis latrans* (Riedman and Estes 1990). In all such cases, whatever killed the otters is perhaps less important than the cause of the weak condition, or whatever made the sea otters spend time on the dangerous shores. However, this is not the case for the recent massive surge in predation on sea otters, which is discussed in more detail below.

Otters living in tropical areas encounter a variety of large predators. **Giant otters** are attacked occasionally by black caiman (*Melanosuchus niger*), which are common in their habitat (personal observation; Davenport and Hajek 2005). **Neotropical otters** living in the same waters are subject to the same hazards. Sometimes, however, the boot is on the other foot, and a group of giant otters will attack a caiman, causing a great deal of damage (Davenport and Hajek 2005). Nicole Duplaix (2004) reported a giant otter cub taken by black caiman in Guyana, and she mentioned anaconda *Eunectes murinus*, jaguar *Panthera onca* and puma *Puma concolor* as potential enemies, but stated that 'there are no predators which prey with any regularity on *Pteronura*' (Duplaix 1980, p 530). The much larger Nile crocodile *Crocodilus niloticus* is likely to constitute a danger to **spotted-necked otters** in Africa (Kruuk and Goudswaard 1990; Procter 1963), but in all of these cases there is no hard evidence.

It is likely that intraspecific predation (i.e. cannibalism) occurs in all otter species. It is quite common amongst carnivores (Packer and Pusey 1984), and for otters direct evidence comes from Simpson and Coxon (2000), who found a small **Eurasian otter** cub in the stomach of a male, and from Mourão and Cavalho (2001), who saw an adult male **giant otter** kill and eat a young cub in the Pantanal of Brazil. However, Eurasian otters, and probably the others, have various behavioural patterns (such as site selection for dens, and family parties avoiding adult males) that help prevent cannibalism of cubs by marauding males (see Chapter 6), and actual death through cannibalism is probably not significant in population terms. Intraspecific aggression may be more important.

Predation on **sea otters** (Fig. 12.14), out in the ocean, has recently taken on an altogether different dimension. The enemies are sharks and killer whales. Sharks, especially the great white *Carcharodon carcharias*, have been known to take sea otters in California since otter carcasses were found with shark bites in 1959, and white shark predation probably always occurred whenever the two used the same waters. 'Attacks on sea otters by white sharks may represent a significant source of natural mortality in the California population' (Riedman and Estes 1990, p 88).

Sea otter carcasses with characteristic shark wounds (caused by serrated teeth, and occasionally

Figure 12.14 Sea otter carrying anvil, California. The animals are highly vulnerable to predation. © Sharon Blaziek.

with pieces of shark tooth broken off) have been found in considerable numbers along the Californian shores. Ames and Morejohn (1980) found that 9–15% of all recorded mortality between 1968 and 1979 was caused by white shark, and along coasts of the Monterey peninsula 36% of all dead sea otters had been killed by sharks. More recently, Estes *et al.* (2003a) noted that, in 1999, 1.6% of the Californian sea otter population had been killed by white sharks, and Kreuder *et al.* (2003) found that 13% of all mortality was due to that predator. One should keep in mind that these figures do not include otters that were swallowed by sharks without further trace: these are only the otters that were killed or fatally wounded, and whose remains were found. In addition, shark-attacked otters may be more likely to sink (and not wash up) than those dying from other causes.

There is evidence that white sharks do not target sea otters as a primary prey, but that their food is mostly elephant and harbor seal; sea otters may be taken merely 'by mistake' (Riedman and Estes 1990). Sharks habitually attack a prey and lacerate it, leave it to die and then return to consume it. The injuries inflicted on sea otters are relatively mild, compared with those on seals, so it is possible that sharks do not follow through an attack on a sea otter, and the victim dies later.

A significant finding is that a high proportion of sea otters as shark victims, 57%, have encephalitis caused by *Toxoplasma gondii*. 'Otters with moderate to severe *T. gondii* encephalitis were 3.7 times more likely to be attacked by sharks than otters without this condition', probably because of aberrant behaviour (Kreuder *et al.* 2003, p 501). This suggests that this type of predation on sea otters may repreent only a proximate cause of death, and the real culprit in this substantial mortality amongst sea otters may be *Toxoplasma*.

The most worrying and dramatic recent impact on sea otters—in fact, the impact on an entire coastal ecosystem—is a sudden surge in predation by killer whales *Orcinus orca*, which must have started in the late 1980s. The result is one of the most widespread and precipitous population declines for a mammalian carnivore in recorded history (Doroff *et al.* 2003).

Predation on otters by killer whales was almost unknown until the 1990s. Riedman and Estes (1990) mentioned a Russian observation of *O. orca* capturing a sea otter, but they classified this as 'probably very rare'. Kenyon (1969) saw killer whales within a few metres of resting or feeding sea otters, when the otters sometimes became quite still in response, but the killer whales never attacked. That was until 1992.

In July 1992, three killer whales had been feeding on salmon in the Prince William Sound. Two of them swam into an inlet, where about 40 sea otters were resting, mostly on rocks exposed by the tide. Time and again the whales rushed at the otters, creating a surge of water that covered the rocks. One of the whales noticed an otter that had been resting on the surface nearby, and which had become very alert, 'persicoping' then diving. The killer whale followed at speed, and a few seconds later the sea otter floated to the surface, dead. The whale bit it again, threw it about, pushed it some 30 metres, then dived with it, and probably ate it underwater. A quarter of an hour later the three whales departed (Hatfield *et al.* 1998, p ???).

This incident was documented together with eight subsequent attacks of killer whales on sea otters. In the same years, the Alaskan population of sea otters, which was still in the process of recovering from the historic overhunting, commenced an abrupt and disastrous decline in numbers. Jim Estes and his colleagues first recorded this around some of the Aleutian islands, where numbers went down at a rate of about 25% per year (Estes *et al.* 1998). They noticed that it was prime adults that were disappearing, and that recruitment seemed normal. There was no evidence of disease, pollution or other causes of mortality, and the staple food of the otters, sea urchins, was increasing. They also saw that otter numbers in one large cove inaccessible to killer whales remained stable, whereas a neighbouring, open-access, lagoon lost 76% of its sea otters over 5 years.

Researchers calculated that, for just one group of Aleutian islands, the rate of loss was 6788 otters per year. With the number of observers in the field and the time they spent there, they estimated that they should actually have witnessed killer whale predation on sea otters just over five times. The number of attacks they saw was six: it fitted. They also calculated, from the food requirements of killer whales, that only 3.7 of these animals would need to change their feeding habits for them to consume some 40,000 sea otters, precipitating the gigantic population decline.

'A pod of five individuals could account for the decline in sea otters' (Williams *et al.* 2004a, p 3373).

The evidence for killer whale predation as the cause of the decline is compelling. Since the mid-1980s and until 2000, sea otter numbers throughout the Aleutian archipelago decreased at a rate of 17.5% per year, with an overall decline of 75% since 1965 (Doroff *et al.* 2003). This has continued into the present millennium, and in 2005 Estes *et al.* (2005) calculated that the Aleutian islands had seen a decline of 95.5% since 1965, with only just over 3000 sea otters left. Further, in Prince William Sound killer whales were seen to take sea otters, and their depredations are confounding the recovery of this species from the *Exxon Valdez* oil spill (Garshelis and Johnson 1999). The process is still going on, and the outlook for the Alaskan (and Russian) sea otters is exceedingly bleak.

Two other aspects of the disaster are important: the further consequences of sea otter decline, and the reasons for this sudden, spectacular, development. One immediate consequence is the change in predation pressure on sea urchins: with the disappearance of sea otters, the 'sea urchin barrens' have become a thing of the past. Jon Watt *et al.* (2000) found that sea otter diet in 1992–1994 showed a large increase in sea urchins compared with the 1970s, and a decrease in kelp forest fish species, changes that were immediately attributable to the decreasing predation pressure from sea otters.

Jim Estes and his colleagues compared the sea bed around the Aleutian island of Adak in 1997 with how it was in 1987. They found that sea urchin size and density changed markedly, resulting in an overall 8-fold increase in sea urchin biomass, whereas kelp density declined by a factor of 12. In 1991, kelp tissue loss to herbivores (mostly sea urchins) was 1.1% per day; in 1997 it was 47.5% per day (Estes *et al.* 1998). The changes were paralleled at the other Aleutian islands: it was a regional phenomenon. In addition, the ecological effects of the disappearance of the sea otter perfectly matched the earlier observed changes (in reverse), when sea otters had returned to areas after many decades of absence.

As always in such *post hoc* analyses, it will not be possible to prove what caused the sea otter disaster, and it is still early for a full appreciation of all the details. However, there is now a great deal of independent evidence pointing to a highly likely explanation, with humans as the culprit. The evidence shows that what happened, and is happening, to the sea otter is only a small aspect of a much larger catastrophe, referred to as the 'sequential megafauna collapse' in the northern Pacific (Springer *et al.* 2003). I will discuss this further in the following chapters; here it will suffice that numbers of fin and sperm whales were reduced by whaling by an order of magnitude, in the 1960s and 1970s. They were a main prey of killer whales, who then changed to smaller fare, especially seals and sea-lions. Killer whale predation was probably the main cause of the spectacular decline of these pinnipeds in the 1970s and 1980s—and then the *Orcas* started with sea otters. This provides examples of top-down control of animal population on a scale previously almost unheard of.

Some conclusions

For otter populations on a large, regional scale, even the most basic parameter (their numbers) is known for only a few areas, in only a couple of species, although we are beginning to gain some insight into changes, and into the causes of what is happening. We now know that genetic diversity has suffered in the sea otter, after the very severe bottlenecks that the populations went through. But other, at times more than decimated, populations (e.g. in Europe) appear to have recovered without a great deal of loss of genetic diversity, despite the fact that gene flow is quite restricted to relatively small areas.

Otters of all species, as far as we know, can be relatively long lived (15–25 years), but in practice their life expectancy is short—only a few years (probably longest for the sea otter). The high mortality amongst these animals (even without taking into account regional, population-wide, disasters) is associated with food availability, with the energy-expensive foraging method that otters employ, and with fish or other prey in short supply. Competition with other fish-eating species, even with other otters, is not very substantial, but fish populations themselves can be constrained, and some important prey species have crashed spectacularly. Moreover, mortality from other sources may be very high, associated with the animals' aquatic habitat, especially chemical pollution, but also diseases such as toxoplasmosis. For sea otters, there is also predation, released probably through human mismanagement of the habitat, now causing near elimination of populations.

Keeping this in mind, one realizes that recruitment to populations is agonizingly slow, as otters (with few exceptions) have small numbers of cubs. The female otter, of whatever species, faces the problem of having to provide for its cub or cubs, from resources that are unusually energy expensive to exploit. Connected with this is the young ones' long dependence on the mother, in most species for about a year and sometimes longer, probably determined by the process of learning to exploit a difficult type of prey.

Despite our lack of knowledge for most otter species, one cannot avoid the recognition that the existence of these animals, their population balance, is decidedly precarious. Otters live on the edge of a precipice.

CHAPTER 13

Syntheses: challenges to otter survival

What is special about otters?

In this chapter I will attempt to tie together some of the strings from previous chapters of this book, summarizing results, spelling out implications, and speculating on their biological significance. The emphasis will be on habitat, foraging and population dynamics, referring to the otters' spatial organization and social behaviour. Inevitably, some points have been made earlier, but perhaps one should not object to that. All aspects of otter biology are so closely interwoven that it is unavoidable that a survey of present knowledge will lead one repeatedly to some of the same road intersections.

In these generalizations there are many aspects of otter biology that illustrate rather general principles in animal ecology or behaviour, such as the energetics of foraging or the function of scent marking. But otter specifics will necessarily come into it. The question will repeatedly arise of how do otters differ from one another, and from other species. These animals obviously have some unusual problems; their numbers are often threatened in unusual ways, hence some answers to questions posed need to be based on comparisons.

Undoubtedly, the main distinction between otters and other similar-sized carnivores is the fact that otters spend much time in water (Fig. 13.1). This relates to their shape, metabolism, locomotion, food and other fundamental aspects of their biology, such as foraging behaviour, social organization, survival and mortality.

Because of this distinction, otters are highly unusual carnivores. They have evolved in habitats that are now threatened more than most. They have a lifestyle that is more risky than that of most other mammals. Some of their important prey species are vanishing, and they are unusually exposed to human-made changes in the environment such as climate change, pollution and over-fishing. Several otter species are subject to fierce exploitation. What

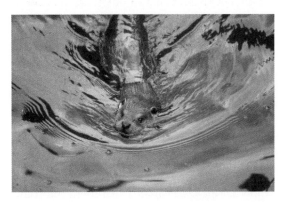

Figure 13.1 Eurasian otter—their ecology revolves around aquatic needs.

are the aspects that are especially significant and different in their natural history?

Substantial ecological and behavioural knowledge exists for only a few species, mostly for the Eurasian otter, the sea otter and the North American river otter. However, a fair amount of hard data is also available for most others, as well as many casual observations. One of the important conclusions that can be drawn is the striking similarity between all otter species, not just in appearance, but also in their natural history, ecology and behaviour. When researching a river otter in Prince William Sound, a neotropical otter in the Pantanal, a spotted-necked otter in Lake Victoria or a Eurasian otter in Shetland, I am hard pushed to note differences between them. Many of the observations and conclusions made on one of these species will therefore probably also be relevant for the others. At the very least, they will provide a template for our understanding, with working hypotheses. Some species, such as sea and giant otters, are from a somewhat different mould, however, and one has to be more careful with comparisons.

A main aspect of the relations between otters and their environment, often critically affecting populations and behaviour, is feeding ecology, as for most other carnivores (Kruuk 2002). Much of the research that I have described here, for all otter species, is relevant to this in one way or another. However, what is becoming more evident in our knowledge of these animals is the significance of predation, not *by* otters, but *on* otters, and, at least in some species and habitats, this has reverberations throughout their ecology.

Evidence from studies of DNA on relatedness and evolution of otters suggests strongly that not only do species all over the world look very similar (except the sea otter), they also are closely related (Koepfli and Wayne 1998; see Chapter 3). There is a concentration of otter species diversity in south-east Asia (starting with *Lutra*), from where they appear to have moved along the waters' edges into Africa (*Aonyx* and *Lutra*), and across the Bering Strait into the New World (*Lontra* and *Enhydra*), then south into Latin America (*Lontra* and *Pteronura*). They did not make it across large stretches of open water, for instance into eastern parts of Indonesia and towards Australia, or to Madagascar, because of their restriction to the waters' edges.

In the process of this dispersion, the animals have changed—albeit not that much. Some have affected distinct morphological specializations, notably the sea otters (with unusually dense fur, and a short tail adapting to a cold and exclusively all-aquatic existence), the giant otter (a large flat tail enabling fast movement in open water), and the long-fingered clawless and small-clawed species, adapted to foraging for crabs. This list could be much longer, as, naturally, all species show differences, yet the overall similarities between the otters are much more striking. Almost any other family or subfamily of Carnivora shows more diversity of shape and size than do the otters. They are the remarkable result of a trend starting in terrestrial mammals, to utilize the vast resources of the aquatic environment.

Habitats

Otters have been depicted as symbols of undamaged nature, of clean water and pure vegetation. However, one learns from comparing the uses of different habitats by otters, by animals of the same species in different places, or by different species, that a more fitting portrait is that of animals for which almost any bank, any shore, will do. Chapter 4 demonstrated that otters occur along large rivers, tiny streams, lakes, sea coasts, some with forests along their shores, others without a tree or shrub in sight; they occur in swamps and in moorland, amongst agriculture, in rice paddies, even along dense housing or industry. Some mild eutrophication of waters by agriculture is probably beneficial, rather than detrimental.

Of course, this is somewhat of an oversimplification. Open, sandy beaches and steep cliffs do not attract otters, and there are other types of coast they avoid. But, generally, detailed studies of ranging behaviour show that the animals are very catholic in their taste for where they live, swim, hunt and rear their offspring along the waters' edges—that is, as long as there is sufficient prey.

However, the large range of habitats used by the animals is somewhat deceptive. Within the range of different landscapes, the actual living space for otters is quite confined. To some extent all the above attributes are only the non-essential tapestry behind the stage where everything happens, and, more accurately, most otters' habitat (except that of sea otters) can be characterized as a narrow strip on either side of the interface between water and land, where food is acquired in the cold, watery inhospitality of one side, and where recovery from this exposure and all other activities take place on the other. The animals' living space differs, therefore, from that of other carnivores in that it necessarily includes both water and land, and is to some extent one-dimensional, linear. We talk about otter ranges in terms of kilometres rather than hectares, even though we know that the width of the strip also counts, and that fishing at depth extracts its energetic toll from otters.

Sea otters are different, in that they have moved almost the entire range of activities offshore. In addition, they are unique amongst otters as they are able to modify their underwater habitat (discussed in Chapter 8 and below).

One may consider the otters' body shape as a compromise between adaptations to the requirements of existence on land and in water, with all the resulting limitations. With their large, webbed feet and long, heavy tails the animals are fairly clumsy on land (compared with, say, foxes, cats or badgers), and therefore vulnerable to terrestrial predators. In water they can move fast, the flat-tailed ones even faster than the others (though less able to manoeuvre). However, otters also cool rapidly in water, probably because really efficient thermo-insulation would interfere even more with movement on land (as in seals). To make up for this cooling in water, the otters' energetic requirements are large, and their food intake has to be high. This renders otters more vulnerable to prey fluctuations, as well as to the cumulative effects of pollutants.

Although the elongated, streamlined shape of otters makes them eminently suited to a semi-aquatic existence in their particular habitats, strictly speaking we should not consider this an adaptation. The overall body shape is a common mustelid feature that otters share with many other members of the family, such as martens, stoats and weasels, species that do not live anywhere near water. Rather than an adaptation, it is likely that the elongated mustelid shape enabled the *Lutrinae* to evolve in their aquatic direction, which would have been an evolutionary impossibility for a canid or felid. Otters have taken body shape several steps further into aquatic use.

One social consequence of the linear habitat of all except the sea otter is that animals sharing an area are bound to meet more often than they would in a two-dimensional space. This could affect competition for resources, as I will discuss below. It also affects otters that have a direct interest in keeping away from others of the same species. An example is the female with small cubs: she has to avoid infanticidal males, a common phenomenon in many Carnivora, including otters (see Chapters 5 and 6). Her strategy, in the case of Eurasian and river otters, often appears to be to establish a 'natal holt' quite far from the waterline, far from the usual otter habitat. This renders her more vulnerable to predation, and to more recent threats such as traffic (many roads follow river courses, and they have to be crossed). Only when the cubs are large enough will families start to use shores and banks.

Another important consequence of habitat linearity is that animals have to move around a great deal (imagine one's own house and garden laid out in a narrow strip, and the effects of that on distances walked when going about daily business). Much of the otters' travel takes place in water, easier because of the animals' morphological adaptations to catching fish, but expensive because of heat loss in this highly conductive medium. It is not surprising, therefore, that any feature of the habitat that enables animals to escape from the consequences of this linearity is favoured by otters, such as shallow coasts (with a wide shelf to catch prey), shortcuts across peninsulas and meanders in rivers, and islands. These are the places where one finds many signs of otters, and where the animals spend a lot of time. Any shortcut through a convolution of coasts is advantageous, enabling a long interface with water to be accessed from places of safety or of low energy expenditure on land. This applies to all except sea otters.

When otters (again excepting *Enhydra*) live along the sea, they are confined to foraging in a narrow strip of water (because of the limitations on depths at which they can fish). However, most of their prey is not so restricted. This does not arise in most freshwater areas, but in the sea, especially in winter and spring of northern latitudes, many potential prey of otters remove themselves to waters too deep for otters to fish efficiently. Migratory prey for carnivores in land-based ecosystems, for instance the ungulates which move through territories of lions and hyenas, can to some extent be followed by the predators in a 'commuting' system, but this option is not open to otters. In consequence, there is a seasonal low in prey abundance, especially along sea coasts, which is associated with high otter mortality.

The linearity of habitat also has consequences for resource exploitation. When a resource patch, such as a river tributary with fish, is rich enough to be used by several otters, complications arise over range-sharing. In the simplest form this can be mere overlap of neighbouring home ranges, but when there are many such rich 'patches' range-sharing becomes difficult. The linearity of the habitat makes confrontation more inevitable than for animals living in easy two-dimensional ranges.

To avoid clashes over food, the otters' options, then, are random dispersion with tolerance between individuals, or some kind of group territorial system. The latter option appears to be common. Many populations consist of quite solitary individuals in single-sex territories. But several species of otter employ territorial arrangements in which several animals share a range, each with its own favourite hunting haunts that it uses as a solitary individual (river, Eurasian, neotropical, marine, Cape clawless otter). Others have gone a step further: they live in tight groups of often related individuals, sometimes cooperating during foraging (some river otters, giant, spotted-necked, smooth, small-clawed otters). We should seek to explain such variations in organization in terms of adaption to environmental characteristics, and an important role here may be played by the Resource Dispersion Hypothesis (RDH) (Carr and Macdonald 1986; Kruuk and Macdonald 1985; Macdonald 1983).

In simple terms, the RDH explains the existence of group territories as an adaptation to resources, especially food, occurring in relatively small, temporal concentrations or 'patches'. One individual needs several such patches within its range in order to have prey always available, but because patches are rich (though temporal) several other individuals will also be able to feed from them without substantial competition. If there are large risks involved in dispersing from a natal range, it may be advantageous for an animal to stay in the home group, and reproduce there or await future opportunities of dispersal. Group size is determined by the number of prey per patch, range size by the distance between patches.

This 'RDH explanation' holds fairly well for group-living animals such as badgers (Kruuk 1978) or red foxes (Macdonald 1983), but does it predict differences in group organization for otters? We suggested earlier (Kruuk and Moorhouse 1991) that the answer for Eurasian otters is a qualified 'yes', as a vague generality: we found group territories along coasts of clumped food resources. However, it would not have been possible to predict the details of otter organization from our knowledge of food dispersion, for instance the existence of separate core areas of individual females along coasts, and the system of males exploiting main rivers and females the tributaries and lakes. Nor would we be able to predict numbers of otters per group range, or even the size of ranges, merely from our knowledge of fish dispersion and productivity. For an appreciation of the extent to which the spatial and social system is adapted to environmental requirements, one needs to know many more details, such as foraging strategies and resources other than food (e.g. fresh water along the sea coasts).

The linear nature of the otters' habitat has other important effects on spatial organization. For instance, some carnivores (e.g. Eurasian badgers, red foxes) have the option of 'central place foraging'. This means using one site or den as a base and exploiting the home range from there, going off in different directions on different occasions, foraging in the various patches at suitable times, but always returning again to the central site. This, probably, is not efficient for otters, except on the relatively rare occasions when they live in a non-linear habitat such as a small lake (Fig. 13.2). They have to move about, basing themselves in different parts of the range at different times, compromising between the need to 'know' the stretch they are fishing and the need to exploit different areas far apart. The core area system described in Chapter 5 is one solution to this need for compromise.

Apart from problems of the consequences of habitat linearity, otters of all species also have to cope with consequences of spending much time in a wet medium. This is energetically costly, because of heat loss, and the otters' land base has to provide facilities for the animals to recover after swimming—safe holts, rocky shelters or couches (e.g. in reed beds), or, when living in a marine habitat, freshwater washing sites along the coast or inside holts, to cleanse the fur. Only sea and marine otters have escaped this constraint. Along many inland waters, dry underground holts will be an impossibility, although the otters' couch building behaviour makes up for this. In fact, when islands or reedbeds are available, many otters appear to prefer to rest above ground even in northern winters, perhaps because their pelts dry better.

There are other, specific, risks attached to an aquatic habitat: it may freeze over, making access impossible or very difficult, or it may dry up. Neither of these risks appears to be a great deterrent to otters; the animals have a range of behavioural patterns to cope with ice, including the use of special entry-holes, or exploiting open patches or running water. Species such as the Cape clawless otter move elsewhere over large distances when rivers dry up, returning with the rains, when the crabs emerge again. Nevertheless, food availability is greatly reduced during such conditions, and it is likely that this

Figure 13.2 Otter habitat—a lake with reed beds (Dinnet Lochs, north-east Scotland).

constitutes a bottleneck for populations, although it has not been demonstrated. Social systems must be seriously disrupted during these events, but the consequences of this are as yet unknown.

When comparing habitats of different species of otter, one notices that they may be largely similar, but that there is variation between them. Variation is noticeable especially when different species occur in the same area, such as sea otters further out at sea than river otters, and smooth otters in a Thai river using mostly the main stem, Eurasian otters mostly the narrower upstream parts and tributaries, and the small-clawed otter often in damp areas along the banks. South American giant otters prefer deep ox-bows, where neotropical otters usually forage along the edges of the main rivers, and along Chilean coasts marine otters use the wild, rocky areas where southern river otters use shallow, sheltered sites. However, there is large overlap between them, and the animals are not very choosy.

One species, however, stands out strikingly amongst the otters in its relationship with the environment. The sea otter is unique in its huge, modifying effect on the vegetation. By affecting populations of herbivorous sea urchins, the presence or absence of sea otters dictates whether there are sea urchin 'barrens' or miles and miles of dense kelp forests. Kelp affects the sea otters' prey, and provides the animals' resting sites. It is a demonstration of the possible role of carnivores in ecosystems, even though such a large-scale effect is rare, and decidedly unique amongst otters. Yet, also for other otters, the ecosystem influence may be stronger than we credit at present. We have seen that it would be quite feasible for Eurasian otters to affect fish prey populations, which in turn may extend an influence to invertebrates and vegetation. We are becoming more aware of top-down effects, and much more research is called for.

Foraging

The first otters were probably fish-eating, *Lutra*-like animals. In later stages of evolution they branched out with crab-catching adaptations (*Aonyx*), and began to exploit echinoderms, molluscs and other invertebrates (*Enhydra*-like fossils). At present, all otters eat fish, as well as crustaceans, amphibians, molluscs, sea urchins and perhaps some other aquatic fauna. Their prey is relatively small; in the case of fish it is usually bottom-living and/or taken during its inactive period. In some otters there is variation in prey between the sexes, and individual otters of a given species may differ in their prey preference. Giant otters females feed their young with smaller fish than they take themselves; Eurasian otters with larger ones. To a large extent, otters of any species take fish and other suitable prey proportional to their availability.

The relationships between otter and prey populations show characteristics that appear to be general to many carnivores, and are equally applicable to predator–prey interactions in a more terrestrial environment. Where otters are unusual is in the energetic cost of their foraging trips, in the expensive requirement of keeping warm whilst fishing, even in tropical waters. To meet this, they need to eat the equivalent of 15–20% of their bodyweight every day, sea otters even 20–30%, and giant otters (in their warm waters) somewhat less (see Chapter 7).

All is well if fish occur in sufficient density for otters to be able to get their food quickly and eat. However, if they have to search for a long period, the expense of foraging increases so rapidly that otter life becomes difficult to sustain. This applies even to sea otters, which are permanently aquatic, but whose deep foraging dives take much energy, over and above what the animals need just to rest on the cold surface of the ocean.

In several areas where Eurasian otters were studied, they appeared to be living uncomfortably close to their energetic limits, and a relatively small reduction in prey availability makes, or would make, such places unsuitable for otters. This curious situation is caused by the negative relationship between foraging profitability and the length of time of cold-water foraging needed each day to stay alive. This is not linear, but approaches a negative exponential (see Fig. 9.12).

To reduce the high energy bill that otters have to face during foraging, they depend on thermo-insulation, especially in cold climates (Fig. 13.3), which is based almost entirely on the excellent qualities of the fur. This, in itself, gets otters into trouble with humans (see Chapter 14), but is not sufficient completely to prevent a considerable heat-drain in cold water. A more efficient option for thermo-insulation, such as blubber, is not available to otters, because of other demands on their mobility on land and in water (see Chapter 10).

Whilst often living close to the point where the curve of daily cost of foraging goes up sharply, otters also have to contend with the fact that their staple food is harder to catch than that of many other carnivores. Compare the skill and energy needed by an

Figure 13.3 High demands on thermo-regulation—North American river otters on ice. © Angie Berchielli.

otter to catch, for example, a trout with what a terrestrial mammal such as a badger needs to get at its earthworms, or a cat to ambush its voles. It is hardly surprising, then, that it takes a young otter more than a year to acquire these skills and efficiency, with initially a great deal of what appears to be 'teaching' by its mother (see Chapter 6). It is highly likely that this long period of dependence affects the breeding interval and lifetime reproductive output (see Chapter 11).

From the many studies on diet it is clear that otters do not take just any fish (Fig. 13.4), frog, crab or mollusc, but are selective. They do not, or only rarely, chase fast-swimming fish, and most otters leave many crabs alone as unrewarding; they concentrate on slow, bottom-living or resting fish species, preferably ones with a high lipid content such as eels or salmonids. Sea otters take those molluscs that are easy to extract, and otters that are crab specialists tend to take their prey in very shallow water, enabling fast processing.

Other efficient foraging strategies include hunting at the appropriate time of day (taking fish when they are inactive), in the sea at the right stage of the tide, hunting in shallows, avoidance of back-tracking, and the repeated use of well known, specific feeding patches, harvesting and reharvesting on an almost daily basis with the fish repopulating the empty niches before the otter returns (see Chapter 9). When watching otters swim, dive and catch their fish, they leave us with little doubt that they have a detailed knowledge of their beat, just like an experienced fisherperson who knows the best pools and individual rocks from which to cast his or her fly.

Detailed knowledge of fishing sites and the kinds of prey likely to be encountered should have important benefits to otters. It appears that the expectation of success is expressed in the otters' success rates per dive: otters tolerate lower success rates per dive where prey is large or the water shallow. Only minimal changes in success rate occur between seasons, despite changes in fish availability. In other words, when otters dive they expect a given return for effort, and their knowledge of feeding areas enables them to maximize this (see Chapter 9).

It is important, therefore, for animals to be able to use the same resource-rich patches again and again, without interference from others. This need has implications for the social and spatial structure of the otter population, affecting even behaviour patterns such as scent marking or sprainting, which all otters (except *Enhydra*) do unusually frequently, compared with other carnivores.

I have argued that, on the one hand, a Eurasian otter in Shetland needs to be able to use different stretches of a sea coast, because various important prey species occur in given sites and seasons. Some fish need sheltered areas, some need exposed sites, and different vegetations of algae play a role. But, on the other hand, once an otter has a range encompassing such a variety of sites, it can allow others to share, without competition for food, provided they can avoid exploiting exactly the same patches (see Chapter 5). This, it appears, is achieved by organization into core areas, and a signalling system with spraints that informs other otters that someone is already exploiting a particular site (see Chapter 6). This enables newcomers to keep out, to the benefit of the animal who was there first as well as to the newcomer itself (the latter because it does not have to waste energy by exploring an already partly emptied site).

One may expect, *a priori*, that food is likely to have its main impact on populations of otters, and on the behaviour of the animals, during periods of shortage or low abundance—hence the increase in sprainting during such periods (see Chapter 6). A first ecological

Figure 13.4 Slow fish preferred—neotropical otter eating a catfish, Pantanal, Brazil.

effect to look for is the choice of prey; during shortages, low-preference food (prey such as rabbits, birds or fish carrion in the case of Eurasian otters) is taken more often. This may be the case also for other otter species. The role of amphibians in the diet, especially frogs, is somewhat ambiguous; they are taken mostly, and in large numbers, during times when fish stocks are low, but that is also the time when frogs themselves are much more available to otters. At least for Eurasian otters, frogs appear to be an excellent prey, of the right size and in the right kind of places (and for some populations frogs are a staple).

Studies on Eurasian otter feeding ecology in fresh water show that the quantity of fish consumed by an otter population may be large in relation to the fish populations themselves. Otters in some areas may eat considerably more per year than the actual total 'standing crop' of fish. Annual fish productivity is high, often larger than the standing crop itself (see Chapter 8), but even if we express otter predation as a proportion of fish productivity, rather than in relation to biomass, it is still very considerable—in our Scottish study area, more than half. This is likely to be true in many areas with good populations of otters, dealing with different species of fish. In the case of sea otter predation, it was clearly demonstrated that this seriously depletes prey populations of sea urchins and shellfish (see Chapter 8).

Under such conditions it is likely that prey numbers will determine numbers of otters, and, indeed, we found that there are more Eurasian otters where there is more fish, in a significant positive correlation. In this complicated scenario, otters may affect fish numbers, but at the same time fish numbers are likely to be determined partly by food conditions in the streams and lakes. Given the high rate of predation, under certain conditions, otters could be competitors for food (fish) with other predators, such as humans. This has been demonstrated only for sea otters and shellfish, but the ingredients are there for such competition between other species with fisheries elsewhere. There is, for instance, suggestive evidence that Eurasian otters possibly affect populations of Atlantic salmon and brown trout (see Chapter 8). They are certainly perceived to damage salmonid and cyprinid populations (see Chapter 14), especially also in fish farms, which is an important factor in conservation management. Further, spotted-necked otters and other species take large numbers of fish from nets in Africa and elsewhere (see Chapter 11).

One interesting complication, in the relationship between populations of otters and fish, is the fact that, at least in species such as the North American river otter and the Eurasian one, male and female otters do not use the same habitat. They live in different waters, and eat different sizes of fish. Males of Eurasian otters are shown to live along more exposed coasts and in larger rivers, feeding on larger prey, but there is, of course, much overlap between the sexes. In winter in mainland Scotland, it is the male Eurasian otters that often take large salmon (themselves also mostly male fish). Strikingly, in the otter 'female areas' in the sheltered bays in Shetland, the main food species, eelpout, showed a large increase in numbers in summer, and we demonstrated that this was closely associated with otters' reproductive success. It is possible, therefore, that fluctuations in numbers of given prey species affect male and female otters differently, but much more data are needed before we begin to understand the implications of this.

Social life

From the dietary variation that I have described, occurring between and within species, it appears that the food of various otter species does not differ all that much. It varies with size of otters, with availability, with region, but, compared with most other carnivores, as a group otters are highly specialized feeders. It is all the more interesting, therefore, that their social organizations vary so greatly, especially as, amongst carnivores, social organization is often strongly affected by feeding ecology (Macdonald 1983).

The standard picture in popular literature of a 'pair' of otters is largely a figment of anthropomorphic imagination. Some otter species are as solitary as can be (at least in some populations), such as the Eurasian, the neotropical and the hairy-nosed. Others live in family groups or even extended families (giant, smooth, small-clawed, spotted-necked); some form temporary attachments of small groups (Cape clawless); others are solitary with some or

Figure 13.5 Sea otter. Predation affects these animals more than was thought hitherto. © Jane Vargas.

most of their males living in groups (river, Cape clawless); and the sea otter often occurs in large, sexually segregated, aggregations. The spectrum of social systems is very rich indeed.

Moreover, in some otters, intraspecific variation is also large. Eurasian otters may usually be solitary, but in high-density areas a system of group territories applies, with females in exclusive core areas, and males quite separate. The large family groups of spotted-necked otters are found in East African lakes, but not in South African streams, where the species is much more solitary. Such systems appear to be variations on a theme of family group, of female with offspring, in a home range that is independent of that of males. Male ranges overlap with females, often with several. In giant otters (and perhaps in other gregarious species), a male is permanently attached to a family group. Group sizes could be determined by the age at which cubs leave the family.

As yet we are far from clear which environmental variables affect the social status of these otters. They must be quite important; for instance, it can hardly be a coincidence that the sea otters' existence in large aggregations is similar to that of other sea mammals such as seals and sea-lions. More specifically, the variation in Eurasian otter society has been linked to resource exploitation, and along Shetland's sea coasts the existence of rich resource patches is linked with group territories, as are the feeding patches of Cape clawless otters. However, a quite different explanation may be needed to explain why North Amercian river otter males live in groups, especially in the sea, and females do not (and neither do Eurasian or neotropical males).

One aspect that group-living otters of different species have in common is that they face a greater threat from predators, and it was recognized as long ago as the middle of last century (Kruuk 1964; Tinbergen 1951) that group living can contribute to an efficient anti-predator defence. Spotted-necked otters crowd together when crossing large stretches of open, crocodile-infested water, packs of male river otters in the sea are exposed to killer whales and sharks as they swim further offshore than females, as do sea otters, whilst giant otter family groups are often faced with the danger of piranhas. Groups of small-clawed otters spend much time in dense, wet vegetation and debris along rivers and in rice paddies, exposing themselves to terrestrial predators, and often large packs of smooth otters in their open rivers and lakes share their waters with dangerous crocodiles. The importance of predators as an environmental force shaping social organization may well have been underestimated, although the recent effects of predation on sea otters (Fig. 13.5) could well rectify this.

Otters have many behavioural mechanisms to maintain and service social organization, and the way in which it is adapted to environmental requirements, although a great deal more research is required for a better understanding. Olfactory communication is an important aspect of this, and there probably are no other carnivores that produce faecal scent-marks (spraints) as often as do some of the otters. One aspect of sprainting has been discussed above, as preventing intraspecific and interspecific competition during foraging. But much more is being communicated by the otters, in spraints, urine, by scratching the ground, and by rubbing cheeks, belly or other body parts on substrate: we know virtually nothing of the content of such messages.

During direct confrontation, smell is also demonstrably important—at least as important as visual signals. But, in general, one can recognize only a few distinct body postures or facial expressions in otters, fewer than in many other carnivores. There are also few different vocalizations, and only the very gregarious giant otter can be described as a quite noisy species. However, all otters frequently use vocal communication within that most basic social unit, between mother and offspring.

Maternal care in otters is intensive over a relatively long period, in no species more than in the sea otter, where mothers back-paddle with a pup on their belly for many months. The long period of dependence in all species is a consequence of the problems young otters face in acquiring their prey. In some species, mothers provision cubs of up to 1.5 years in age. This long period of dependence is an important factor in the low reproductive rate of otters.

Populations

The population dynamics of otters present a somewhat bleak picture. Animal ecologists often categorize species, on the basis of their lifetime reproductive strategies, as 'r selected' or 'k selected', terms derived from population dynamics terminology. R-selected species produce many offspring within a short time, but invest little parental care in them; consequently many of their young die quickly, and the parents themselves tend to be relatively short lived (e.g. rabbits). K-selected species have few young, look after them for a long time and invest a great deal of energy in that; they tend to be longer lived (e.g. elephants). With few exceptions, otters clearly fall into the k-selected end of the spectrum, with large maternal investment resulting in only one or a few cubs. Their problem is that the adults themselves live for only a short time, compared with many other k-selected species.

One expects carnivores of the size of otters to have a long life, after surviving the vicissitudes of early independence, and indeed the potential longevity of most species is 15–25 years (in captivity). However, for North American river and Eurasian otters in the wild, the mean life expectancy at the age of 1 year is only some 3–4 years, in areas that have been studied (see Chapter 12). So far, there is little to suggest that other species fare any better, except the sea otter, with a mean life expectancy of 7–8 years.

This implies that an average *Lutra* or *Lontra* female, even if she breeds every year, will produce hardly more than two litters, in some areas with a mean of fewer than two cubs per litter. Sea otters have only one cub at a time. Some individuals, of course, will live much longer than the average age, but that does not alter the general picture that otters have very little room for manoeuvre with such small numbers of offspring. A small increase in adult mortality, or a decrease in recruitment of cubs, will seriously affect the viability of a whole population.

Unusually for mammals, annual mortality in Eurasian otters in Britain increases with age (see Chapter 11), and this may also be the case in other otter species, except in sea otters. There is as yet no good explanation for this; a likely possibility is that it is a consequence of the high metabolic rate of these animals, the price of their high-expense foraging.

An alternative is that pollutants are accumulating with age, although there is no evidence for this.

The pattern of recruitment and mortality in otters can usefully be compared with that in another well studied mustelid of comparable size, the Eurasian badger *Meles meles* (Cheeseman *et al*. 1987). The adult mortality rate of these badgers is similarly high, but does not increase with age as in otters, and most deaths occur in young animals. Females produce more cubs per year and per lifetime than do otters, and this is then followed by high cub mortality. This implies that badger recruitment is more easily adjusted to food availability, or to adult population density, than the recruitment in otter populations.

In otters, both adult mortality and cub recruitment are affected by fish availability, at least in some species and populations, but probably everywhere (see Chapter 8). However, these two aspects of otter populations are not influenced by fish numbers in the same way. In Eurasian otters, adult mortality occurs especially during the annual periods of fish shortage (see Chapter 12), for instance in northern latitudes at times when waters are coldest (so foraging is most costly). Cub recruitment, at least along Shetland sea coasts, is correlated with prey abundance during the annual high season for fish, in midsummer. Whatever affects fish numbers and productivity in their environment is likely, along one of these pathways, also to have implications higher up the ladder in the aquatic community, in the populations of otters.

Other environmental effects on adult otter mortality, either direct, or indirect by reducing fish populations, may be caused by pollution, which affects aquatic habitats more than most others. Effects of pollution have to be considered in a population context, jointly with other mortality factors such as food shortage, and this has been attempted to some extent in the Shetland studies. One possible scenario in Shetland was that otter numbers were limited by food, and that older animals, with high levels of mercury in their bodies, would suffer the highest mortality rate. Diseases may also affect survival.

A typical characteristic of populations of otters is that the animals always occur in small numbers; this is true for all species, except for *Enhydra*, which may occur in 'rafts' of several hundred animals. Species such as the fox or the Eurasian badger can attain densities of up to some 30 per km^2 in Britain (0.1–0.3 per hectare) (Cheeseman *et al*. 1987; Macdonald 1985), but for *Lutra lutra* no more than some 0.8 adults were found along 1 km of rich Shetland coast, or 0.3 per km of stream in a 'good' area in mainland Scotland. However, such figures for Eurasian otters translate into actual densities of up to 0.5 animals per hectare of fresh water, or 0.1 per hectare of suitable coastal water, that is, the actual densities per area of habitat are of the same order of magnitude as those of other, similarly sized carnivores (see Chapter 11). Crucially, in any one region there is less suitable habitat for otters than for those other terrestrial carnivores. Otter populations, despite densities per area of suitable habitat that compare well with other carnivores, cannot fall back on substantial numbers if something goes wrong.

Recent studies of the genetic variability of otter populations have concentrated on the Eurasian and the sea otter (see Chapter 11). There is relatively little genetic exchange between populations of Eurasian otters more than 100 km apart, and within populations the genetic variability is relatively low in that species. This is unrelated to recent population bottlenecks (e.g. the pollution disasters), so it may well be a general feature for other otter species. Some populations, such as the one in Shetland, are clearly genetically distinct, and even morphologically different. Sea otters have lost much of their genetic diversity because of severe population reduction in the recent past (see Chapter 12), and also in that species there is only restricted gene flow between neighbouring populations (see Chapter 11). Small genetic variability of otters may well render them more vulnerable to disease, as in the case of the African cheetah (*Acinonyx jubatus*) and other species (Wayne *et al*. 1986), and, at least for sea otters, disease, especially toxoplasmosis, has become a demonstrably important factor in populations (see Chapter 12).

Noticeably, in recent studies, more emphasis is being placed on factors other than food resources that may limit numbers of otters in populations. Disease is one such factor, but what has caught most recent attention is predation on otters. Sea otters, especially diseased ones, are being preyed upon by sharks (see Chapter 12), which leads to population declines, and the most dramatic increase in killer whale predation upon that species is part of a general 'megafauna collapse' (Springer *et al.* 2003). I have argued above that predators may also be an important selection pressure favouring group formation, in otters of many species. 'Top-down' factors may play a more important role than was recognized in the past, when we concentrated on the 'bottom-up' effects of resources.

Life at the edge of a precipice

Considering the threats to otters, and their problems as outlined in the previous sections, it is in some ways surprising that otter populations do still exist. Some species are more under threat than others (Fig. 13.6), but many of the comments made would be valid for almost all.

We are looking at animals that live in a linear habitat of which there is not much around, where they have to cover large distances to satisfy their needs, exposing them to hazards such as traffic. They are highly specialized foragers on difficult prey types; much learning is needed to acquire the prey, and foraging itself is energetically so expensive that large quantities of prey have to be caught in a short time. The underlying relationships are not linear, and daily foraging costs in cold water increase very fast with decreasing prey availability. Toxic substances are more likely to accumulate in aquatic prey, and hence into the predators, than in other habitats, and concentrations of environmental pollutants such as mercury in otters increase with age. Genetic diversity is low, and this may affect susceptibility to disease; everywhere otter populations are becoming more exposed to contact with people and domestic animals, as possible vectors of disease. The animals are sought after for their fur, they compete for fish with humankind, and some are prone to falling victim to predators. Populations have high mortality and low reproductive rates.

It is not surprising, then, that many otter populations are seriously vulnerable, that they have taken some severe knocks in past decades, and very recently. Somehow, most of them usually cope with these dangers, but there is little doubt that otters have more odds stacked against them than many other carnivores. Extrapolating from what we know

Figure 13.6 Giant otters are gregarious and boisterous, but vulnerable. © Nicole Duplaix.

at the moment, it is likely that only a small increase in one of the water contaminants or prevalence of a disease, a slight decrease in prey population, a severe winter or a drought, or an increase in predation, may have effects much more dramatic than for many other similar-sized species. All of this is a consequence of their extreme feeding specialization.

Further research

Despite interesting results from the many studies on otters, we are left with the main message that very little is yet known about these animals. Several questions need to be addressed urgently, to cover important conservation problems or because of more academic interest, but usually for both reasons. For some species we know next to nothing about ecology and behaviour, but even for those that have had some attention, we still know little about actual numbers over larger areas, about population sizes, and about changes in areas where previous estimates have been made.

For all, the genetic composition of populations needs further study, to determine their variability and differences between regions. We need to know what are the barriers between populations, and whether there are differences that should be kept in mind when considering transplantations. Given the limited genetic diversity within populations, one wants to know about dangers associated with inbreeding. DNA techniques are opening new doors here, as well as providing new means of assessing otter numbers and social organizations.

The processes of population change still need further elucidation. For instance, what is the cause of the increase in Eurasian otter mortality with age? How and why is the oestrus cycle synchronized with the seasons in some regions, and apparently random elsewhere? What is the profitability of foraging in various areas, taking into account success rates and prey types, as well as water temperatures? Much more information is required on populations of relevant fish species, their population dynamics, fluctuations and responses to predation. The characteristics of fish populations will be crucial to otters everywhere, and a dedicated, interdisciplinary approach to the role of fish populations is much needed.

More information is required on intraspecific as well as interspecific variation in the spatial organization of otters, and underlying environmental differences. For instance, how is *L. lutra* organized in countries in south-east Asia, where it lives in totally different vegetation types, where it has to compete with other species of otters, and where it may be subject to much higher predation? Such questions of intraspecific differences need to be answered in order to establish the flexibility of a species in the face of environmental change. What is the relationship between predation on North American river otters and their organization of male packs and solitary females? Similar questions should be asked for other species of otters; as just one example, what causes the spotted-necked otter to live in large, diurnally active, packs in Lake Victoria, and as nocturnal, solitary individuals in streams of South Africa?

Perhaps the most immediate need is for an understanding of the relationship between sea otters and killer whales, and how this can possibly be managed. There are other direct conservation problems: how do Eurasian otters respond to transplantation (reintroduction), how do populations respond to increases or decreases in their food supply, what are the pollutant levels in otters from stable, or increasing or decreasing populations? One could add many more such questions, but even the most basic problems in the ecology and behaviour of all but two or three species of otter are still open. A great deal of fascinating research has yet to be done to serve as a solid basis for conservation management.

CHAPTER 14

Otters, people and conservation

A changing world

Several of the issues that are relevant to conservation have come up in previous chapters, relating to the ecology of otters. Here, I want to provide an overview, for a background to conservation management of some of these most appealing and interesting carnivores, in their own environment.

The fate of otters in the world would seem a puny concern, at a time when we are producing a massive change in climate, when the size of human populations is sky-rocketing, when entire faunas dwindle and pollution kills. Yet, one should take note of, and guide, what happens to these animals. Otters live on fringes, and they are amongst the mammals that are most sensitive to some of these changes. Their vicissitudes show the sites for red signal flags and, for our own sake and that of all the natural environment, we should observe these. Despite our concerns because of world-shocking changes, we should still look after the smaller issues, and make certain that populations of such wonderful creatures do not slip through our fingers into oblivion.

The existence of otters touches on the most significant environmental issues affecting the world today, and several of these will impact on the chances of survival for these animals. Climate change is the most relevant and all embracing, though also the least predictable. Increases in temperature, averaging 0.6°C over the last century and projected to accelerate rapidly (Root *et al.* 2003), have been shown as the cause of long-term changes in the composition of fish communities (Genner *et al.* 2004). They have a significant effect on other species (ranging from molluscs to mammals) worldwide, causing geographical range shifts averaging 6.1 km per decade (Parmesan and Yohe 2003). The extent and effects of changes in rainfall, cloud cover and other aspects of climate are more difficult to project, but are likely to be profound. There is little doubt, therefore, that climate change will seriously affect otters of all species.

Not all such effects are necessarily deleterious. An increase in water temperatures will benefit otters, in that less energy is required to compensate for heat loss, when swimming. Otters will need fewer prey, and a decrease in ice cover in northern latitudes could enable Eurasian and river otters to extend their geographical ranges northwards. Further south, however, some waters will dry up and habitats will become unsuitable, especially where the water needs of human populations escalate.

Burgeoning human populations will directly affect otters in many areas. The world in 2005 is inhabited by 6.5 billion people, an increase of 67% since 1970, and according to UN figures, numbers are expected to reach 8.9 billion in 2050, peaking soon thereafter. Present-day spectacular increases are evident especially in the tropics, and in countries such as China and India this is augmented by an upward change in wealth. Because people affect otters directly or indirectly through hunting (acquiring otter products), fishing, habitat change and pollution, there is every reason to be concerned about long-term upward trends.

Coastal, wetland and riparian habitats are being altered just about everywhere, affected by people in many different ways. It is estimated that since 1900 more than half of the world's wetlands have disappeared (www.waterconserve.info [2005]). The most obvious disasters are those where housing or industry fills up coasts and takes the place of wild vegetation, but the alterations take many different forms. Sometimes entire rivers, lakes or even seas (Aral Sea) dry up because of increased use of water by people and agriculture. In addition, there are more subtle effects, not included in these statistics, such as drainage of nearby lands, removal of vast areas of mangroves, and replacement of rivers and streams by hydroelectric dams. Deforestation of land bordering on water is taking place everywhere. From previous chapters it will have become clear that some of these changes are highly significant to otters, whereas others matter less.

Then we are faced with pollution, which may poison through the food chain, diminish the prey base, or remove animals' thermo-insulation. Despite all publicity, despite the many clean-ups and legislation, Rachel Carson's *Silent Spring* is still with us. Increases in human populations and intensification of industry and agriculture have gone hand in hand with increased eutrophication and pollution of the environment; pollution, especially aquatic, but also aerial and terrestrial, potentially affects many wild carnivores in several different ways.

Many of the harmful compounds that have affected wildlife in the past have been taken out of circulation, at least in some countries, but others are still present, and the problem is likely to persist forever. Chemicals are added to the otters' environment directly, in outflows from industry and conurbations, in run-off from agricultural land, and in rainfall from polluted skies. Various organochlorines are involved, PCBs and many others, as well as heavy metals, especially mercury, all of them with potentially lethal effects when they accucmulate in mammals' bodies. Organochlorine insecticides also remove the insect food base for important prey species of otters, such as salmonid fish and eels. A rather less subtle variety of pollution is produced by foundering oil-tankers: spilled oil damages thermo-insulation as well as poisoning animals. All predictions suggest that oil spills are likely to become more frequent.

Another spin-off from our civilizations to wild carnivores is the incidence of disease. Some lethal or seriously debilitating pathogens can be transmitted from one mammal species to another, and some of our domestic animals carry them: distemper, rabies, tuberculosis, toxoplasmosis, brucellosis and many others. Contact rate between wild and domestic carnivores is bound to be increasing, and it does affect otters. In turn, otters and other small carnivores are carriers of serious human diseases apart from the above; for instance, as found in markets in Vietnam, Laos and China, otters are among the species suspected as a reservoir of the recent, seriously worrying, virus causing severe acute respiratory syndrome (SARS) (Bell *et al.* 2004).

The impact of humankind on entire, huge ecosystems, involving complete faunas of large mammals as well as invertebrates and vegetation, is brutally evident along the Pacific rim of North America. **Sea otters** are involved in a big way, after having escaped the effects of unrestricted hunting earlier on. In the past two decades, sequential near-extinctions, one species of sea mammal after another, have been documented. Unrestrained fishing and whale hunting set it all off, natural predation by killer whales has changed in consequence, and the continued existence of several species, including the sea otter, is now in serious jeopardy (Estes 2002).

Otter vulnerabilities

How are the different species of otter likely to be affected by the most important challenges that are being thrown at them now, at the beginning of the twenty-first century? How are they vulnerable, and what are the specific threats?

Earlier sections in this book have demonstrated that otters inhabit often fragile ecosystems with both aquatic and terrestrial characteristics. However, otters are not particularly choosy, and even areas that have been seriously affected by our civilization may be inhabited by them. They do not necessarily need clean water, or trees along banks, or places for dens. Otters are not necessarily the 'ambassadors for a clean environment', a role that is often ascribed to them by conservationists—they can take a great deal of human impact. However nasty such human effects are to our perception of the environment, otters often cope remarkably well.

Several species (river, Eurasian, neotropical, small-clawed, Cape clawless, spotted-necked) occur in urban situations, where streams or rivers pass through cities, and the giant oil terminal in Shetland has **Eurasian otters** swimming between the huge ships, leaving their cubs between the boulders of the jetties, often within a few metres of vehicles and noisy people with hard hats. Some otters actually sleep in pump houses at the oil terminal.

Radio-tagged otters elsewhere in Scotland frequently sleep in or under human-made structures: rubble, concrete blocks on banks, or in an old car in the stream close to a farm. **River otters** in California often scavenge in and around fishing boats, and are a familiar sight in several harbours. **Sea otters** are remarkably tame in some very touristy places, and may obtain some of their food (octopus) from empty drink cans.

Dense bank vegetation, reed beds and marshes are often used by otters of different species, and may therefore be important in conservation management. Pristine condition of habitat may not be that important to otters, yet vulnerabilities arise because of its near-linear shape. A single otter needs a very long stretch indeed, one that is likely to bring it into contact with potential dangers such as traffic, much more than the small, simple patch of more terrestrial carnivores.

The habitat factor that has emerged as *sine qua non* is the presence of sufficient biomass of potential prey (i.e. fish or, for some species, shellfish or crustaceans). This has to be accessible in such abundance that otters can acquire their daily ration quickly (to avoid getting cold); therefore, any suitable otter habitat needs high prey densities. Otters need unusually large amounts of food, and the consequences of this single requirement, in terms of their survival and conservation, are profound and wide ranging.

First, it means that their habitat needs large prey populations, and often some eutrophication can be beneficial. Second, one of the broader consequences of the difficult foraging technique is that it takes young otters a long time to become proficient and independent. It follows from this that otters have a slow reproductive rate, which makes populations vulnerable. This is a vulnerability to even a relatively small increase in mortality, to any kind of exploitation by people, or an increase in natural predation. Third, the large quantities of prey required by otters, predators at the apex of the aquatic feeding pyramid, also mean that they are particularly prone to accumulate dangerous compounds, released in the waters directly or through aerial pollution, by agriculture or industry. Some of these compounds may affect otter mortality directly, or they may affect fertility, or they may affect prey populations.

Another direct threat from people to various otter species arises mostly from the animals' efficient mechanism of thermo-insulation during foraging: their fur. To people, fur is an attractive commodity, for adornment or for protection against cold. Amongst local hunters as much as in the organized fur industry, otter skins are highly prized (and especially the skins of the one species that is always in water, the sea otter).

Maintenance of this precious otter fur, in most species that forage in the sea, has the consequent requirement of needing frequent washing in fresh water. Because of this, the presence of suitable freshwater sources along sea coasts is a vital habitat requirement, and one that frequently determines whether otters can live along a given coast or not.

Finally, one of the animals' special vulnerabilities arises because otters compete with us for their aquatic prey—this is one of the reasons for their persecution.

Otter nuisance

Frequently, death of otters at the hands of people is brought on to the animals by their own activities: in some places they affect fisheries, or they take domestic fowl. Although, especially in Europe and North America, otters may have achieved remarkable popularity, in the not too distant past, say more 50 years ago, they were considered as vermin, no better than foxes, raccoons or polecats. They still are considered to be vermin in many developing countries. In ecological terms, otter nuisance is a manifestation of direct, interspecific competition between predators: otters prey on the same species as humans do, and they interfere with fishing nets and fish farms.

Now, in a more tolerant age and with few otters left, conservationists tend to deny what are seen as the otters' bad deeds, but perhaps it should be acknowledged that the view of countryfolk of old was not totally off the mark, at least not everywhere. In our own recent research in Scotland we recorded **Eurasian otters** eating many salmon, often large ones, and in some rivers they consumed almost half of the productivity of fish populations, as documented in Chapter 8. More recently, researchers in western Scotland confirmed that otters took 28% of tagged adult salmon within a few weeks (Reynolds 2003). This leaves some people concerned that the salmon populations in Scottish rivers, which have seriously declined for unknown reasons, are being delivered a final blow by the otters.

In Austria the government pays compensation for damage by Eurasian otters to trout and carp populations in fish ponds and stocked streams. There has been a large increase in this since the recovery of the otter populations in the 1980s and 1990s, and in 1995 the level of compensation had reached the equivalent of almost US$200,000 per year (Bodner 1998). Similarly, in the USA, the sport-fishing public in Missouri is worried about the alleged dramatic increase in damage by **river otters** to the inhabitants of fish ponds and streams, following a successful otter reintroduction programme (Hamilton 2004). Within 1 year, the government received over 500 complaints.

Fisherfolk in Europe have suffered damage to nets, caused by otters, and I have seen this elsewhere. In Thailand, where my Thai colleagues and I used gill nets in rivers, Eurasian otters frequently tore up the nets (they were actually seen doing so) and took much of the catch, as did the **spotted-necked otter** with gill nets during my work on Lake Victoria. In Rwanda, this last species took an estimated 15% of the catch from the nets of local fisherfolk (Lejeune 1989, 1990), and there are similar reports from other species. Not infrequently the otters are caught in fish nets themselves, and drown. Little wonder that people whose livelihood depends on fish catches show less sympathy for the animals than do the urban viewers of spectacular television documentaries. Persecution of otters, therefore, is not based solely on prejudice (although this does not imply that I approve of it; nor do I think that it is effective in preventing damage).

Anywhere in the world, fish farms may occasionally be visited by otters, if there are chinks in the armour of the ponds or of the floating fish cages, such as holes in the protective netting, or absence of anti-predator wiring, and substantial numbers of fish may be taken. Some types of fish farm are easy to protect, and are ignored by the animals. In Shetland, for instance, where a large, well protected fish farm with floating fish cages was established right in the centre of our study area, we frequently saw **Eurasian otters** swim closely past the cages full of salmon, never taking any overt interest in them. But in South Africa (Southern Cape province) I watched a **Cape clawless otter** for about half an hour attempting, unsuccessfully, to get into a floating fish cage similar to those in Shetland; it was walking on the boardways and biting at the netting.

Kranz (2000) confirmed the widespread occurrence of predation by **Eurasian otters** in central European fish farms; especially in Austria and the Czech Republic the species is generally perceived as a pest. Adamek *et al.* (2003) documented serious problems in Czech fish farms, with otters taking substantial numbers of large carp, up to 11 kg in weight, as well as perch and zander. Often the otters ate only the viscera, leaving 63–73% of the fish body mass on the bank. Farmed fish also died from stress caused by otters hunting them under the ice in winter.

Elsewhere in Europe depredations by otters in fish farms are a cause for concern, with significant quantities of salmonid fish being taken in Finland, especially in winter (Ludwig *et al.* 2002), often as surplus kills left uneaten. Annual losses were estimated at US$75,000 in the 1980s, but are likely to have increased substantially since then. Similar observations of otter damage in fish farms have been made in Poland (Kloskowski 2000), France (Leblanc 2003) and the north of England (Morgan 2003), and no doubt fish farms in other continents will have records to match. Both the **smooth** and the **small-clawed otter** have been mentioned as taking fish and shrimps from aquaculture projects in Malaysia (Foster-Turley 1992).

Otters occasionally lower their popularity ratings by taking poultry: ducks, hens and geese. My own three ducks, free-roaming in the garden in Scotland, were taken from the pond one snowy March, by a Eurasian otter. They disappeared one at a time at night, over a couple of weeks, carried off and eaten in the marsh about 1 km away, with good tracking snow to record it all. A large Eurasian male otter regularly raided the hen-house of one of my then PhD students on Skye, Paul Yoxon, who studied the animals along the coasts there, and on the Scottish mainland one of our radio-tagged otters, followed by Leon Durbin, often foraged from a collection of captive ducks (the owner blamed mink). Shetland farmers frequently lose hens and ducks to otters, and a Scottish newspaper (*Press and Journal*, 23 October 1996) recounted the loss of several scores of hens, ducks and geese around crofts on the Outer Hebrides island of North Uist (headline: 'Killer otters should be shot').

Fanshawe *et al.* (2003) concluding that along Californian coasts, **sea otters** and fisheries for abalone are incompatible: sea otters take such a high proportion of harvestable sizes of these molluscs that not enough is left for people to fish. In Alaska, the increasing populations of sea otters in the 1970s caused serious disruption and the end of many clam fisheries (Johnson 1982a,b; Wendell *et al.* 1986).

The otter-loving public often doubts or denies such damage, a response that, I believe, is short sighted. Conservation management policies for otters will have to take account of problems with people's use of resources, and of the perception of problems. Maintaining that they do not exist inevitably leads to loss of credibility of observers and conservationists, and therefore loss of effectiveness.

Exploitation of otters

When advertising the attractions for eco-tourists of an island, shore or lake, the presence of otters of any species is a certain asset: the animals now have a charisma that moves anyone with a love of nature, even the most experienced naturalists. Visitors to an area may not encounter the animal, but the knowledge of its presence alone is sufficient to attract, and otters are an asset in television documentaries. However, not so long ago, and to some extent today, the animals had or have more practical uses. The most important of these applications is their fur, of which more below, but there are other benefits that people derive from otters.

The species that is displayed most often in zoos all over the world is the **small-clawed otter** (Foster-Turley and Engfer 1988), which is relatively easy to breed and takes well to captivity. Playful and diurnal, it makes a highly attractive exhibit in a suitable pool. Other species are kept occasionally; in several western European countries otter parks have become fashionable, where mostly the local species, Eurasian otters, are kept.

A rather less peaceful entertainment, involving especially **Eurasian otters**, was the sport of otter hunting, with specially bred and trained otter hounds. It was popular in England until the 1960s, and similarly common in many other countries of Europe, with much cruelty involved in the process. The otter hunt was celebrated in a well known 1844 painting by Sir Edwin Landseer. In Britain otter hunting was outlawed in 1981, and in virtually all countries in western Europe the animals are now protected.

In Asia, trained **smooth otters** are commonly used to help with fishing, especially in Bangladesh (Feeroz 2004; Hendrichs 1975). In the Sundarbans, small fishing boats go out at night, carrying three people and special nets, and three otters in individual harnesses on a lead. The otters surround and chase fish into the nets. In the study by Feeroz (2004), as many as 256 different otters were involved in this artisanal fishing, the animals caught wild or specially bred.

In many countries in Asia, otters are caught and marketed for food, or for medicine, and the African trade in 'bush meat' often involves otters. In Europe otters were also on the menu, and from Germany I was sent a (no doubt useful) recipe for Otter *aux fines herbes* (Kruuk 2002). One of the reasons for the **Eurasian otter**'s popularity as food was the fact that it was classified as an honorary fish by the Roman Catholic Church, and could be eaten on fast days (Fridays and the six weeks before Easter). A splendid large painting by the Flemish painter Frans Snyders (1579–1657) in the Louvre shows a fish stall in the market, with an otter in pride of place.

At least as important is the present-day use of otter body parts in traditional medicine, in countries of south-east Asia and in China. A dried otter penis fetches US$40–50 (presumably without the baculum), and other parts are used as well. In India the blood of the **smooth otter** is used against epilepsy: it is collected in a clean vessel, a cloth is soaked in it and dried. When needed, a patient soaks the cloth in a glass of water and takes the fluid three times daily for 3 days. If fresh blood is available, then only once per day for 3 days will suffice (Nagulu *et al.* 1999).

However interesting, all such usage of the animals by people pales into commercial insignificance compared with the trade in their skins. Otter fur is highly prized and, although now outlawed in many western countries, there still is a roaring business elsewhere. Staggering numbers of otters have been caught over the years, different species in different countries, and in many places to this day, they die a nasty death to feed this trade.

The best known example, with its dramatic consequences for the species, is the slaughter of **sea otters** in the north Pacific ocean, remarkably well documented in 'Otter skins, Boston ships, and China goods' (Gibson 1992). The disastrous trade came to a halt in 1911, but not before the species had been all but wiped out. In the 18th and 19th centuries, a sea otter skin was worth more than that of almost any other animal; after his expedition in 1804–1806, the explorer William Clark wrote: 'The fur of them were more butiful than any fur I have ever seen, it is the richest and I think the most delightful fur in the world at least I cannot form an idea of any more so, it is deep thick silky in the extream and strong' (Gibson 1992, p 6). The fur from northern coasts (Kurils and Aleutians) was better than that from California.

After a Russian expedition in 1741 'discovered' the sea otter, a gold-rush for their skins started, and until 1800 Russian ships shot and collected tens of thousands of animals. In the late eighteenth century, Spanish ships joined in the fur trade along the Californian coast, and further north both British and American traders moved into the fray. Unlike the Russians, crews on the British and American ships did not capture sea otters themselves, but traded with the Indian population who did the hunting for them. Often barbaric in their dealings with the locals, these traders had a massive influence on the demise of the Indian tribes and their culture (Gibson 1992). The sea otter skins were transported to Canton and sold to the Chinese, for excellent profits.

The American traders alone sold 158,070 sea otter skins from 1804 to 1837 (an average of 4790 per year), although in the last decade of this the supply of skins had dwindled to a trickle. The efforts of the British and Russians came on top of this, and it is known that numbers taken before 1800 were much higher. However, by about the 1840s there were few sea otters left, and the immense fur bonanza was reduced to just a regular harvest, with sea otter numbers falling steadily until they were fully protected in 1911 (Gibson 1992).

Skins of other otter species were valued far less than that of the sea otter, yet they featured large in the nineteenth century fur trade between America and China. In the period 1808–1837, American ships sold 286,166 **river otter** skins on the Canton market, an average of almost 10,000 per year, but their value was considered far inferior to that of the sea otter fur. Today, river otters are still being trapped in large numbers in most parts of North America; for instance, the number of their skins traded in USA and Canada in 1978 was 47,000 (Kruuk 2002), and more than 50,000 in more recent years. The otter harvest in Louisiana alone sometimes exceeds 10,000 animals per year, usually surpassing that in any other state (Official Website of Nebraska, www.nebraska.gov [July 2005]). In Missouri, and in several other states, river otters were eliminated by fur trappers from all but very few locations (Erickson and Hamilton 1988), before being reintroduced.

Such capture and trade of river otters is within the law, but through the Convention on International Trade in Endangered Species (CITES) the import and export of skins of protected otter species is now illegal. It may be outlawed, yet in 1977 one single New York dealer smuggled, amongst many other furs, the skins of 15,470 **neotropical** and 271 **giant otters** into the country (Eltringham 1984).

There were several large sources of South American otter skins. From Peru, Schenck (1997) reported that, between 1946 and 1971, 24,282 **giant otter** skins were exported, almost 1000 per year, with a peak of 2248 at the beginning of the period. In the 9 years before 1969, 19,925 skins were exported from the Brazilian Amazon, more than 2000 per year, and from Colombia more than 1000 in 1965. Each skin was worth about the same as that of a jaguar. **Neotropical otter** skins were traded much more often; Peru alone officially exported 113,718 skins between 1959 and 1972, more than 8000 per year, with 14,000 in 1970 alone (and this was presumed to be only about half the real figure; Smith 1981).

In Europe, except in Russia and other eastern countries, trapping of **Eurasian otters** for their fur has virtually ceased, but up to the Second World War there was a substantial market for otter skins, as now in North America. The species is frequently trapped in Russia and in Asian countries, however. In the international fur trade more than half of all otter skins are bought by Russia, China, Korea and Japan. In a recent case of illicit trade in otter skins between India and Nepal, 665 otter skins (probably all or mostly **smooth otter**, originating in India) were impounded

Figure 14.1 Skin trade is a major threat. A display in a shop in Qing Hai, China, near the Tibet border, August 2005, showing seven skins of smooth otter, and leopard and snow leopard. © Chris Wood (Zhejiang University, China).

Figure 14.2 An effective, traditional trap for Eurasian otter in Shetland, called an 'otter house', long out of use, but many are still there. The wooden trap door has gone.

in October 2003 (www.careforthewild.org newsstory October 2005). In and around Tibet, skins of smooth otters are used in ceremonial dress, and in 2005 in one market 'very few of 40-odd shops did not display at least 3 or 4 full [smooth] otter skins, some up to 10 or 15' (C. Wood, personal communication, 2005; Fig. 14.1).

The fur trade still targets probably all species of otter, despite legal protection in many countries. What is important, though, is whether this affects numbers in populations, and we know little about that. The trade was clearly highly detrimental to the sea otter, and there is strong suggestive evidence that the giant otter is also affected (Schenck 1997); both of these species are relatively easy to kill. Whether the same is true for other species has not been demonstrated. It is likely that this kind of mortality is density dependent: hunters and trappers will desist when the effort involved becomes too great. But whether this is the case or not, fur trapping does constitute a source of mortality, that in these slowly reproducing species could be fatal for a population when it is hit simultaneously with other adverse conditions, and is unacceptable to many because of the cruelty involved.

To catch an otter

Conservation management of otters occasionally has to involve dealing with otters that cause damage, or with people who interfere with the animals, or with translocations of otters. One needs to know the means of persecution of today and yesterday, and there are times when catching the animals for conservation purposes, or for research, is unavoidable. For this reason it can be important to know the different ways of live-trapping the animals, and which of these methods are legally permitted. Where otters are fully protected, any kind of trapping or interference with them, for whatever purpose, needs to be done under licence.

The most impressive old otter traps—historic monuments to the role of otters in the fur trade in Europe—are in Shetland. Scores of unique and ingenious 'otter houses', as they are called locally, were constructed by the Shetlanders in the very distant past, and can still be found there, more or less intact: well built, stone structures at strategic places just above the high tide mark (Fig. 14.2). They were made to catch **Eurasian otters** by the construction of what is almost a small artificial cave, the entrance with a wooden trapdoor, with inside a sprainting stone or couch-site and a trip release, and a 'lid' in the ceiling to kill and remove the animal. I talked to several old Shetlanders who had caught otters with them. Many of the structures must be centuries old, and all are now abandoned by the people who used them. However, they are still popular with the wild otters of today, which sleep and spraint inside the otter houses as they have done over the ages. Interestingly, these Shetland otter houses are almost the same as the constructions that I found used by Inuit in Greenland for catching arctic foxes: a remnant, perhaps, of Viking civilization.

Otters of all species are relatively easy to catch for killing, usually because they follow obvious paths and return to specific sprainting sites. On mainland Scotland the traditional method of catching otters was the leg-hold or 'gin' trap, with its sharp teeth, now outlawed in Britain and many other countries because of its cruelty, but still in common use in North America. The animals are highly susceptible to being caught by this type of trap, or by a neck snare, when it is set on one of their well worn runs. The nastiest type of otter trap that I met was in Norway, where a fisherman showed me a spring-loaded underwater trap, with huge spikes to go through the body of the animal. It is set off by the otter on touching a piece of string under water, in the diving approach close to a holt or sprait site (Fig. 14.3).

Shooting, after flushing otters with dogs from their couch or lying-up place, also accounted for considerable numbers of otters killed in Scotland by gamekeepers. In north-west Scotland, I met one keeper who had shot 93 otters in 15 years, along a 20-km coast of a peninsula (Kruuk and Hewson 1978). This cull stopped 3 years before we began to study the animals there, when the population appeared to be thriving. Even now otters are still shot along some Scottish rivers, despite full legal protection (see Fig. 12.9), and, of course, also anywhere else in the world many otters are simply shot. Not surprisingly, we only occasionally find the odd victim, because the evidence is so easy to dispose of.

In North America the **river otter**, and in many parts of the world various other species of otter, are trapped with leg-hold traps, which appear to be by far the most efficient method of catching—often extremely cruel. Other countries where leg-hold traps are still legal now require the jaws of the trap to be padded or smoothed, so the animal is less likely to injure itself once caught. When caught in the original, basic version of the leg-hold trap, otters will often respond to the severe pain by biting off parts of their leg.

To catch otters for research purposes, or to acquire them for relocation programmes, they need to be live-trapped by less traumatic methods, and this has proved to be more difficult. **Sea otters** can be netted from boats, often by divers with large, hand-held nets, but otters of the other species are often highly adept at avoiding capture. The most efficient method to date, for catching otters for research, is still capture with a leg-hold trap, the jaws modified to lessen the risk of injury, and the trapping mechanism connected to a radio-transmitter to tell the operator that it has sprung, so the otter spends only a short time in restraint (Blundell *et al.* 1999). Even with such precautions, serious injuries to legs and teeth do occur occasionally, however, and in several countries (including Britain) the leg-hold trap is totally outlawed, whether padded or not.

A research alternative is the 'Hancock trap', described in detail by Mitchell-Jones *et al.* (1984), in which an otter stepping on a treadle releases a heavily sprung steel bar with a chain-link net that covers the animal. It is cumbersome and less effective (Blundell *et al.* 1999), otters caught are prone seriously to injure their teeth, and the traps can potentially break the leg of a larger animal or person. There are also various types of box-trap, large cages with a treadle to spring the door ('Tomahawk' type), baited with fish. These are quite effective for some species (e.g. **Cape clawless otters**; Somers and Nel 2004), but in my own studies **Eurasian otters** refused to enter them, and also were not attracted to bait. I used large, wooden, walk-through box-traps, tunnels with a trapdoor at both ends and a central dangling piece of wire connected with a mouse-trap on top to set off the doors (Kruuk 1995). The trap was set on some otter passage and made attractive by fresh sprait inside. It often took a long time to catch an animal, and I left these traps *in situ* for weeks on end, but they were relatively safe. Once caught in any of

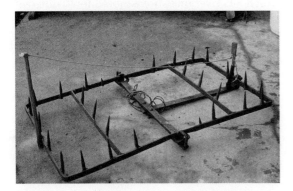

Figure 14.3 Underwater spring trap for Eurasian otters, as still used in Norway. Set close to a landing site, the jaws of the trap are activated when an otter swims into the string.

these trap types, the otter is anaesthetized with an injection of a proprietary anaesthetic (Blundell et al. 1999), often administered by blow-pipe.

With all types of trap the problem arises that otters are, in anthropomorphic terms, highly strung animals that not infrequently go berserk when captured, and damage or kill themselves even in the largest of cage-traps. There is a distinct risk to the animal that can be minimized, but never eliminated completely.

To study otters in the field, radio-tracking is often essential, depending of course on the questions asked. Details of methodology and radio-tracking procedures are beyond the scope of this book, but, as one specific point for the use of this technique with otters, it should be mentioned that to attach a radio-transmitter to an otter it is necessary to implant it surgically, intraperitoneally, into the animal (e.g. Blundell et al. 2000). This is a relatively minor operation for a suitably qualified scientist or veterinarian. Attaching the transmitter in any other way (e.g. on a harness) causes much and continuous distress to the otter, and the otter's body shape renders a collar quite impossible.

Reintroductions of otters, and conservation genetics

In countries where things have gone totally wrong for otters, when pollution has overcome them or they have been trapped out, one hopes that some day the otters will return, after conditions have improved. They may come back naturally, through range expansion from neighbouring regions, because, as we have shown, some individuals move over enormous distances. Alternatively, conservation managers may decide to give the otters a helping hand.

Many parts of the USA and Europe were left without otters, after the disastrous 1950s, 1960s and 1970s. Since then, an astonishing number of reintroductions has been organized by various wildlife authorities: a recent summary accounted for some 3987 **river otters** that were moved to new areas in North America (Breitenmoser et al. 2001), 1037 **sea otters** (Breitenmoser et al. 2001; Riedman and Estes 1990), and in western Europe at least 120 **Eurasian otters** (Breitenmoser et al. 2001; Fernandez-Moran et al. 2002; S. Broekhuizen, personal communication). People want the animals back in their natural habitat, and quickly.

Achieving a reintroduction can be a major management tool; it may also be controversial, and does not always succeed. The first targeted reintroductions of otters were those of the **sea otter**, on five occasions in the 1950s, when animals were captured from the remnant population in Amchitka and transported to the Alaskan mainland coast. Most of them died from cold when still in captivity and during transport, because their fur had become soiled and lost its insulating properties, and the operations failed (Kenyon 1969). But, after 1965, with the authorities having learned from the previous fiasco, more than 400 sea otters were moved and well cared for, and the species became established again in south-eastern Alaska, increasing to over 5000 sea otters in the late 1980s at an annual rate of 17.6% (Riedman and Estes 1990).

Many more sea otter reintroductions followed to coasts further south, most of them successful. A prerequisite of success is sufficient numbers in the start-up population (at least 25–30 annually, over 3–5 years; Jameson et al. 1982), and of course appropriate habitat. Yet some of the reintroductions failed for unknown reasons, such as one in Oregon in 1970–1971 involving 93 sea otters (Riedman and Estes 1990).

The numerous reintroductions of **river otters** throughout the USA have been highly successful. The most spectacular was in Missouri in the 1980s, in which 845 river otters were moved to 40 different sites over a period of 8 years. Almost all of these animals originated downstream of Missouri, in Louisiana (which has been the source of animals for most otter reintroductions in the USA). The Missouri reintroductions were followed up by intensive research, by radio-tracking many of the animals, and observing them and their signs in the field (Reading and Clark 1996). The success of the operation was due largely to strict adherence to rules: selecting suitable and undisturbed habitat, releasing at least 20 otters at any one site, with female-biased sex ratios, selecting release sites close enough to one another to permit mixing of populations, in any one site releasing all animals together to enable optimal spacing between them, and of course prohibiting all trapping nearby (Erickson and Hamilton 1988).

Amongst the numerous other river otter reintroductions in North America, the 1982 project in Pennsylvania has been particularly well researched. A total of 82 otters from Louisiana were moved into five streams and rivers, many followed by radio-tracking, and an overall evaluation took place afterwards (Serfass *et al.* 1999). Where river otters were present in the state only in a small corner before, the species is now present again more or less throughout, and otter trapping is now illegal (though animals still get caught in traps set for beavers, raccoons and mink).

The **Eurasian otter** has been successfully reintroduced into southern Sweden (Sjöasen 1996), eastern England (East Anglia; Reading and Clark 1996) and north-eastern Spain (Fernandez-Moran *et al.* 2002), and a further reintroduction in the Netherlands is still ongoing (with very small numbers; S. Broekhuizen, personal communication). It is likely that more will follow: reintroductions are popular, because they constitute active, direct and positive management.

Strict guidelines for reintroduction procedures have been issued by the International Union for the Conservation of Nature (IUCN) and various authors, to encourage success, and to prevent mistakes (IUCN/SSC 1998). Yet there are voices objecting, mostly on the grounds that such management involves trapping and suffering, and people feel that nature should be left to correct itself, with otters eventually re-establishing without our help. An important argument against interference relates to the genetic composition of otter populations: introduced otters originate from populations with a genetic composition different from what the local population was, or should be. This is a major concern of conservation genetics.

Intensive research on the molecular genetics of otters has demonstrated that, in the natural condition and without the interference of humankind, gene flow between populations of Eurasian, river or sea otters is present, but limited (Cronin *et al.* 1996; Dallas *et al.* 2002; Randi *et al.* 2003; Serfass *et al.* 1998). In Europe, Eurasian otter populations in the Iberian peninsula, in Denmark and in south-west England are genetically distinct, and even within one country such as Britain (Dallas *et al.* 2002) or Germany (Effenberger and Suchentrunk 1999) one can recognize genetic differences between populations. The same holds for river otter populations from different parts of the USA, even to the extent of being called different subspecies by some. Three different subspecies of sea otter have been recognized. When reintroductions are carried out, such distinctions between populations are sometimes forgotten.

From a conservation management viewpoint, this matters. Storfer (1999) warns that animals with a different genetic make-up may not show the necessary adaptations to local conditions, for instance in characteristics such as litter size. Even if such adaptation is not genetic, acquired characteristics, such as responses to hazards or prey acquisition, which animals often learn from their parents or peers, may be decidedly maladaptive in a strange environment. In **sea otters** it was demonstrated that, in a translocated population that came from two different sources, animals from one reproduced better than those from the other, suggesting important and possibly genetic effects of origin on the fitness of translocated otters (Bodkin *et al.* 1999).

There is a case, therefore, to consider seriously the source of otters for reintroductions (genetic composition and habitat similarity) or, alternatively, to let population recovery run its course with natural immigration from neighbouring source populations. John Dallas concluded from his studies of genetic variability in Eurasian otters that special areas of conservation, as sources of recruitment for populations elsewhere, should be located no more than 100–150 km apart (Dallas *et al.* 2002).

Sufficient genetic diversity is an important requirement for any population, and provides a clear argument for using sufficiently large numbers of otters for a start-up population, when reintroducing. **Sea otters** suffered a large loss of diversity after their populations came through a bottleneck created by the disastrous persecutions from the fur trade (Larson *et al.* 2002b). The results from present-day research suggest a need to monitor the genetic diversity of sea otter populations, if and when they emerge from their recent ordeals.

Conservation needs of ecosystems

The general message for conservation management from research on otter ecology is that, to maintain populations, animals have to be protected in their context, as part of an ecosystem, as part of a food web. One has to think in terms of entire coasts, wetlands or river catchments. Nevertheless, public interest tends to focus on species such as otters, rather than on, for example, the fish or sea urchins they eat, or the reed beds they use.

In Europe and North America, otters have become flagship species, which attract compassion. Most people in these regions will never see an otter, yet they are fascinated by them, and are prepared to pay for the otters' survival. A telephone survey in Britain established that the average person is willing to contribute almost £12 (US$21) in tax for the conservation of otters (White *et al.* 1997), an amount that puts the puny expenditure by government in this direction to shame. This would provide for a large government effort towards the conservation of wetlands, in the widest sense. Although public sentiment in other countries, especially in the developing world, is unlikely to match this, such quantification of public concern should provide impetus for both national and international conservation organizations.

If one thinks in terms of wetland conservation as a practical application of the extensive observations on otter food, foraging, populations and pollutants, this would cover the vital importance of sufficient populations of prey species. Repeatedly, throughout the research in the Americas, in Shetland and elsewhere, in seas, rivers and lakes in Scotland, Africa and Asia, we found observations consistent with the idea that otters are food limited. Our model suggested that even a relatively small decline in the food supply can have disastrous effects.

Translating this into practical suggestions for conservation: when facing diminishing fish populations, waters could be made more suitable for otters by improving the conditions for fish or, for instance, by establishing and stocking fish ponds (as suggested for Austria by Foster-Turley *et al.* 1990), or by providing compensation to existing fish farms for damage by otters. The trophic status of streams can be improved (e.g. by liming the more acid streams as in Scandinavia or, better still, addressing the causes of acid rain). Anything that adversely affects fish populations or, for sea otters, clam populations should be discouraged, including intensive fishing, and especially pollution. Where otters use specific aspects of the habitat for their fishing (e.g. rough banks, and riffles in rivers for **Eurasian otters** [Carss *et al.* 1990], or semi-aquatic scrub for the **spotted-necked otter** [Kruuk and Goudswaard 1990]), the management of such parameters of food availability would be important. Obviously, the effects need to be monitored closely.

Some of the methods that could be used to 'promote' otters would be highly artificial, and therefore controversial. Maintaining specially stocked fish ponds for otters amounts to otter farming. In addition, there could be important side-effects of putative 'improvements' for otters to other aspects of the environment. For instance, a species such as the **Cape clawless otter** may benefit from eutrophication of streams, as this would encourage crabs, and it may be that the introduction of American crayfish into Europe has boosted **Eurasian otter** numbers in Spain and Portugal, and perhaps **Cape clawless otters** in Kenya. However, the wider effects of such changes on vegetation, and on other species of animals, can be profound, and need to be taken into account.

Other results from recent research on habitat usage by otters are relevant to otter conservation. Trees along river banks look attractive to us, and they may even benefit salmonids and other prey fish. However, they appear to have little effect on otter dispersion or movement, despite the fact that otters spraint near them (Jenkins 1981, and others). Conclusions on habitat preferences of otters have often been based on spraint distribution, but we now know that this may lead to false deductions. In contrast, radio-tracking showed that islands in rivers or lakes, even artificial ones, are used by animals even from a long distance away, so they may constitute important elements in the habitat, and should be protected. Favourite bank vegetation for resting and sleeping includes reed beds, thick bushes such as bramble, gorse, and in Scotland often rhododendron. If one is going to design 'otter havens', these habitat preferences could be highly relevant. Constructing artificial holts may be less effective, because in most habitats otters sleep above ground.

An observation on habitat use that has recurred frequently for several otter species is that they do not appear to avoid houses, industry, roads, campsites or much of the mess left by people (Fig. 14.4). Artificial

Figure 14.4 The threat of oil spills at sea: an oil tanker loading in Shetland. In the foreground is a boom to protect any spillage from spreading.

embankments and protective piles of concrete blocks attract otters. Such developments may be undesirable for various environmental reasons, but, as far as otter conservation is concerned, it is difficult to argue that these riparian developments are detrimental—often, otters will tolerate them or take to them with alacrity.

One crucial habitat issue has been identified along sea coasts. Otters of many species cannot do without fresh water in the form of small streams and pools, or holts going down to groundwater level. They need frequent freshwater baths to be able to survive the consequences of sea-water foraging. Development such as land drainage, which reduces access to fresh water, is likely to have an immediate effect on otter numbers, and, for conservation purposes, coasts that are well provided with pools and/or peaty soils should be protected adequately. At present, many of the really 'good' otter coasts in any of those areas have no protection of any sort, often because visually they are not particularly spectacular. Along some coasts, otters of several different species could benefit from small, artificial, freshwater pools, which could increase numbers effectively.

These points do not detract from the fact that a conservation policy is needed for entire wetlands. The above comments are merely suggestions to pursue the restricted aim of maximizing numbers of otters; probably, we now have the substantial knowledge needed to follow such a course. However, merely maximizing numbers of otters, of whatever species, is not necessarily a desirable management option. Some types of wetland, such as acid and nutrient-poor bogs, are inherently unsuitable as otter habitats, as there are few fish or other prey. In theory it should be possible to attract the animals there, with moderate agricultural eutrophication that would bring invertebrates and fish. Otters could follow—but we would have lost a whole, fascinating, oligotrophic community. In order to manage an area—a wetland or a river catchment or a coast—one has to decide first what kind of community is desired, and otters are not necessarily part of that.

Conflicts occur between the conservation of otters and that of other species, especially with regard to predation. This may be predation by otters on declining species of fish, such as various species of salmon in North America, or Atlantic salmon in Scotland where in some rivers, after severe declines of fish populations from other causes, Eurasian otters consume a high proportion of salmon numbers and productivity. But, most dramatically, conflicts arise when predators on the otters themselves arrive on the scene.

As an example, in eastern and southern Africa there are many places where **Cape clawless** and, to a lesser extent, **spotted-necked otters** are quite abundant in rivers exposed to agriculture and human disturbance. However, along rivers in national parks such as Kruger or Serengeti, otters are few, seen only rarely, and one finds only the occasional track of a Cape clawless otter on the many sandy places, and no sign of the spotted-necked. The likely explanation for the difference compared with more agricultural areas is the presence of many crocodiles in the protected rivers, and their virtual absence elsewhere, because of poaching. Conservation of crocodiles is at odds with the protection of otters.

More serious and urgent is the conflict between **sea otter** conservation and that of their main predators, the killer whale and, to a lesser extent, the white shark. All three are flagship species, and at the time of writing the sea otter seems to be heading for extinction along several coasts, largely because of the activities of the other two (with killer whales themselves seriously affected by human activities). Rarely has the need for management of an entire ecosystem been more urgent, but because of our uncertainty on a course of action, with the enormous areas involved, with the politics and economics of whaling and fisheries management, and the slow response of the populations concerned to any changes, the outlook for sea otters is bleak.

On a smaller scale, questions need to be addressed such as how many fish one can harvest before affecting numbers of top predators, how nutrient input from agriculture and forestry affects fish populations (through plankton and invertebrates), how organochlorines, mercury and other pollutants affect the food web. We need to know much more about these problems and several others before we can feel some confidence that we are managing rationally. I hope that at least some conservation agencies will direct funding towards these ends, because rational management of wetlands is vitally important.

Wherever humankind is a prominent element of the habitat, various otter species, especially in Latin America and Asia, are exploited for their fur, usually illegally. This may cause numbers to be much lower than they could be in any of the areas concerned, but to my knowledge only in the case of the **giant otter** is there good evidence that poaching may cause local extinctions (Schenck 1997). In any case, whether the exploitation is legal or not, the nature of these areas makes policing virtually impossible. However, instead of persecuting poachers, an important alternative to conservation managers is to use the presence of otters as an attraction for eco-tourism, as has been successfully achieved in the Amazon of Peru and Brazil, and in the Pantanal. In these areas, both giant and **neotropical otters** are high on the list of 'must sees' of many visitors, who make a substantial contribution to the local economy, and this, as well as the presence of tourists themselves, tends to discourage poaching. There is a great deal of potential for this in many other tropical countries.

Other human assaults on entire habitats are of concern, not just to otters but to the entire ecosystem, to all animals and plants therein. An oil spill such as that of the *Exxon Valdez* in Prince William Sound in 1989 killed **sea otters** and **river otters** in large numbers. Over the years after the spill the populations recovered—as they always will—but the suffering that was caused was horrendous, and the remains of the oil will be there in the substrate along the shores, probably for ever. It is vitally important that stricter regulations for oil tankers are enforced, worldwide.

The organochlorine pollution of the twentieth century, costing the lives of untold numbers of otters in Europe and North America, has more or less disappeared; otters have returned in many places, and their numbers are increasing still. Some compounds, however, such as PCBs, are still in the environment, especially in the seas. These and others, such as

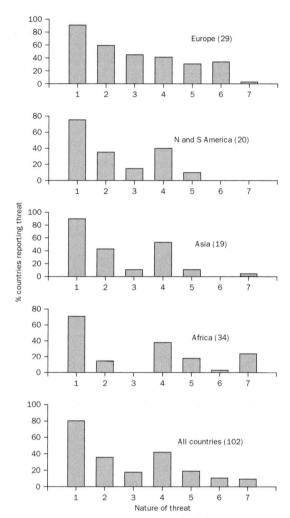

Figure 14.5 Threats to national otter populations, as recorded by an IUCN worldwide survey of observers (Foster-Turley *et al.* 1990). All species of otter were combined; the numbers of countries are given in parentheses. 1, Habitat deterioration; 2, pollution; 3, accidents; 4, persecution; 5, availability of fish; 6, recreation; 7, disturbance. Overall, habitat deterioration was of most concern, followed by persecution and pollution. In Europe, road accidents are perceived to be more important than elsewhere.

mercury, are concentrated in otters' bodies, although the damage they cause to populations is as yet an unknown factor. The threat from new, unrecognized, pollutants is still there and, in the meantime, the all-embracing, global change in climate is going on unabated.

Management of almost all types of habitat by people is now an unavoidable necessity. Especially in the case of otters, because of the size of areas used by these animals, a strong human influence, including agriculture or fishing, will almost necessarily have to be included in any management plan. It is possible, however, to accommodate this next to an impressive diversity of wild fauna, and I believe that it is one of our more important duties as research scientists to advise on how it can be done.

The general picture of today's concerns about otters worldwide shows a change since the 1990s. Figure 14.5 shows what conservationists perceived as the most severe threats to otters in different continents, based on data from Foster-Turley *et al.* (1990), and before much was known about feeding ecology, disease and predation. Then, habitat change and persecution were seen as the main sources of threat. These hazards are still with us, but others, as discussed in previous chapters, have increased in significance.

Conservation of otter species

The International Union for the Conservation of Nature (IUCN 2001) has assembled data on the present status of many vertebrates, including the otters. It recognizes a species' status as *extinct* (no reasonable doubt that the last individual has died), *endangered* (very high risk of extinction in the wild in the next few years), *vulnerable* (a high risk of extinction in the wild), *near threatened* (close to the vulnerable category, or likely to be in the near future), *least concern* (widespread and abundant) or *data deficient* (inadequate information on abundance and distribution).

On this scale, the otters are listed as:

Eurasian otter *Lutra lutra*	Near threatened
River otter *Lontra canadensis*	Least concern
Sea otter *Enhydra lutris*	Endangered
Neotropical otter *Lontra longicaudis*	Data deficient
Southern river otter *L. provocax*	Endangered
Marine otter *L. felina*	Endangered
Giant otter *Pteronura brasiliensis*	Endangered
Smooth otter *Lutrogale perspicillata*	Vulnerable
Hairy-nosed otter *Lutra sumatrana*	Data deficient
Small-clawed otter *Aonyx cinereus*	Least concern
Cape clawless otter *A. capensis*	Least concern
Congo clawless otter *A. congicus*	Data deficient
Spotted-necked otter *Lutra maculicollis*	Vulnerable

Three of the 13 otter species are *data deficient*, that is, we know so little about them that we cannot even assess whether they are abundant or almost extinct. Almost half are *endangered* or *vulnerable*, with a very high or a high risk of extinction in the wild in the foreseeable future. The two species that occur in the human population centres of the west, the Eurasian otter and the North American river otter, which have faced large threats from chemical pollution and persecution in the past, now appear to be relatively safe.

Much of the ecological research has been done on 'safe' species such as these two, and the Cape clawless otter. In the case of the sea otter, most research and assessment was done at the time when it appeared to be safe and increasing, before the very recent collapse of the species, so we know a great deal about these animals. A fair amount is now also known about the conspicuously endangered species.

The most threatened species are the endangered sea otter, giant, southern river and marine otters. The last two have an uncomfortably small geographical range in South America, occurring in areas where persecution is difficult to prevent. The giant otter is now recovering from previous slaughter by poachers, and has become a popular target for ecotourism, but is often demonstrably disturbed by the visitors in loud motorboats in the Amazon and Pantanal areas.

Figure 14.6 Giant otter splashing in the Amazon. © Nicole Duplaix.

We now have at least some knowledge of these and some other species, but for all otters there still is a need for surveys, and for studies of populations, and we need to learn how to interpret the results of the fieldwork. Almost all surveys are based on the presence or absence of faeces, spraints, and the use of this method may well be inescapable, despite its limitations. Although the sources of error are conservative (i.e. they give rise to underestimates rather than overestimates of otter presence), they do mean that at present we have no way of translating spraint densities into otter densities. We still are a long way from understanding the conservation status of otters in many regions, but often this may not be as bad as it appears. New techniques based on DNA analysis will bring substantial progress in this field.

Conservationists often claim declines, even where sound evidence is lacking. Sometimes detailed figures on numbers of otters are presented where, in fact, these are no more than guesses. This is regrettable, because it reduces the confidence of the public and policy makers. There is a need for openness about our ignorance, as well as for critical assessment of conclusions that fieldworkers can draw from their observations.

Above all, scientists who work in the field should communicate their research to a wide public, and include their legitimate doubts and concerns. The questions we ask about otter survival, about the ecology and behaviour of these animals, are of interest to many. Getting people's attention for wetland ecosystems and the otters themselves is not just a question of pretty pictures and simple alarm calls from conservationists. It is a matter of intelligent engagement of naturalists, of fishing people and of anyone involved in land management.

Despite all the problems that otters face, be it the sea or river otters along the Pacific coasts, Eurasian otters along the rivers in Europe, smooth otters in a national park in India, giant otters in Amazonia (Fig. 14.6) or spotted-necks in a lake in Africa, it is possible for many people to watch and study them. With their beautiful adaptations others are able to exist between a rock and a wet place, and, if one can understand how they manage, this will enrich us, and it will help the otters' survival. Between them, scientists, naturalists and conservationists have every incentive to ensure that the classical image—of a small head in a ring of water—will always be with us.

References

Adamek Z, Kortan D, Lepic P and Andreji J (2003). Impacts of otter (*Lutra lutra* L.) predation on fishponds. *Aquaculture International*, **11**, 389–396.

Adrian MI and Delibes M (1987). Food habits of the otter (*Lutra lutra*) in two habitats of the Doñana National Park, SW Spain. *Journal of Zoology, London*, **212**, 399–406.

Alarcon GG and Simoes-Lopes PC (2003). Preserved versus degraded coastal environments: a case study of the neotropical otter in the environmental protection area of Anhatomirim, southern Brazil. *IUCN Otter Specialist Group Bulletin*, **20**, 6–18.

Alarcon GG and Simoes-Lopes PC (2004). The neotropical otter *Lontra longicaudis* feeding habits in a marine coastal area, southern Brazil. *IUCN Otter Specialist Group Bulletin*, **21**, 24–30.

Albertson RC, Markert JA, Danley PD and Kocher TD (1999). Phylogeny of a rapidly evolving clade: the cichlid fishes of Lake Malawi, East Africa. *Proceedings of the National Academy of Sciences of the USA*, **96**, 5107–5110.

Ames JA and Morejohn GV (1980). Evidence of white shark, *Carcharodon carcharias*, attacks on sea otters, *Enhydra lutris*. *California Fish and Game*, **66**, 196–209.

Andrews E and Crawford AK (1986). *Otter survey of Wales 1984–1985*. Vincent Wildlife Trust, London, UK.

Angelici FM, Politano E, Bogudue AJ and Luiselli L (2005). Distribution and habitat of otters (*Aonyx capensis* and *Lutra maculicollis*) in southern Nigeria. *Italian Journal of Zoology*, **72**, 223–227.

Anon. (2003a). Forellenfang in Schweizer Fliessgewässern. www.fischnetz.ch

Anon. (2003b). Suspected cat link to sea otter disease. *Marine Pollution Bulletin*, **46**, 272–272.

Anoop KR and Hussain SA (2004). Factors affecting habitat selection by smooth-coated otters (*Lutra perspicillata*) in Kerala, India. *Journal of Zoology, London*, **263**, 417–423.

Anoop KR and Hussain SA (2005). Food and feeding habits of smooth-coated otters (*Lutrogale perspicillata*) and their significance to the fish population of Kerala, India. *Journal of Zoology, London*, **266**, 15–23.

Ansorge H, Schipke R and Zinke O (1997). Population structure of the otter, *Lutra lutra*. Parameters and model for a Central European region. *Zeitschrift für Säugetierkunde*, **62**, 143–151.

Anthony RG, Miles AK, Estes JA and Isaacs FB (1999). Productivity, diets, and environmental contaminants in nesting Bald Eagles from the Aleutian Archipelago. *Environmental Toxicology and Chemistry*, **18**, 2054–2062.

Arden-Clarke C (1983). *Population density and social organisation of the Cape clawless otter, Aonyx capensis Schinz, in the Tsitsikama Coastal National Park*. MSc Thesis, University of Pretoria, South Africa.

Arden-Clarke CHG (1986). Population density, home range and spatial organization of the Cape clawless otter, *Aonyx capensis*, in a marine habitat. *Journal of Zoology, London*, **209**, 201–211.

Aued MB, Chéhebar C, Porro G, Macdonald DW and Cassini MH (2003). Environmental correlates of the distribution of southern river otters *Lontra provocax* at different ecological scales. *Oryx*, **37**, 413–421.

Bacon PJ, Ball FG and Blackwell PG (1991). A model for territory and group formation in a heterogenous habitat. *Journal of Theoretical Biology*, **148**, 445–468.

Baker JR, Jones AM, Jones TP and Watson HC (1981). Otter *Lutra lutra* L. mortality and marine oil pollution. *Biological Conservation*, **20**, 311–321.

Baranga J (1995). The distribution and conservation status of otters in Uganda. *Habitat*, **11**, 29–32.

Barel CDN, Dorit R, Greenwood PH et al. (1985). Destruction of fisheries in Africa's lakes. *Nature* **315**, 19–20.

Bartholomew GA (1977). Energy metabolism. In MS Gordon (ed.), *Animal physiology: principles and adaptations*, pp. 57–110. Macmillan, New York, USA.

Beach FA and Gilmore RN (1949). Response of male dogs to urine from females in heat. *Journal of Mammalogy*, **30**, 391–392.

Beckel AL (1990). Foraging success rates of North American river otter, *Lutra canadensis*, hunting alone and hunting in pairs. *Canadian Field-Naturalist*, **104**, 586–588.

Begg CM, Begg KS, Du Toit JT and Mills MGL (2003). Scent-marking behaviour of the honey badger, *Mellivora capensis* (Mustelidae), in the southern Kalahari. *Animal Behaviour*, **66**, 917–929.

Beja PR (1991). Diet of otters (*Lutra lutra*) in closely associated freshwater, brackish and marine habitats in southwest Portugal. *Journal of Zoology, London*, **225**, 141–152.

Beja PR (1992). Effects of freshwater availability on the summer distribution of otters *Lutra lutra* on the southwest coast of Portugal. *Ecography*, **15**, 273–278.

Beja PR (1995a). *Patterns of availability and use of resources by otters* (Lutra lutra L.) *in southwest Portugal*. PhD Thesis, University of Aberdeen, Aberdeen, UK.

Beja PR (1995b). Structure and seasonal fluctuations of rocky littoral fish assemblages in south-western

Portugal: implications for otter prey availability. *Journal of the Marine Biological Association, UK*, **75**, 833–847.

Beja PR (1996a). An analysis of otter *Lutra lutra* predation on introduced American crayfish *Procambarus clarkii* in Iberian streams. *Journal of Applied Ecology*, **33**, 1156–1170.

Beja PR (1996b). Temporal and spatial patterns of rest-site use by four female otters *Lutra lutra* along the south-west coast of Portugal. *Journal of Zoology, London*, **239**, 741–753.

Beja PR (1997). Predation by marine-feeding otters (*Lutra lutra*) in south-west Portugal in relation to fluctuating food resources. *Journal of Zoology, London*, **242**, 503–518.

Bekoff M (1989). Behavioural development of terrestrial carnivores. In JL Gittleman (ed.), *Carnivore behavior, ecology, and evolution*, pp. 89–124. Cornell University Press, Ithaca, NY, USA.

Bell D, Robertson S and Hunter PR (2004). Animal origins of the SARS coronavirus: possible links with the international trade in small carnivores. *Philosophical Transactions of the Royal Society, London, Series B*, **359**, 1107–1114.

Ben-David M, Bowyer RT and Faro JB (1995). Niche separation by mink and river otters: co-existence in a marine environment. *Oikos*, **75**, 41–48.

Ben-David M, Bowyer RT, Duffy LK, Roby DD and Schell DM (1998). Social behavior and ecosystem processes: river otter latrines and nutrient dynamics of terrestrial vegetation. *Ecology*, **79**, 2567–2571.

Ben-David M, Williams TM and Ormseth OA (2000). Effects of oiling on exercise physiology and diving behavior of river otters: a captive study. *Canadian Journal of Zoology*, **78**, 1380–1390.

Bergan JF (1990). Kleptoparasitism of a river otter, *Lutra canadensis*, by a bobcat, *Felis rufus*, in South Carolina (Mammalia: Carnivora). *Brimleyana*, **16**, 63–65.

Bergheim A and Hesthagen T (1990). Production of juvenile Atlantic salmon, *Salmo salar* L., within different sections of a small enriched Norwegian river. *Journal of Fish Biology*, **36**, 545–562.

Berry RJ (2000). *Orkney nature*. Academic Press, London, UK.

Berry RJ and Johnston JL (1980). *The natural history of Shetland*. Collins, London, UK.

Bininda-Emonds ORP, Gittleman JL and Purvis A (1999). Building large trees by combining phylogenetic information: a complete phylogeny of the extant Carnivora (Mammalia). *Biological Reviews*, **74**, 143–175.

Blundell (2001). *Social organization and spatial relationships in coastal river otters: assessing form and function of social groups, sex-biased dispersal, and gene flow*. PhD Thesis, University of Alaska, Fairbanks, Alaska, USA.

Blundell GM, Kern JW, Bowyer RT and Duffy LK (1999). Capturing river otters: a comparison of Hancock and leg-hold traps. *Wildlife Society Bulletin*, **27**, 184–192.

Blundell GM, Bowyer RT, Ben-David M, Dean TA and Jewett SC (2000). Effects of food resources on spacing behavior of river otters: does forage abundance control home-range size? In JH Eiler, DJ Alcorn and MR Neuman (eds), *Proceedings of the Fifteenth International Symposium on Biotelemetry*, pp. 325–333. International Society of Biotelemetry, Wageningen, Netherlands.

Blundell GM, Maier JAK and Debevec EM (2001). Linear home ranges: effects of smoothing, sample size, and autocorrelations on kernel estimates. *Ecological Monograph*, **71**, 469–489.

Blundell GM, Ben-David M, Groves P, Bowyer RT and Geffen E (2002). Characteristics of sex-biased dispersal and gene flow in coastal river otters: implications for natural recolonization of extirpated populations. *Molecular Ecology*, **11**, 289–303.

Blundell GM, Ben-David M, Groves P, Bowyer RT and Geffen E (2004). Kinship and sociality in coastal river otters: are they related? *Behavioural Ecology*, **15**, 705–714.

Bodkin JL, Ballachey BE, Cronin MA and Scribner KT (1999). Population demographics and genetic diversity in remnant and translocated populations of sea otters. *Conservation Biology*, **13**, 1378–1385.

Bodkin JL, Burdin AM and Ryazanov DA (2000). Age- and sex-specific mortality and population structure in sea otters. *Marine Mammal Science*, **16**, 201–219.

Bodkin JL, Esslinger GG and Monson DH (2004). Foraging depths of sea otters and implications to coastal marine communities. *Marine Mammal Science*, **20**, 305–321.

Bodner, M. (1998). Damage to fish ponds as a result of otter (*Lutra lutra*) predation. *BOKU Report on Wildlife Reserves and Game Management (Vienna)*, **14**, 106–117.

Bohlin T, Hamrin S, Heggberget TG, Rasmussen G and Saltveit SJ (1989). Electrofishing—theory and practice with special emphasis on salmonids. *Hydrobiologia*, **173**, 9–43.

Bonesi L and Macdonald DW (2004a). Differential habitat use promotes sustainable coexistence between the specialist otter and the generalist mink. *Oikos*, **106**, 509–519.

Bonesi L and Macdonald DW (2004b). Impact of released Eurasian otters on a population of American mink: a test using an experimental approach. *Oikos*, **106**, 9–18.

Bonesi L, Chanin P and Macdonald DW (2004). Competition between Eurasian otter *Lutra lutra* and American mink *Mustela vison* probed by niche shift. *Oikos*, **106**, 19–26.

Borgwardt N and Culik BM (1999). Asian small-clawed otters (*Amblonyx cinerea*): resting and swimming metabolic rates. *Journal of Comparative Physiology B*, **169**, 100–106.

Bowyer RT, Testa JW, Faro JB, Schwartz CC and Browning JB (1994). Changes in diets of river otters in Prince William Sound, Alaska: effects of the *Exxon Valdez* oil spill. *Canadian Journal of Zoology*, **72**, 970–976.

Bowyer RT, Testa JW and Faro JB (1995). Habitat selection and home ranges of river otters in a marine environment: effects of the *Exxon Valdez* oil spill. *Journal of Mammalogy*, **76**, 1–11.

Bowyer RT, Blundell GM, Ben-David M, Jewett SC, Dean TA and Duffy LK (2003). Effects of the *Exxon Valdez* oil spill on river otters: injury and recovery of a sentinel species. *Wildlife Monographs*, **153**, 1–53.

Bradbury JW and Vehrencamp SL (1976). Social organisation and foraging in Emballonurid bats. II. A model for the determination of group size. *Behavioral Ecology and Sociobiology*, **2**, 1–17.

Brandstätter F (2005). Maximum age of giant otters in captivity. *Friends of the Giant Otter*, **12**, 5.

Breathnach S and Fairley JS (1993). The diet of otters *Lutra lutra* (L.) in the Clare River system. *Biology and Environment Proceedings of the Royal Irish Academy*, **93B**, 151–158.

Breen PA, Carson TA, Foster JB and Stewart EA (1982). Changes in subtidal community structure associated with British sea otter transplants. *Marine Ecological Programme Series*, **7**, 13–20.

Breitenmoser U, Breitenmoser-Würsten C, Carbyn LN and Funk SM (2001). Assessment of carnivore reintroductions. In JL Gittleman, SM Funk, DW Macdonald and RK Wayne (eds), *Carnivore conservation*, pp. 241–281. Cambridge University Press, Cambridge, UK.

Broekhuizen S (1989). Belasting van otters met zware metalen en PCBs. *De Levende Natuur*, **90**, 43–47.

Brzezinski M, Jedrzejewski W and Jedrzejewska B (1993). Diet of otters (*Lutra lutra*) inhabiting small rivers in the Bialowieza National Park, eastern Poland. *Journal of Zoology, London*, **230**, 495–501.

Bunyan PJ, Stanley PI, Blundell CA, Wardall GL and Tarrant KA (1975). The investigation of pesticide and wildlife incidents. *Reports of the Pest Infestation Control Laboratory 1971–1973*, Ministry of Agriculture, Guildford, UK.

Butler JRA and du Toit JT (1994). Diet and conservation status of Cape clawless otters in eastern Zimbabwe. *South African Journal of Wildlife Research*, **24**, 41–47.

Buxton A (1946). *Fisherman naturalist*. Collins, London, UK.

Calkins DG (1978). Feeding behavior and major prey speices of the sea otter, *Enhydra lutris*, in Montague Strait, Prince William Sound, Alaska. *Fisheries Bulletin*, **76**, 125–131.

Carr GM and Macdonald DW (1986). The sociability of solitary foragers: a model based on resource dispersion. *Animal Behaviour*, **34**, 1540–1549.

Carr RA, Jones MM and Russ ER (1974). Anomalous mercury in near-bottom water of a mid-Atlantic rift valley. *Nature, London*, **251**, 249.

Carss DN (1995). Foraging behaviour and feeding ecology of the otter *Lutra lutra*: a selective review. *Hystrix*, **7**, 179–194.

Carss DN and Elston DA (1996). Errors associated with otter *Lutra lutra* faecal analysis. II. Estimating prey size distribution from bones recovered in spraints. *Journal of Zoology, London*, **238**, 319–332.

Carss DN and Nelson KC (1998). Cyprinid prey remains in otter *Lutra lutra* faeces: some words of caution. *Journal of Zoology, London*, **245**, 238–244.

Carss DN and Parkinson SG (1996). Errors associated with otter *Lutra lutra* faecal analysis. I. Assessing general diet from spraints. *Journal of Zoology, London*, **238**, 301–317.

Carss DN, Kruuk H and Conroy JWH (1990). Predation on adult Atlantic Salmon, *Salmo salar*, by otters *Lutra lutra* within the River Dee system, Aberdeenshire, Scotland. *Journal of Fish Biology*, **37**, 935–944.

Carss DN, Elston DA and Morley HS (1998a). The effects of otter (*Lutra lutra*) activity on spraint production and composition: implications for models which estimate prey-size distribution. *Journal of Zoology, London*, **244**, 295–302.

Carss DN, Nelson, KC, Bacon PJ and Kruuk H (1998b). Otter (*Lutra lutra*) prey selection in relation to fish abundance and community structure in two different freshwater habitats. *Symposia of the Zoological Society of London*, **71**, 191–213.

Carss DN, Elston DN, Nelson KC and Kruuk H (1999). Spatial and temporal trends in unexploited yellow eel stocks in two shallow lakes and associated streams. *Journal of Fish Biology*, **55**, 636–654.

Carter SK and Rosas FCW (1997). Biology and conservation of the giant otter *Pteronura brasiliensis*. *Mammal Review* **27**, 1–26.

Carugati C and Perrin MR (2003a). Habitat, prey and area requirements of otters (*Aonyx capensis*) and (*Lutra maculicollis*) in the Natal Drakensberg, South Africa. *Proceedings of the Eighth International Otter Colloquium, Trebon, Czech Republic*, pp. 18–23.

Carugati C and Perrin MR (2003b). Requirements of the cape clawless otter (*Aonyx capensis*) and the spotted-necked otter (*Lutra maculicollis*) in the Natal Drakensberg, South Africa. *Proceedings of the Eighth International Otter Colloquium, Trebon, Czech Republic*, pp. 12–17.

Castilla JC (1982). Nuevas observaciones sobre conducta, ecologia y densidad de *Lutra felina* (Molina 1782)

(Carnivora: Mustelidae) en Chile. *Museo Nacional de Historia Natural (Santiago, Chile), Publicación Ocasional*, **38**, 187–206.

Castonguay M, Hodson PV, Moriarty C, Drinkwater KF and Jessop BM (1994). Is there a role of ocean environment in American and European eel decline? *Fisheries Oceanography*, **3**, 197–203.

Caughley G (1977). *Analysis of vertebrate populations*. Wiley, London, UK.

Chanin PRF (1981). The diet of the otter and its relations with the feral mink in two areas of south-west England. *Acta Theriologica*, **26**, 83–95.

Chanin PRF (1985). *The natural history of otters*. Croom Helm, London, UK.

Chanin PRF (1991). Otter *Lutra lutra*. In GB Corbet and S Harris (eds), *The Handbook of British Mammals*, Blackwell Scientific Publications, Oxford, UK.

Chanin P (2003). Monitoring the otter *Lutra lutra*. *Conserving Natura 2000 Rivers Monitoring Series*, **10**, 1–43.

Chanin PRF and Jefferies DJ (1978). The decline of the otter *Lutra lutra* L. in Britain: an analysis of hunting records and discussion of causes. *Biological Journal of the Linnaean Society*, **10**, 305–328.

Cheeseman CL, Wilesmith JW, Ryan J and Mallinson PJ (1987). Badger population dynamics in a high-density area. *Symposia of the Zoological Society of London*, **58**, 279–294.

Chéhebar CE (1985). A survey of the southern river otter *Lutra provocax* Thomas in Nahuel Huapi National Park, Argentina. *Biological Conservation*, **32**, 299–307.

Clark JD (1981). *Pollution trends in the river otter of Georgia*. MSc Thesis, University of Georgia, Athens, USA.

Clavero M, Prenda J and Delibes M (2003). Trophic diversity of the otter (*Lutra lutra* L.) in temperate and Mediterranean freshwater habitats. *Journal of Biogeography*, **30**, 761–769.

Clode D and Macdonald DW (1995). Evidence for food competition between mink (*Mustela vison*) and otter (*Lutra lutra*) on Scottish islands. *Journal of Zoology, London*, **237**, 435–444.

Clutton-Brock TH (1991) *The evolution of parental care*. Princeton University Press, Princeton, NJ, USA.

Clutton-Brock TH and Harvey PH (1977). Primate ecology and social organization. *Journal of Zoology, London*, **183**, 1–39.

Clutton-Brock TH, Guinness FE and Albon SD (1982). *Red deer: behaviour and ecology of two sexes*. Edinburgh University Press, Edinburgh, UK.

Conroy JWH and French DD (1987). The use of spraints to monitor populations of otters (*Lutra lutra* L.). *Symposia of the Zoological Society of London*, **58**, 247–262.

Conroy JWH and Jenkins D (1986). Ecology of otters in northern Scotland VI. Diving times and hunting success of otters at Dinnet Lochs, Aberdeenshire and in Yell Sound, Shetland. *Journal of Zoology, London*, **209**, 341–346.

Conroy J and Kruuk H (1995). Changes in otter numbers in Shetland between 1988 and 1993. *Oryx*, **29**, 197–204.

Conroy JWH, Watt J, Webb, JB and Jones A (1993). A guide to the identification of prey remains in otter spraints. *Occasional Publications of the Mammal Society*, **16**, 1–52.

Cooke SJ, Bunt CM, Hamilton SJ et al. (2005). Threats, conservation strategies, and prognosis for suckers (Catostomidae) in North America: insights from regional case studies of a diverse family of non-game fishes. *Biological Conservation*, **121**, 317–331.

Corbet GB and Harris S (1991). *The handbook of British mammals*. Blackwell Scientific Publications, Oxford, UK.

Corbett LK (1979). *Feeding ecology and social organisation of wild cats* (Felis sylvestris) *and domestic cats* (Felis catus) *in Scotland*. PhD Thesis, University of Aberdeen, UK.

Correia AM (2001). Seasonal and interspecific evaluation of predation by mammals and birds on the introduced red swamp crayfish *Procambarus clarkii* (Crustacea, Cambaridae) in a freshwater marsh (Portugal). *Journal of Zoology, London*, **255**, 533–541.

Costa DP (1982). Energy, nitrogen, and electrolyte flux and sea-water drinking in the sea otter *Enhydra lutris*. *Physiological Zoology*, **55**, 35–44.

Costa DP and Kooyman GL (1982). Oxygen consumption, thermo-regulation, and the effect of fur oiling and washing on the sea otter *Enhydra lutris*. *Canadian Journal of Zoology*, **60**, 2761–2767.

Cotgreave P (1995). Relative importance of avian groups in the diets of British and Irish predators. *Bird Study*, **42**, 246–252.

Crawford A (2003). *Fourth otter survey of England 2000–2002*. Environment Agency, Bristol, UK.

Crawford AK, Jones A, Evans D and McNulty J (1979). *Otter survey of Wales 1977–1978*. Society for the Promotion of Nature Conservation, Nettleham, Lincoln, UK.

Creel S and Creel NM (1998). Six ecological factors that may limit African wild dogs, *Lycaon pictus*. *Animal Conservation*, **1**, 1–9.

Creel SR and Creel NM (1991). Energetics, reproductive suppression and obligate communal breeding in carnivores. *Behavioral Ecology and Sociobiology*, **28**, 263–270.

Cronin MA, Bodkin J, Ballachey B, Estes J and Patton JC (1996). Mitochondrial-DNA variation among subspecies and populations of sea otters (*Enhydra lutris*). *Journal of Mammalogy*, **77**, 546–557.

Dallas JF and Piertney SB (1998). Microsatellite primers for the Eurasian otter. *Molecular Ecology*, **7**, 1248–1251.

Dallas JF, Bacon PJ, Carss DJ et al. (1999). Genetic diversion in the Eurasian otter, *Lutra lutra*, in Scotland. Evidence from microsatellite polymorphism. *Biological Journal of the Linnean Society*, **68**, 73–86.

Dallas JF, Marshall F, Piertney SB, Bacon PJ and Racey PA (2002). Spatially restricted gene flow and reduced microsatellite polymorphism in the Eurasian otter *Lutra lutra* in Britain. *Conservation Genetics*, **3**, 15–29.

Dallas JF, Coxon KE, Sykes T et al. (2003). Similar estimates of population genetic composition and sex ratio derived from carcasses and faeces of Eurasian otter *Lutra lutra*. *Molecular Ecology*, **12**, 275–282.

Danilov PI and Tumanov IL (1975). The reproductive cycles of some Mustelidae species. *Byulleten Moskovskogo Obshchestva Ispytatelei (Oto. Biol.)*, **80**, 35–47.

Davenport L and Hajek F (2005). Violent encounters observed between cayman and giant otter. *Friends of the Giant Otter*, **12**, 6–8.

Davies JM (1981). Survey of trace elements in fish and shellfish landed at Scottish ports, 1975–76. *Scottish Fisheries Research Reports*, **19**, 1–28.

Dayan T and Simberloff D (1994). Character displacement, sexual dimorphism, and morphological variation among British and Irish mustelids. *Ecology*, **75**, 1063–1073.

De Silva PK (1997). Seasonal variation of the food and feeding habits of the Eurasian otter *Lutra lutra* (Carnivora: Mustelidae) in Sri Lanka. *Journal of the South Asian Natural History Society*, **2**, 205–216.

Dean TA, Bodkin JL, Fukuyama AK et al. (2002). Food limitation and the recovery of sea otters following the 'Exxon Valdez' oil spill. *Marine Ecology Progress Series*, **241**, 255–270.

Defler TR (1983). Associations of the giant river otter (*Pteronura brasiliensis*) with fresh-water dolphins (*Inia geoffrensis*). *Journal of Mammalogy*, **64**, 692.

Delibes M and Adrian I (1987). Effects of crayfish introduction on otter *Lutra lutra* food in the Doñana National Park, S.W. Spain. *Biological Conservation*, **42**, 153–159.

Delibes M, Ferreras P and Blazquez MC (2000). Why the Eurasian otter leaves a pond? An observational test of some predictions on prey depletion. *Terre et la Vie*, **55**, 57–65.

Desai JH (1974). Observations on the breeding habits of the Indian smooth otter *Lutrogale perspicillata* in captivity. *International Zoo Yearbook*, **14**, 123–124.

Doroff AM and DeGange AR (1994). Sea otter, *Enhydra lutris*, prey composition and foraging success in the northern Kodiak Archipelago. *Fishery Bulletin*, **92**, 704–710.

Doroff AM, Estes JA, Tinker T, Burn DM and Evans TJ (2003). Sea otter population declines in the Aleutian Archipelago. *Journal of Mammalogy*, **84**, 55–64.

Dubuc LJ, Krohn WB and Owen RB (1990). Predicting occurrence of river otters by habitat on Mount Desert Island, Maine. *Journal of Wildlife Management*, **54**, 594–599.

Duggins DO, Simenstad CA and Estes JA (1989). Magnification of secondary production by kelp detritus in coastal marine ecosystems. *Science*, **245**, 170–173.

Dunbar IF (1977). Olfactory preferences in dogs: the response of male and female beagles to conspecific odors. *Behavioral Biology*, **20**, 471–481.

Dunstone N (1993). *The mink*. Poyser, London, UK.

Duplaix N (1978). Synopsis of the status and ecology of the giant otter in Suriname. In N Duplaix (ed.), *Proceedings of the First Working Meeting of the Otter Specialist Group, Paramaribo, Suriname, 1977*, pp. 48–54. IUCN, Gland, Switzerland.

Duplaix N (1980). Observations on the ecology and behavior of the giant river otter *Pteronura brasiliensis* in Suriname. *Revue d'Ecology: La Terre et la Vie*, **34**, 495–620.

Duplaix N (1982). *Contribution à l'écologie et à l'ethologie de Pteronura brasiliensis Gmelin 1788 (Carnivora, Lutrinae): implications evolutives*. PhD Thesis, Université de Paris-sud, Paris, France.

Duplaix N (2004). Giant otter cub killed by black caiman. *Friends of the Giant Otter*, **10**, 4.

Duplaix-Hall N (1975). River otters in captivity. In RD Martin (ed.) *Breeding endangered species in captivity*, pp. 315–327. Academic Press, London, UK.

Durbin L (1993). *Food and habitat utilization of otters (Lutra lutra L.) in a riparian habitat—the River Don in north-east Scotland*. PhD Thesis, University of Aberdeen, UK.

Durbin LS (1996). Some changes in the habitat use of a free-ranging female otter *Lutra lutra* during breeding. *Journal of Zoology, London*, **240**, 761–764.

Durbin LS (1997). Composition of salmonid species in the estimated diet of otters (*Lutra lutra*) and in electrofishing catches. *Journal of Zoology, London*, **243**, 821–825.

Durbin LS (1998). Habitat selection by five otters *Lutra lutra* in rivers of northern Scotland. *Journal of Zoology*, **245**, 85–92.

Ebensperger LA and Botto-Mahan C (1997). Use of habitat, size of prey, and food-niche relationships of two sympatric otters in southernmost Chile. *Journal of Mammalogy*, **78**, 222–227.

Ebensperger LA and Castilla JC (1992). Habitat selection in land by the marine otter, *Lutra felina*, at Pan-de-Azucar-Island Azucar, Chile. *Revista Chilena de Historia Natural*, **65**, 429–434.

Effenberger S and Suchentrunk F (1999). RFLP analysis of the mitochondrial DNA of otters *Lutra lutra* from Europe—implications for conservation of a flagship species. *Conservation Biology*, **90**, 229–234.

Egglishaw HW (1970). Production of salmon and trout in a stream in Scotland. *Journal of Fish Biology*, **2**, 117–136.

Elliott JE, Henny CJ, Harris ML, Wilson LK and Norstrom RJ (1999). Chlorinated hydrocarbons in livers of american mink (*Mustela vison*) and river otter (*Lutra canadensis*) from the Columbia and Fraser River basins, 1990–1992. *Environmental Monitoring and Assessment*, **57**, 229–252.

Elliott JM (1984). Growth, size, biomass and production of young migratory trout *Salmo trutta* in a Lake District stream, 1966–83. *Journal of Animal Ecology*, **53**, 979–994.

Elmhirst R (1938). Food of the otter in the marine littoral zone. *Scottish Naturalist*, **1938**, 99–102.

Eltringham SK (1984). *Wildlife Resources and economic development*. John Wiley & Sons, Chichester, UK.

Emmerson W and Philip S (2004). Diets of Cape clawless otters at two South African coastal localities. *African Zoology*, **39**, 201–210.

Erickson DW and Hamilton DA (1988). Approaches to river otter restoration in Missouri. *Transactions of the North American Wildlife and Natural Resources Conference*, **53**, 404–413.

Erlinge S (1967). Home range of the otter *Lutra lutra* in Southern Sweden. *Oikos*, **18**, 186–209.

Erlinge S (1968). Territoriality of the otter *Lutra lutra* L. *Oikos*, **19**, 81–98.

Estes JA (1980). *Enhydra lutris*. Mammalian Species, American Society of Mammalogists, **133**, 1–8.

Estes JA (1989). Adaptations for aquatic living by carnivores. In JL Gittleman (ed.), *Carnivore behavior, ecology and evolution*, pp. 242–282, Cornell University Press, Ithaca, NY, USA.

Estes JA (1990). Growth and equilibrium in sea otter populations. *Journal of Animal Ecology*, **59**, 385–401.

Estes JA (2002). From killer whales to kelp: food web complexity in kelp forest ecosystems. *Wild Earth*, **12**, 24–28.

Estes JA and Harrold C (1988). Sea otters, sea urchins and kelp beds: some questions of scale. In GR VanBlaricom and JA Estes (eds), *The community ecology of sea otters*, pp. 116–150. Springer, Berlin, Germany.

Estes JA and Jameson RJ (1988). A double-survey estimate for sighting probability of sea otters in California. *Journal of Wildlife Management*, **52**, 70–76.

Estes JA and Palmisano JF (1974). Sea otters: their role in structuring nearshore communities. *Science*, **185**, 1058–1060.

Estes JA and VanBlaricom GR (1985). Sea-otters and shell-fisheries. In JR Beddington, RJH Beverton and DM Lavigne (eds), *Marine mammals and fisheries*, pp. 187–235. Allen and Unwin, London, UK.

Estes JA, Jameson RJ and Johnson AM (1981). Food selection and some foraging tactics of sea otters. In JA Chapman and D Pursley (eds), *Proceedings of the Worldwide Furbearers Conference*, pp. 606–641. University of Maryland Press, Baltimore, MD, USA.

Estes JA, Jameson RJ and Rhode EB (1982). Activity and prey selection in the sea otter: influence of population status on community structure. *American Naturalist*, **120**, 242–258.

Estes JA, Duggins DO and Rathbun GB (1989). The ecology of extinctions in kelp forest communities. *Conservation Biology*, **3**, 252–264.

Estes JA, Tinker MT, Williams TM and Doak DF (1998). Killer whale predation on sea otters linking oceanic and nearshore systems. *Science*, **282**, 473–476.

Estes JA, Crooks K and Holt R (2001). Ecological role of predators. In S Levin (ed.), *Encyclopedia of biodiversity*, Vol. 4, pp. 857–878. Academic Press, San Diego, CA, USA.

Estes JA, Hatfield BB, Ralls K and Ames J (2003a). Causes of mortality in California sea otters during periods of population growth and decline. *Marine Mammal Science*, **19**, 198–216.

Estes JA, Riedman ML, Staedler MM, Tinker MT and Lyon BE (2003b). Individual variation in prey selection by sea otters: patterns, causes and implications. *Journal of Animal Ecology*, **72**, 144–155.

Estes JA, Tinker MT, Doroff AM and Burn DM (2005). Continuing sea otter population declines in the Aleutian archipelago. *Marine Mammal Science*, **21**, 169–172.

Estes RD (1967). Predators and scavengers. *Natural History*, **76**, 20–29.

Estes RD (1969). Territorial behaviour of the wildebeest (*Connochaetes taurinus* Burchell 1823). *Zeitschrift für Tierpsychologie*, **26**, 284–370.

Ewer R (1973). *The carnivores*. Weidenfeld and Nicolson, London, UK.

Fagan R (1981). *Animal play behavior*. Oxford University Press, New York, USA.

Fangstrom B, Athanasiadou M, Grandjean P, Weihe P and Bergman A (2002). Hydroxylated PCB metabolites and PCBs in serum from pregnant Faroese women. *Environmental Health Perspectives*, **110**, 895–899.

Fanshawe S, VanBlaricom GR and Shelly AA (2003). Restored top carnivores as detriments to the performanace of marine protected areas intended for fishery sustainability: abalones and sea otters. *Conservation Biology*, **17**, 273–283.

Feeroz MM (2004). Otters (*Lutra* spp.) of the Sundarbans: status, distribution and use of otters in fishing. *Abstracts* of the Ninth Otter Colloquium, University of Frostburg, MD, USA.

Fernandez-Moran J, Saavedra D and Manteca-Vilanova X (2002). Reintroduction of the Eurasian otter in northeastern Spain: trapping, handling and medical management. Journal of Zoo and Wildlife Medicine, **33**, 222–227.

Fish FE (1994). Association of propulsive swimming mode with behavior in river otters (Lutra canadensis). Journal of Mammalogy, **75**, 989–997.

Foster-Turley P (1992). Conservation aspects of the ecology of Asian small-clawed and smooth otters on the Malay peninsula. IUCN Otter Specialist Group Bulletin, **7**, 26–29.

Foster-Turley P and Engfer S (1988). The species survival plan for the Asian small-clawed otter Aonyx cinerea. International Zoo Yearbook, **27**, 79–84.

Foster-Turley P, Macdonald SM and Mason CF (1990). Otters, an action plan for conservation. International Union for the Conservation of Nature, Gland, Switzerland.

Franklin IR (1980). Evolutionary changes in small populations. In ME Soulé and BA Wilcox (eds), Conservation biology. An evolutionary ecological perspective, pp. 135–139. Sinauer, Sunderland, MA, USA.

Fraser NHC, Metcalfe NB and Thorpe JE (1993). Temperature-dependent switch between diurnal and nocturnal foraging in Salmon. Proceedings of the Royal Society of London, B, **252**, 135–139.

Furuya Y (1977). Otters in Padas Bay, Sabah, Malaysia (Amblonyx cinerea). Journal of the Mammal Society of Japan, **7**, 39–43.

Garrott RA, Eberhardt LL and Burn DM (1993). Mortality of sea otters in PrinceWilliam Sound following the Exxon Valdez oil spill. Marine Mammal Science, **9**, 343–359.

Garshelis DL (1983). Ecology of sea otters in Prince William Sound, Alaska. PhD Thesis, University of Minnesota, MN, USA.

Garshelis DL (1997). Sea otter mortality estimated from carcasses collected after the Exxon Valdez oil spill. Conservation Biology, **11**, 905–916.

Garshelis DL and Garshelis JA (1984). Movements and management of sea otters in Alaska. Journal of Wildlife Management, **48**, 665–678.

Garshelis DL and Johnson CB (1999). Otter eating orcas. Science, **283**, 175.

Garshelis DL and Johnson CB (2001). Sea otter population dynamics and the Exxon Valdez oil spill: disentangling the confounding effects. Journal of Applied Ecology, **38**, 19–35.

Garshelis DL, Johnson AM and Garshelis JA (1984). Social organization of sea otters in Prince William Sound, Alaska. Canadian Journal of Zoology, **62**, 2648–2658.

Garshelis DL, Garshelis JA and Kimker AT (1986). Sea otter time budgets and prey relationships in Alaska. Journal of Wildlife Management, **50**, 637–647.

Genner MJ, Sims, DW, Wearmouth VJ et al. (2004). Regional climatic warming drives long-term community changes of British marine fish. Proceedings of the Royal Society of London, B, **271**, 655–661.

Gibson JR (1992). Otter skins, Boston ships, and China goods. University of Washington Press, Seattle, USA.

Gittleman JL (1986). Carnivore life history patterns: allometric, phylogenetic, and ecological associations. American Naturalist, **127**, 744–771.

Gora M, Carpaneto GM and Ottion P (2003). Spatial distribution and of the Neotropical otter Lontra longicaudis in the Ibera Lake (northern Argentina). Acta Theriologica, **48**, 495–504.

Gorbics CS and Bodkin JL (2001). Stock structure of sea otters (Enhydra lutris kenyoni) in Alaska. Marine Mammal Science, **17,** 632–647.

Gorman ML and Mills MGL (1984). Scent marking strategies in hyaenas (Mammalia). Journal of Zoology, London, **202**, 535–547.

Gorman ML and Trowbridge BJ (1989). The role of odor in the social lives of carnivores. In JG Gittleman (ed.), Carnivore behavior, ecology, and evolution, pp. 57–88. Cornell University Press, Ithaca, NY, USA.

Gorman ML, Kruuk H, Jones C, McLaren G and Conroy JWH (1998a). The demography of European otters Lutra lutra. Symposia of the Zoological Society of London, **71**, 107–118.

Gorman, ML, Mills MG, Raath JP and Speakman JR (1998b). High hunting costs make African wild dogs vulnerable to kleptoparasitism by hyaenas. Nature, **391**, 479–481.

Gosling LM (1982). A reassessment of the function of scent marking in territories. Zeitschrift für Tierpsychologie, **60**, 89–118.

Gourvelou E, Papageorgiou N and Neophytou C (2000). Diet of the otter Lutra lutra in lake Kerkini and stream Milli-Aggistro, Greece. Acta Theriologica, **45**, 35–44.

Grant TR (1983). Body temperatures of free-ranging platypuses, Ornithorhynchus anatinus (Montremata) with observations on their use of burrows. Australian Journal of Zoology, **31**, 117–122.

Grant TR and Dawson TJ (1978). Temperature regulation in the platypus, Ornithorhynchus anatinus. Physiological Zoology, **51**, 1–6, 315–332.

Green J (1977). Sensory perception in hunting otters, Lutra lutra L. Otters, Journal of the Otter Trust, **1977**, 13–16.

Green J and Green R (1980). Otter survey of Scotland, 1977–1979. Vincent Wildlife Trust, London, UK.

Green J and Green R (1987). *Otter survey of Scotland, 1984–1985*. Vincent Wildlife Trust, London, UK.

Green J, Green R and Jefferies DJ (1984). A radio-tracking survey of otters *Lutra lutra* on a Perthshire river system. *Lutra*, **27**, 85–145.

Green R and Green J (1997). *Otter survey of Scotland, 1991–94*. Vincent Wildlife Trust, London, UK.

Groenendijk J, Hajek F, Duplaix N et al. (2005). Surveying and monitoring distribution and population trends of the giant otter (*Pteronura brasiliensis*). *Habitat*, **16**, 6–60.

Gutleb AC, Schenck C and Staib E (1997). Giant otter (*Pteronura brasiliensis*) at risk? Total mercury and methylmercury levels in fish and otter scats, Peru. *Ambio*, **26**, 511–514.

Halbrook RS, Woolf A, Hubert GF, Ross S and Braselton WE (1996). Contaminant concentrations in Illinois mink and otter *Ecotoxicology*, **5**, 103–114.

Hall KRL and Schaller GB (1964). Tool-using behavior of the California sea-otter. *Journal of Mammalogy*, **45**, 287–298.

Hamilton DA (2004). River otter restoration, research and population management in the spotlight of a hostile angling public in Missouri. *Abstracts of the Ninth Otter Colloquium*, University of Frostburg, MD, USA.

Hamilton WJ and Eadie WR (1964). Reproduction in the otter, *Lutra canadensis*. *Journal of Mammalogy*, **45**, 242–252.

Hansen H (2003). Food habits of the North American river otter (*Lontra canadensis*). *River Otter Journal*, **12**, 1–5.

Haque MN and Vijayan VS (1995). Food habits of the smooth otter (*Lutra perspicillata*) in Keolado National Park, Bharatput, Rajasthan (India). *Mammalia*, **59**, 345–348.

Harper RJ (1981). Sites of three otter (*Lutra lutra*) breeding holts in fresh-water habitats. *Journal of Zoology, London*, **195**, 554–556.

Harris CJ (1968). *Otters: a study of recent Lutrinae*. Weidenfeld and Nicholson, London, UK.

Harris S, Morris P, Wray P and Yalden D (1995). *A review of British mammals: population estimates and conservation status of British mammals other than cetaceans*. Joint Nature Conservancy Councils, Peterborough, UK.

Harrison K, Crimmen O, Travers R, Maikweki J and Mutoro D (1989). Balancing the scales in Lake Victoria. *Biologist*, **36**, 189–191.

Harvey JT (1989). Assessment of errors associated with harbor seal (*Phoca vitulina*) fecal sampling. *Journal of Zoology, London*, **219**, 101–111.

Harvie-Brown JA and Buckley TE (1892). *A vertebrate fauna of Argyll and the Inner Hebrides*. Douglas, Edinburgh, UK.

Hatfield BB, Marks D, Tinker MT, Nolan K and Pierce J (1998). Attacks on sea otters by killer whales. *Marine Mammal Science*, **14**, 888–894.

Hauer S, Ansorge H and Zinke O (2000). A long-term analysis of the age structure of otters (*Lutra lutra*) from eastern Germany. *Zeitschrift für Säugetierkunde*, **65**, 360–368.

Hauer S, Ansorge H and Zinke O (2002a). Mortality patterns of otters (*Lutra lutra*) from eastern Germany. *Journal of Zoology, London*, **256**, 361–368.

Hauer S, Ansorge H and Zinke O (2002b). Reproductive performance of otters *Lutra lutra* (Linnaeus, 1758) in Eastern Germany: low reproduction in a long-term strategy. *Biological Journal of the Linnaean Society*, **77**, 329–340.

Heggberget TM (1984). Age determination in the European otter *Lutra lutra*. *Zeitschrift für Säugetierkunde*, **49**, 299–305.

Heggberget TM (1993). Marine-feeding otters (*Lutra lutra*) in Norway: seasonal variation in prey and reproductive timing. *Journal of the Marine Biological Association UK*, **73**, 297–312.

Heggberget TM (1995). Food resources and feeding ecology of marine feeding otters (*Lutra lutra*). In HR Skjoldal, C Hopkins, KE Erikstad and HP Leinas (eds), *Ecology of fjords and coastal waters*, pp. 609–618. Elsevier Science, London, UK.

Heggberget TM and Christensen H (1994). Reproductive timing in Eurasian otters on the coast of Norway. *Ecography*, **17**, 339–348.

Heggberget TM and Moseid K-E (1994). Prey selection in coastal Eurasian otters *Lutra lutra*. *Ecography*, **17**, 331–338.

Heggberget TM and Myrberget S (1979). The otter *Lutra lutra* population in Norway 1970–1977. *Fauna*, **32**, 89–95.

Heggenes J, Krog OMW, Lindas OR, Dokk JG and Bremner T (1993). Homeostatic behavioural responses in a changing environment: brown trout (*Salmo trutta*) become nocturnal in winter. *Journal of Applied Ecology*, **62**, 295–308.

Helder J and De Andrade HK (1997). Food and feeding habits of the neotropical river otter *Lontra longicaudis* (Carnivora, Mustelidae). *Mammalia*, **61**, 193–203.

Hendrichs H (1975). The status of the tiger *Panthera tigris* (Linne, 1758) in the Sundarbans Mangrove Forest (Bay of Bengal). *Säugetierkundige Mitteilungen*, **23**, 161–199.

Herfst M (1984). Habitat and food of the otter *Lutra lutra* in Shetland. *Lutra*, **27**, 57–70.

Hewson R (1969). Couch building by otters *Lutra lutra*. *Journal of Zoology, London*, **195**, 554–556.

Hillegaart V, Ostman J and Sandegren F (1981). Area utilization and marking behaviour among two captive otter (*Lutra lutra* L.) pairs. *Abstract, Second International Otter Colloquium, Norwich, September 1981.*

Houston AI and McNamara JM (1994). Models of diving and data from otters: comments on Nolet *et al*. (1993). *Journal of Animal Ecology*, **63**, 1004–1006.

Humphrey SR and Zinn TL (1982). Seasonal habitat use by river otters and Everglades mink in Florida. *Journal of Wildlife Management*, **46**, 375–381.

Hung C-M, Li S-H and Lee L-L (2004). Faecal DNA typing to determine the abundance and spatial organisation of otters (*Lutra lutra*) along two stream systems in Kinmen. *Animal Conservation*, **7**, 301–311.

Hussain SA (1996). Group size, group structure and breeding in smooth-coated otter *Lutra perspicillata* Geoffroy (Carnivora, Mustelidae) in National Chambal Sanctuary, India. *Mammalia*, **60**, 289–297.

Hussain SA and Choudhury BC (1995). Seasonal movement, home range, and habitat use by smooth coated otters in National Chambal Sanctuary, India. *Habitat*, **11**, 45–55.

Hussain SA and Choudhury BC (1997). Distribution and status of the smooth-coated otter *Lutra perspicillata* in National Chambal Sanctuary, India. *Biological Conservation*, **80**, 199–206.

Hussain SA and Choudhury BC (1998). Feeding ecology of the smooth-coated otter *Lutra perspicillata* in the National Chambal Sancutary, India. *Symposia of the Zooogical Society of London*, **71**, 229–249.

Irving L and Hart S (1957). The metabolism and insulation of seals as bare-skinned mammals in cold water. *Canadian Journal of Zoology*, **35**, 497–511.

IUCN (2001). Red Data 2001. www.redlist.org/search/search-basic.html

IUCN/SSC Reintroduction Specialist Group (1998). *IUCN guidelines for re-introductions*. IUCN, Gland, Switzerland.

Iverson JA and Krog J (1973). Heat production and body surface area in seals and sea otters. *Norwegian Journal of Zoology*, **21**, 51–54.

Ivlev VS (1966). The biological productivity of waters. *Journal of the Fisheries Research Board Canada*, **23**, 1727–1759.

Jackson JBC, Kirby MX, Berger WH *et al.* (2001). Historical overfishing and the recent collapse of coastal ecosystems. *Science* **293**, 629–638.

Jacobsen L and Hansen HM (1992). Analysis of otter (*Lutra lutra* L.) spraints to estimate prey proportions and size of prey: a comparison of methods through feeding experiments. *Journal of Zoology, London*, **238**, 167–180.

Jacques H, Alary F and Parnell R (in press). *Aonyx congicus*. In J Kingdon, T Butynski and D Kappold (eds.). The Mammals of Africa Vol. 5 Carnivora, Pholidota, Perissodactyla. Academic Press, Amsterdam.

Jameson RJ (1989). Movements, home range and territories of male sea otters in central California. *Marine Mammal Science*, **5**, 159–172.

Jameson RJ and Johnson AM (1993). Reproductive characteristics of female sea otters. *Marine Mammal Science*, **9**, 156–167.

Jameson RJ, Kenyon KW, Johnson AM and Wight HM (1982). History and status of translocated sea otter populations in North America. *Wildlife Society Bulletin*, **10**, 100–107.

Jarman PJ (1974). The social organization of antelope in relation to their ecology. *Behaviour*, **48**, 215–266.

Jedrzejewska B, Sidorovich VE, Pikulik MM and Jedrzejewski W (2001). Feeding habits of the otter and the American mink in Bialowieza Primeval Forest (Poland) compared to other Eurasian populations. *Ecography*, **24**, 165–180.

Jefferies DJ (1969). Causes of badger mortality in eastern counties of England. *Journal of Zoology, London*, **157**, 429–436.

Jefferies DJ (1989). The changing otter population of Britain 1700–1989. *Biological Journal of the Linnaean Society*, **38**, 61–69.

Jefferies DJ, Green J and Green R (1984). *Commercial fish and crustacean traps: a serious cause of otter* (Lutra lutra) *mortality in Britain and Europe*. Vincent Wildlife Trust, London, UK.

Jenkins D (1980). Ecology of otters in northern Scotland I. Otter (*Lutra lutra*) breeding and dispersion in mid-Deeside, Aberdeenshire in 1974–1979. *Journal of Animal Ecology*, **49**, 713–735.

Jenkins D (1981). Ecology of otters in northern Scotland IV. A model scheme for *Lutra lutra* L. in a freshwater system in Aberdeenshire. *Biological Conservation*, **20**, 123–132.

Jenkins D and Burrows GO (1980). Ecology of otters in northern Scotland III. The use of faeces as indicators of otter (*Lutra lutra*) density and distribution. *Journal of Animal Ecology*, **49**, 755–774.

Jenkins D and Harper RJ (1980). Ecology of otters in northern Scotland II. Analysis of otter (*Lutra lutra*) and mink (*Mustela vison*) faeces from Deeside, NE Scotland, in 1977–78. *Journal of Animal Ecology*, **49**, 737–754.

Jenkins M (2003). Prospects for biodiversity. *Science* **302**, 1175–1177.

Jensen S, Kihlstrom JE, Olsson M, Lundberg C and Ordberg J (1977). Effects of PCB and DDT on mink (*Mustela vison*) during the reproductive season. *Ambio*, **6**, 239.

Johnson AM (1982a). Status of Alaska sea otter populations and developing conflicts with fisheries. *Transactions of the North American Wildlife and Natural Resources Conference*, **47**, 293–299.

Johnson AM (1982b). The sea otter, *Enhydra lutris*. *Mammals of the Sea, FAO Fisheries Series 5*, **4**, 521–525.

Johnson SA and Berkley KA (1999). Restoring river otters in Indiana. *Wildlife Society Bulletin*, **27**, 419–427.

Johnston JL (1999). *A naturalist's Shetland*. Poyser, London, UK.

Jolly GM (1969). The treatment of errors in aerial counts of wildlife populations. *East African Agricultural and Forestry Journal*, **34**, 50–55.

Jones T and Jones D (2004). *Otter survey of Wales 2002*. Environment Agency, Bristol, UK.

Kain JM (1979). A view of the genus Laminaria. *Oceanography and Marine Biology*, **17**, 101–161.

Kanchanasaka B (2001). Tracks and other signs of the hairy-nosed otter (*Lutra sumatrana*). *IUCN Otter Specialist Group Bulletin*, **18**, 57–62.

Kanchanasaka B (2004). Status and distribution of the hairy-nosed otter (*Lutra sumatrana*). *Poster, Ninth International Otter Colloquium*, Frostburg, MD, USA.

Kauhala K (1996). Distributional history of the American mink (*Mustela vison*) in Finland with reference to trends in otter (*Lutra lutra*) populations. *Annales Zoologici Fennici*, **33**, 283–291.

Kenyon KW (1969). The sea otter in the Eastern Pacific Ocean. *North American Fauna, United States Department of the Interior*, **68**, 1–352.

Kilmer FH (1972). A new species of sea otter from the Pleistocene of northwestern California. *Bulletin of the South California Academy of Science*, **71**, 150–157.

Kimber KR and Kollias GV (2000). Infectious and parasitic diseases and contaminant-related problems of North American river otters (*Lontra canadensis*): a review. *Journal of Zoo and Wildlife Medicin*, **31**, 452–472.

King JM (1983). Abundance, biomass and diversity of benthic macroinvertebrates in a Western Cape river, South Africa. *Transactions of the Royal Society of South Africa*, **45**, 11–34.

Kingdon J (1997). *The Kingdon field guide to African mammals*. Academic Press, London, UK.

Kinlaw A (2005). High mortality of nearctic river otters on a Florida, USA Interstate highway during an extreme drought. *IUCN Otter Specialist Group Bulletin*, **21**, 76–88.

Kitching JA (1941). Studies in sub-littoral ecology III. Laminaria forest on the west coast of Scotland: a study of zonation in relation to wave action and illumination. *Biological Bulletin, Marine Biological Laboratory, Woods Hole, Massachusetts*, **80**, 324–337.

Kleiber M (1975). *The fire of life: an introduction to animal energetics*. Wiley, New York, USA.

Kloskowski J (2000). Selective predation by otters *Lutra lutra* on common carp *Cyprinus carpio* at farmed fisheries. *Mammalia*, **64**, 287–294.

Koeman JH, van de Ven WSM, de Goeij JJM, Tijoe PS and van Haaften JL (1975). Mercury and selenium in marine mammals and birds. *Science of the Total Environment*, **3**, 279–287.

Koepfli K-P and Wayne RK (1998). Phylogenetic relationships of otters (Carnivora: Mustelidae) based on mitochondrial cytochrome *b* sequences. *Journal of Zoology, London*, **246**, 401–416.

Koop JH and Gibson RN (1991). Distribution and movements of intertidal butterfish *Pholis gunnellus*. *Journal of the Marine Biological Association, United Kingdom*, **71**, 127–136.

Kramer DL (1988). The behavioral ecology of air breathing by aquatic animals. *Canadian Journal of Zoology*, **66**, 89–94.

Kranz A (1996). Variability and seasonality in spraiting behaviour of otters *Lutra lutra* on a highland river in Central Europe. *Lutra*, **39**, 33–44.

Kranz A (2000). Otters (*Lutra lutra*) increasing in Central Europe: from the threat of extinction to locally perceived overpopulation? *Mammalia*, **64**, 357–368.

Krebs JR (1978). Optimal foraging decision rules for predators. In JR Krebs and NB Davies (eds), *Behavioural ecology: an evolutionary approach*, pp. 23–63, Sinauer, London, UK.

Kreuder C, Miller MA, Jessup DA et al. (2003). Patterns of mortality in southern sea otters (*Enhydra lutris nereis*) from 1998–2001. *Journal of Wildlife Diseases*, **39**, 495–509.

Kruuk H (1964). Predators and anti-predator behaviour of the black-headed gull, *Larus ridibundus*. *Behaviour (Supplement)*, **11**, 1–129.

Kruuk H (1972). *The spotted hyena, a study in predation and social behavior*. University of Chicago Press, Chicago, IL, USA.

Kruuk H (1975a). *Hyaena*. Oxford University Press, Oxford, UK.

Kruuk H (1975b). Functional aspects of social hunting by carnivores. In G Baerends, C Beer and A Manning (eds). *Function and evolution in behaviour*, pp. 119–141. Clarendon Press, Oxford, UK.

Kruuk H (1978). Foraging and spatial organisation of the European badger, *Meles meles* L. *Behavioral Ecology and Sociobiology*, **4**, 75–89.

Kruuk H (1986). Interactions between Felidae and their prey species: a review. In SD Miller and DD Everell (eds), *Cats of the world: biology, conservation and management*, pp. 353–373. National Wildlife Federation, Washington, DC, USA.

Kruuk H (1989). *The social badger*. Oxford University Press, Oxford, UK.

Kruuk H (1992). Scent marking by otters (*Lutra lutra*): signalling the use of resources. *Behavioral Ecology*, **3**, 133–140.

Kruuk H (1993). The diving behaviour of the platypus (*Ornithorhynchus anatinus*) in waters with different trophic status. *Journal of Applied Ecology*, **30**, 592–598.

Kruuk H (1995). *Wild otters: predation and populations*. Oxford University Press, Oxford, UK.

Kruuk H (2002). *Hunter and hunted*. Cambridge University Press, Cambridge, UK.

Kruuk H (2003). *Niko's nature*. Oxford University Press, Oxford, UK.

Kruuk H and Balharry D (1990). Effects of seawater on thermal insulation of the otter, *Lutra lutra* L. *Journal of Zoology, London*, **220**, 405–415.

Kruuk H and Carss DN (1996). Costs and benefits of fishing by a semi-aquatic carnivore, the otter *Lutra lutra* L. In S Greenstreet and M tasker (eds), *Aquatic predators and their prey*, pp. 10–17. Blackwell Scientific Publications, Oxford, UK.

Kruuk H and Conroy JWH (1987). Surveying otter *Lutra lutra* populations: a discussion of problems with spraints. *Biological Conservation*, **41**, 179–183.

Kruuk H and Conroy JWH (1991). Mortality of otters *Lutra lutra* in Shetland. *Journal of Applied Ecology*, **28**, 83–94.

Kruuk H and Conroy JWH (1996). Concentrations of some organochlorines in otters (*Lutra lutra* L) in Scotland: implications for populations. *Environmental Pollution*, **92**, 165–171.

Kruuk H and Goudswaard PC (1990). Effects of changes in fish populations in Lake Victoria on the food of otters (*Lutra maculicollis* Schinz and *Aonyx capensis* Lichtenstein). *African Journal of Ecology*, **28**, 332–329.

Kruuk H and Hewson R (1978). Spacing and foraging of otters (*Lutra lutra*) in a marine habitat. *Journal of Zoology, London*, **185**, 205–212.

Kruuk H and Macdonald D (1985). Group territories of carnivores: empires and enclaves. In RM Sibley and RH Smith (eds), *Behavioural ecology: ecological consequences of adaptive behaviour*, pp. 521–536. Blackwell Scientific Publications, Oxford, UK.

Kruuk H and Moorhouse A (1990). Seasonal and spatial differences in food selection by otters *Lutra lutra* in Shetland. *Journal of Zoology, London*, **221**, 621–637.

Kruuk H and Moorhouse A (1991). The spatial organization of otters (*Lutra lutra* L.) in Shetland. *Journal of Zoology, London*, **224**, 41–57.

Kruuk H and Parish T (1981). Feeding specialization by the European badger *Meles meles* in Scotland. *Journal of Animal Ecology*, **50**, 773–788.

Kruuk H, Gorman M and Parish T (1980). The use of 65–Zn for estimating populations of carnivores. *Oikos*, **34**, 206–208.

Kruuk H, Glimmerveen U and Ouwerkerk E (1985). The effects of depth on otter foraging in the sea. *Institute of Terrestrial Ecology, Annual Report*, **1984**, 112–115.

Kruuk H, Conroy JWH, Glimmerveen U and Ouwerkerk E (1986). The use of spraints to survey populations of otters (*Lutra lutra*). *Biological Conservation*, **35**, 187–194.

Kruuk H, Conroy JWH and Moorhouse A (1987). Seasonal reproduction, mortality and food of otters *Lutra lutra* L. in Shetland. *Symposia of the Zoological Society of London*, **58**, 263–278.

Kruuk H, Nolet B and French D (1988). Fluctuations in numbers and activity of inshore demersal fishes in Shetland. *Journal of the Marine Biological Association of the United Kingdom*, **68**, 601–617.

Kruuk H, Moorhouse A, Conroy JWH, Durbin L and Frears S (1989). An estimate of numbers and habitat preference of otters *Lutra lutra* in Shetland, UK. *Biological Conservation*, **49**, 241–254.

Kruuk H, Wansink D and Moorhouse A (1990). Feeding patches and diving success of otters (*Lutra lutra* L.) in Shetland. *Oikos* **57**, 68–72.

Kruuk H, Conroy JWH and Moorhouse A (1991). Recruitment to a population of otters (*Lutra lutra*) in Shetland, in relation to fish abundance. *Journal of Applied Ecology*, **28**, 95–101.

Kruuk H, Carss DN, Conroy JWH and Durbin L (1993a). Otter (*Lutra lutra* L.) numbers and fish productivity in rivers in N.E. Scotland. *Symposia of the Zoological Society of London*, **65**, 171–191.

Kruuk H, Kanchanasaka B, O'Sullivan S and Wanghongsa S (1993b). Kingfishers *Halcyon capensis* and *Alcedo atthis* and pond-heron *Ardeola bacchus* associating with otters *Lutra perspicillata*. *Natural History Bulletin of the Siam Society*, **41**, 67–68.

Kruuk H, Kanchanasaka B, O'Sullivan S and Wanghongsa S (1993c). Identification of tracks and other sign of three species of otter, *Lutra lutra, L. perspicillata* and *Aonyx cinerea* in Thailand. *Natural History Bulletin of the Siam Society*, **41**, 23–30.

Kruuk H, Kanchanasaka B, O'Sullivan S and Wanghongsa S (1994a). Niche separation in three sympatric otters *Lutra perspicillata, L. lutra* and *Aonyx cinerea*. *Biological Conservation*, **69**, 115–120.

Kruuk H, Balharry E and Taylor PT (1994b). The effect of water temperature on oxygen consumption of the Eurasian otter *Lutra lutra*. *Physiological Zoology*, **67**, 1174–1185.

Kruuk H, Conroy JWH and Webb A (1997a). Concentrations of mercury in otters (*Lutra lutra* L) in Scotland in relation to rainfall. *Environmental Pollution*, **96**, 13–18.

Kruuk H, Taylor PT and Mom GAT (1997b). Body temperature and foraging behaviour of the Eurasian otter (*Lutra lutra*) in relation to water temperature. *Journal of Zoology, London*, **241**, 689–697.

Kruuk H, Jones C, McLaren GW, Gorman ML and Conroy JWH (1997c). Changes in age composition in populations of the Eurasian otter *Lutra lutra* in Scotland. *Journal of Zoology, London*, **243**, 853–857.

Kruuk H, Carss DN, Conroy JWH and Gaywood MJ (1998). Habitat use and conservation of otters (*Lutra lutra*) in Britain: a review. *Symposia of the Zoological Society of London*, **71**, 119–133.

Kvitek RG and Oliver JS (1992). Influence of sea otters on soft-bottom prey communities in southeast Alaska. *Marine Ecology—Progress Series*, **82**, 103–113.

Kvitek RG, DeGange AR and Beitler MK (1991). Paralytic shellfish poisoning toxins mediate feeding behavior of sea otters. *Limnology and Oceanography*, **36**, 393–404.

Kvitek RG, Bowlby CE and Staedler M (1993). Diet and foraging Behavior of sea otters in southeast Alaska. *Marine Mammal Science*, **9**, 168–181.

Kyne MJ, Small CM and Fairley JS (1989). The food of otters *Lutra lutra* in the Irish Midlands and a comparison with that of mink *Mustela vison* in the same region. *Proceedings of the Royal Irish Academy (B)*, **89**, 33–46.

Lack D (1954). *The natural regulation of animal numbers*. Oxford University Press, Oxford, UK.

Laidler PE (1984). *The behavioural ecology of the giant otter in Guyana*. PhD Thesis, University of Cambridge, UK.

Laidre KL, Jameson RJ and DeMaster DP (2001). An estimation of carrying capacity for sea otters along the California coast. *Marine Mammal Science*, **17**, 294–309.

Lanszki J and Molnar T (2003). Diet of otters living in three different habitats in Hungary. *Folia Zoologica*, **52**, 378–388.

Lanszki J, Körmendi S, Hancz C and Martin TG (2001). Examination of some factors affecting selection of fish prey by otters (*Lutra lutra*) living by eutrophic fish ponds. *Journal of Zoology, London*, **255**, 97–103.

Larivière S (1998). *Lontra felina. Mammalian Species*, American Society of Mammalogists, **575**, 1–5.

Larivière S (1999a). *Lontra provocax. Mammalian Species*, American Society of Mammalogists, **610**, 1–4.

Larivière S (1999b). *Lontra longicaudis. Mammalian Species*, American Society of Mammalogists, **609**, 1–5.

Larivière S (2001a). *Aonyx congicus. Mammalian Species*, American Society of Mammalogists, **650**, 1–3.

Larivière S (2001b). *Aonyx capensis. Mammalian Species*, American Society of Mammalogists, **671**, 1–6.

Larivière S (2002). *Lutra maculicollis. Mammalian Species*, American Society of Mammalogists, **712**, 1–6.

Larivière S (2003). *Amblonyx cinereus. Mammalian Species*, American Society of Mammalogists, **720**, 1–5.

Larivière S and Walton LR (1998). *Lontra candensis. Mammalian Species*, American Society of Mammalogists, **587**, 1–8.

Larsen DN (1983). *Habitats, movements and foods of river otters in coastal south-eastern Alaska*. MSc Thesis, University of Alaska, Fairbanks, AL, USA.

Larson S, Jameson R, Bodkin J, Staedler M and Bentzen P (2002a). Microsatellite DNA and mitochondrial DNA variation in remnant and translocated sea otter (*Enhydra lutris*) populations. *Journal of Mammalogy*, **83**, 893–906.

Larson S, Jameson R, Etnier M, Fleming M and Bentzen P (2002b). Loss of genetic diversity in sea otters (*Enhydra lutris*) associated with the fur trade of the 18th and 19th centuries. *Molecular Ecology*, **11**, 1899–1903.

Latour PB (1988). *The individual within the group territorial system of the European badger (Meles meles L.)*. PhD Thesis, University of Aberdeen, UK.

Le Cren ED (1951). The length–weight relationship and seasonal cycle in gonad weight and condition in the perch *Perca fluviatilis*. *Journal of Animal Ecology*, **20**, 201–219.

Leblanc F (2003). Protecting fish farms from predation by the Eurasian otter (*Lutra lutra*) in the Limousin region of Central France: first results. *IUCN Otter Specialist Bulletin*, **20**, 45–48.

Lejeune A (1989). Les loutres, *Lutra maculicollis* Lichtenstein, et la pêche artisanale au Rwanda. *Revue Zoologique Africaine*, **103**, 215–223.

Lejeune A (1990). Alimentary behavior of the spotted necked otter (*Hydrictis maculicollis*) at Lake Muhazi (Rwanda). *Mammalia*, **54**, 33–45.

Lejeune A and Frank V (1989). Distribution of *Lutra maculicollis* in Rwanda: ecological constraints. *IUCN Otter Specialist Group Bulletin*, **5**, 8–16.

Lenton EJ, Chanin PRF and Jefferies DJ (1980). *Otter survey of England, 1977–79*. Nature Conservancy Council, London, UK.

Lever C (1978). The not so innocuous mink. *New Scientist*, **78**, 481–484.

Leyhausen P (1956). Verhaltensstudien an Katzen. *Zeitschrift für Tierpsychologie (Suppl)*, **2**, 1–120.

Libois RM and Rosoux R (1989). Ecologie de la loutre (*Lutra lutra*) dans le Marais Poitevin, I. Étude de la consommation d'anguilles. *Vie et Milieu*, **39**, 191–197.

Libois RM and Rosoux R (1991). Ecology of the otter (*Lutra lutra*) in Marais Poitevin area. 2. General analysis of feeding habits. *Mammalia*, **55**, 35–47.

Liers EE (1951). Notes on the river otter (*Lutra canadensis*). *Journal of Mammalogy*, **32**, 1–9.

Ligthart MF, Nel JAJ and Avenant NL (1994). Diet of Cape clawless otters in part of the Breede River system. *South African Journal of Wildlife Research*, **24**, 38–39.

Lindquist O and Rohde H (1985). Atmospheric mercury—a review. *Tellus*, **37B**, 136–159.

Loudon A and Racey P (eds) (1987). Reproductive energetics in mammals. *Symposia of the Zoological Society of London*, **57**, 1–371.

Loughlin TR (1980). Home range and territoriality of sea otters near Monterey, California. *Journal of Wildlife Management*, **44**, 576–582.

Lovett L, Kruuk H and Lambin X (1997). Factors influencing use of freshwater pools by otters, *Lutra lutra*, in a marine environment. *Journal of Zoology, London*, **243**, 825–831.

Lowery RS and Mendes AJ (1977). *Procambarius clarkii* in Lake Naivasha, Kenya, and its effects on established and potential fisheries. *Aquaculture*, **11**, 111–121.

Ludwig GX, Hokka V, Sulkava R and Ylonen H (2002). Otter *Lutra lutra* predation on farmed and free-living salmonids in boreal freshwater habitats. *Wildlife Biology*, **8**, 193–199.

MacArthur RA (1979). Seasonal patterns of body temperature and activity in free-ranging muskrats (*Ondatra zibethicus*). *Canadian Journal of Zoology*, **57**, 25–33.

MacArthur RA (1984). Aquatic thermoregulation in the muskrat (*Ondatra zibethicus*): energy demands of swimming and diving. *Canadian Journal of Zoology*, **68**, 241–248.

MacArthur RA and Dyck AP (1990). Aquatic thermoregulation of captive and free-ranging beavers (*Castor canadensis*). *Canadian Journal of Zoology*, **68**, 2409–2416.

Macaskill B (1992). *On the swirl of the tide*. Jonathan Cape, London, UK.

Macdonald DW (1980a). Patterns of scent marking with urine and faeces among carnivore communities. *Symposia of the Zoological Society of London*, **45**, 107–139.

Macdonald DW (1980b). Social factors affecting reproduction amongst red foxes. In E. Zimen (ed.), *The red fox: symposium on behaviour and ecology*, pp. 123–175. Junk, The Hague, Netherlands.

Macdonald DW (1983). The ecology of carnivore social behaviour. *Nature, London*, **301**, 379–384.

Macdonald DW (1985). The carnivores: order Carnivora. In RE Brown and DW Macdonald (eds), *Social odours in mammals*, pp. 619–722. Clarendon Press, Oxford, UK.

Macdonald DW and Moehlman PD (1983). Cooperation, altruism, and restraint in the reproduction of carnivores. In P Bateson and P Klopfer (eds), *Perspectives in ethology*, Vol. 5, pp. 433–467. Plenum Press, New York, USA.

Macdonald DW and Sillero-Zubiri C (eds) (2004). *Biology and conservation of wild canids*. Oxford University Press, Oxford, UK.

Macdonald SM and Mason CF (1987). Seasonal marking in an otter population. *Acta Theriologica*, **32**, 449–462.

Macdonald SM and Mason CF (1992). A note on *Lontra longicaudis* in Costa Rica. *IUCN Otter Specialist Group Bulletin*, **7**, 37–38.

Maitland PS (1972). *Key to the freshwater fishes of the British Isles*. Freshwater Biological Association, Ambleside, UK.

Marburger JE, Johnson WF, Gross TS and Douglas DR (2002). Residual organochlorine pesticides in soils and fish from wetland restoration areas in central Florida, USA. *Wetlands*, **22**, 705–711.

Mason CF (1989). Water pollution and otter distribution: a review. *Lutra*, **32**, 97–131.

Mason CF and Macdonald SM (1980). The winter diet of otters (*Lutra lutra*) on a Scottish sea loch. *Journal of Zoology, London*, **192**, 558–561.

Mason CF and Macdonald SM (1986). *Otters, conservation and ecology*. Cambridge University Press, Cambridge, UK.

Mason CF and Macdonald SM (1987). The use of spraints for surveying otter *Lutra lutra* populations: an evaluation. *Biological Conservation*, **41**, 167–177.

Mason CF and Macdonald SM (2004). Growth in otter (*Lutra lutra*) populations in the UK as shown by long-term monitoring. *Ambio*, **33**, 148–152.

Mason CF and Madsen AB (1993). Organochlorine pesticide residues and PCBs in Danish otters (*Lutra lutra*). *Science of the Total Environment*, **73**, 73–81.

Maxwell G (1960). *Ring of bright water*. Longmans, Green, London, UK.

Mayfield S and Branch GM (2000). Interrelations among rock lobsters, sea urchins, and juvenile abalone: implications for community management. *Canadian Journal of Fisheries and Aquatic Sciences*, **57**, 2175–2185.

Mayfield S, De Beer E and Branch GM (2001). Prey preference and the consumption of sea urchins and juvenile abalone by captive rock lobsters (*Jasus lalandii*). *Marine and Freshwater Research*, **52**, 773–780.

McCleneghan K and Ames JA (1976). A unique method of prey capture by the sea otter, *Enhydra lutra* Linnaeus. *Journal of Mammalogy*, **57**, 410–412.

McShane LJ, Estes JA, Riedman ML and Staedler MM (1995). Repertoire, structure, and individual variation of vocalizations in the sea otter. *Journal of Mammalogy*, **76**, 414–427.

Mead RA (1989). The physiology and evolution of delayed implantation in Carnivores. In JL Gittleman (ed.), *Carnivore behavior, ecology and evolution*, pp. 437–464, Cornell University Press, Ithaca, NY, USA.

Medina G (1995). Acitivity budget and social behaviour of marine otter (*Lutra felina*) in southern Chile. *Proceedings of the International Otter Colloquium*, **6**, 62–64.

Medina G (1996). Conservation and status of *Lutra provocax* in Chile. *Pacific Conservation Biology*, **2**, 414–419.

Medina G (1997). A comparison of the diet and distribution of southern river otter (*Lutra provocax*) and mink (*Mustela vison*) in southern Chile. *Journal of Zoology, London*, **242**, 291–297.

Medina G (1998). Seasonal variations and changes in the diet of southern river otter in different freshwater habitats in Chile. *Acta Theriologica*, **43**, 285–292.

Medina-Vogel G, Kaufman VS, Monsalve R and Gomez V (2003). The influence of riparian vegetation, woody debris, stream morphology and human activity on the use of rivers by southern river otters in *Lontra provocax* in Chile. *Oryx*, **37**, 422–430.

Medina-Vogel G, Rodriguez CD, Alvarez RE and Bartheld JL (2004). Feeding ecology of the marine otter (*Lutra felina*) in a rocky seashore of the South of Chile. *Marine Mammal Science*, **20**, 134–144.

Melquist WE and Dronkert AE (1987). River otter. In M Novak, JA Baker, ME Obbard and B Malloch (eds), *Wild furbearer management and conservation in North America*, pp. 627–641, Ministry of Natural Resources, Ontario, Canada.

Melquist WE and Hornocker MG (1979). Methods and techniques for studying and censusing river otter populations. *University of Idaho Forest, Wildlife and Range Experimental Station, Technical Report 8*.

Melquist WE and Hornocker MG (1983). Ecology of river otters in west central Idaho. *Wildlife Monographs*, **83**, 1–60.

Melquist WE, Whitman JS and Hornocker MG (1981). Resource partitioning and coexistence of sympatric mink and river otter populations. In JA Chapman and D Pursley (eds), *Proceedings of the Worldwide Furbearer Conference, Frostburg, MD 1981*, pp. 187–220.

Mills MGL (1982). Factors affecting group size and territory size of the brown hyaena, *Hyaena brunnea*, in the southern Kalahari. *Journal of Zoology, London*, **198**, 39–51.

Mills MGL (1990). *Kalahari hyaenas: the behavioural ecology of two species*. Unwin Hyman, London, UK.

Mitchell-Jones AJ, Jefferies DJ, Twelves J, Green J and Green R (1984). A practical system of tracking otters *Lutra lutra* using radio-telemetry and 65-Zn. *Lutra*, **27**, 71–74.

Monnett C and Rotterman LM (2000). Survival rates of sea otter pups in Alaska and California. *Marine Mammal Science*, **16**, 794–810.

Monson DH and DeGange AR (1995). Reproduction, preweaning survival, and survival of adult sea otters at Kodiak Island, Alaska. *Canadian Journal of Zoology*, **73**, 1161–1169.

Monson DH, Estes JA, Bodkin JL and Siniff DB (2000). Life history plasticity and population regulation in sea otters. *Oikos*, **90**, 457–468.

Moorhouse A (1988). *Distribution of holts and their utilisation by the European otter* (Lutra lutra *L.*) *in a marine environment*. MSc Thesis, University of Aberdeen, UK.

Moors PJ (1980). Sexual dimorphism in the body size of mustelids (Carnivora): the roles of food habits and breeding systems. *Oikos*, **34**, 147–158.

Morgan CA (2003). *Studies on a reinforced population of the Eurasian otter* (Lutra lutra). PhD Thesis, University of York, UK.

Moriarty C (1978). *Eels, a natural and unnatural history*. David and Charles, London, UK.

Morris PA (1972). A review of mammalian age determination methods. *Mammal Review*, **2**, 69–104.

Morrison P, Rosemann M and Estes JA (1974). Metabolism and thermo-regulation in the sea-otter. *Physiological Zoology*, **47**, 218–229.

Mortensen E (1977). Population, survival, growth and production of trout *Salmo trutta* in a small Danish stream. *Oikos*, **28**, 9–15.

Mourão G and Cavalho L (2001). Cannibalism among giant otters (*Pteronura brasiliensis*). *Mammalia*, **65**, 225–227.

Muanis M (2004). Giant otter diet analysis in the Pantanal. *Friends of the Giant Otter*, **10**, 2.

Muanis M and Waldemarin H (2004). Diet of neotropical otter, *Lontra longicaudis*, and giant otter *Pteronura brasiliensis*, in Pantanal, Brazil. *Abstracts of the Ninth International Otter Colloquium*. Frostburg State University, MD, USA.

Murphy KP and Fairley JS (1985). Food and spraiting places of otters on the west coast of Ireland. *Irish Naturalists Journal*, **21**, 469–508.

Muus BJ and Dahlstrom P (1974). *Collins guide to the sea fishes of Britain and north-western Europe*. Collins, London, UK.

Myhre R and Myrberget S (1975). Diet of wolverine (*Gulo gulo*) in Norway. *Journal of Mammalogy*, **56**, 752–757.

Nagulu V, Vasudeva RV and Srinivasulu C (1999). Curative property of otter blood—a belief. *IUCN Otter Specialist Group Bulletin*, **16**, 44.

Nakhasathien S and Stewart-Cox B (1990). *Nomination of the Thung Yai-Huai Kha Khaeng Wildlife Sanctuary as UNESCO World Heritage Site*. Royal Forest Department, Bangkok, Thailand.

Neal E (1986). *The natural history of badgers*. Croom Helm, London, UK.

Nelson K and Kruuk H (1997). The prey of otters: calorific content of eels (*Anguilla anguilla*) and other fish, frogs

(*Rana temporaria*) and toads (*Bufo bufo*). *IUCN Otter Specialist Group Bulletin*, **14**, 75–80.

Newby TC (1975). A sea otter (*Enhydra lutris*) food dive record. *Murrelet*, **56**, 7.

Newman RM and Waters TF (1989). Differences in brown trout (*Salmo trutta*) production among contiguous sections of an entire stream. *Canadian Journal of Fisheries and Aquatic Science*, **46**, 203–213.

Newton I (1988). Determination of critical pollutant levels in wild populations, with examples from organochlorine insecticides in birds of prey. *Environmental Pollution*, **55**, 29–40.

Newton I, Wyllie I and Asher A (1993). Long-term trends in organochlorine and mercury residues in some predatory birds in Britain. *Environmental Pollution*, **79**, 143–151.

Nguyen XD (2001). New information about the hairy-nosed otter (*Lutra sumatrana*) in Vietnam. *IUCN Otter Specialist Group Bulletin*, **18**, 64–70.

Nguyen XD, Pham TA and Le HT (2003). *Estimate of otter number and assessment of status of otter populations in U Minh Thuong National Park, Kien Giang Province, Vietnam*. Internal Report, U Minh Thuong Nature Reserve Conservation and Community Development Project, Hanoi, Vietnam.

Nolet BA and Kruuk H (1989). Grooming and resting of otters *Lutra lutra* in a marine habitat. *Journal of Zoology, London*, **218**, 433–440.

Nolet BA and Kruuk H (1994). Hunting yield and daily food intake of a lactating otter (*Lutra lutra*) in Shetland. *Journal of Zoology, London*, **233**, 326–331.

Nolet BA, Wansink DEH and Kruuk H (1993). Diving of otters (*Lutra lutra*) in a marine habitat: use of depths by a single-prey loader. *Journal of Animal Ecology*, **62**, 22–32.

Noll JM (1988). *Home range, movements and natal denning of river otters* (Lutra canadensis) *at Kelp Bay, Barranoff Island, Alaska*. MSc Thesis, University of Alaska, Fairbanks, AL, USA.

Norton-Griffiths M (1973). Counting the Serengeti migratory wildebeest using two-stage sampling. *East African Wildlife Journal*, **11**, 135–149.

O'Connor DJ and Nielson SW (1981). Environmental survey of methylmercury levels in wild mink (*Mustela vison*) and otter (*Lutra canadensis*) from the north eastern United States and experimental pathology of methylmercurialism in the otter. In JA Chapman and D Pursley (eds), *Proceedings of the World Furbearer Conference*, Frostburg, Maryland, USA, pp. 1728–1745.

Oftedal OT and Gittleman JL (1989). Patterns of energ output during reproduction in carnivores. In JL Gittleman (ed.), *Carnivore behavior, ecology, and evolution*, pp. 355–378, Cornell University Press, Ithaca, NY, USA.

Olsson ML, Reutergårdh L and Sandegren F (1981). Var är uttern? *Sveriges Natur*, **6**, 234–240.

Osterhaus ADME and Vedder EJ (1988). Identification of virus causing recent seal deaths. *Nature, London*, **335**, 20.

Ostfeld RS (1982). Foraging strategies and prey switching in the sea otter. *Oecologia*, **53**, 170–178.

Ostfeld RS (1991). Measuring diving success of otters. *Oikos*, **60**, 258–260.

Ostfeld RS, Ebensperger L, Klosterman LL and Castilla JC (1989). Foraging, activity budget and social behaviour of the South American marine otter *Lutra felina* (Molina 1782). *National Geographic Research*, **5**, 422–438.

Packer C and Pusey AE (1984). Infanticide in carnivores. In G Hausfater and SB Hardy (eds), *Infanticide: comparative and evolutionary perspectives*, pp. 31–42. Aldine, New York, USA.

Paddack MJ and Estes JA (2000). Kelp forest fish populations in marine reserves and adjacent exploited areas of central Califorña. *Ecological Applications*, **10**, 855–870.

Paine RT and Vadas RL (1969). The effects of grazing by sea urchins, *Strongylocentrotus* spp., on benthic algal populations. *Limnology and Oceanography*, **14**, 710–719.

Palomares F and Caro TM (1999). Interspecific killing among mammalian carnivores. *American Naturalist* **153**, 492–508.

Pardini R (1998). Feeding ecology of the neotropical otter *Lontra longicaudis* in an Atlantic forest stream, south Brazil. *Journal of Zoology, London*, **245**, 385–391.

Pardini R and Trajano E (1999). Use of shelters by the neotropical river otter (*Lontra longicaudis*) in an Atlantic forest stream, southeastern Brazil. *Journal of Mammalogy*, **80**, 600–610.

Parish T and Kruuk H (1982). The uses of radio-location combined with other techniques in studies of badger ecology in Scotland. *Symposia of the Zoological Society of London*, **49**, 291–299.

Parmesan C and Yohe G (2003). A globally coherent fingerprint of climate change impacts across natural systems. *Nature*, **421**, 37–42.

Parrera A (1993). The neotropical river otter *Lutra longicaudis* in Ibera lagoon, Argentina. *IUCN Otter Specialist Group Bulletin*, **8**, 13–16.

Parrera A (1996). Estimating river otter *Lutra longicaudis* population in Ibera lagoon using a direct sightings methodology. *IUCN Otter Specialist Group Bulletin*, **12**, 32–33.

Payne SF and Jameson RJ (1984). Early behavioral development of the sea otter, *Enhydra lutris*. *Journal of Mammalogy*, **65**, 527–531.

Pechlaner H and Thaler E (1983). Beitrag zur Fortpflanzungsbiologie des europäischen Fischotters (*Lutra lutra* L.). *Zoologische Garten NF, Jena*, **53**, 49–58.

Perrin MR and Carranza ID (2000a). Activity patterns of the spotted-necked otter in the Natal Drakensberg, South Africa. *South African Journal of Wildlife Research* **30**, 1–7.

Perrin MR and Carranza ID (2000b). Habitat use by spotted-necked otters in the KwaZulu-Natal Drakensberg, South Africa. *South African Journal of Wildlife Research*, **30**, 8–14.

Perrin MR and Carugati C (2000). Food habits of Cape clawless otter and spotted-necked otter in Drakensberg. *South African Journal of Wildlife Research*, **30**, 85–92.

Perrin MR, Carranza ID and Linn IJ (2000). Use of space by the spotted-necked otter in the KwaZulu-Natal Drakensberg, South Africa. *South African Journal of Wildlife Research*, **30**, 15–21.

Pertoldi C, Hansen MM, Loeschcke V, Madsen AB, Jacobsen L and Baagoe H (2001). Genetic consequences of population decline in the European otter (*Lutra lutra*): an assessment of microsatellite DNA variation in Danish otters from 1883 to 1993. *Proceedings of the Royal Society of London B*, **268**, 1775–1781.

Pfeiffer P and Culik BM (1998). Energy metyabolism of underwater swimming in river-otters (*Lutra lutra* L.). *Journal of Comparative Physiology B*, **168**, 143–148.

Philcox CK, Grogan AL and Macdonald DW (1999). Patterns of otter *Lutra lutra* road mortality in Britain. *Journal of Applied Ecology*, **36**, 748–762.

Pimm SL, Gittleman JL, McCracken GF and Gilpin ME (1989). Plausible alternatives to bottlenecks to explain reduced genetic diversity. *Trends in Ecology and Evolution*, **4**, 176–178.

Pond C (1985). Body mass and natural diet as determinants of the number and volume of adipocytes in eutherian mammals. *Journal of Morphology*, **185**, 183–193.

Postanowicz R (2004). Marine otter (*Lontra felina*). www.lioncrusher.com/animal.asp?animal=166

Powell RA (1979). Mustelid spacing patterns: variations on a theme by *Mustela*. *Zeitschrift für Tierpsychologie*, **50**, 153–165.

Power ME (1990). Effects of fish in river food webs. *Science*, **250**, 811–814.

Procter J (1963). A contribution to the natural history of the spotted-necked otter (*Lutra maculicollis* Lichtenstein) in Tanganyika. *East African Wildlife Journal*, **1**, 93–102.

Purves MG, Kruuk H and Nel JAJ (1994). Crabs *Potamonautes perlatus* in the diet of otter *Aonyx capensis* and water mongoose *Atilax paludosus* in a freshwater habitat in South Africa. *Zeitschrift für Säugetierkunde*, **59**, 332–341.

Quadros J and Monteiro ELA (2001). Diet of the neotropical otter, *L. longicaudis*, in an Atlantic Forest area, Santa Catarina State, southern Brazil. *Studies of Neotropical Fauna and Environment*, **36**, 15–21.

Ralls K and Harvey PH (1985). Geographic variation in size and sexual dimorphism of North American weasels. *Biological Journal of the Linnaean Society*, **25**, 119–167.

Ralls K, Hatfield BB and Siniff DB (1995). Foraging patterns of California sea otters as indicated by telemetry. *Canadian Journal of Zoology*, **73**, 523–531.

Ralls K, Eagle TC and Siniff DB (1996). Movement and spatial use patterns of California sea otters. *Canadian Journal of Zoology*, **74**, 1841–1849.

Randi E, Davoli F, Pierpaoli M, Pertoldi C, Madsen AB and Loeschcke V (2003). Genetic structure in otter (*Lutra lutra*) populations in Europe: implications for conservation. *Animal Conservation*, **6**, 93–100.

Randolph PA, Randolph JC, Mattingly K and Foster MM (1977). Energy costs of reproduction in the cotton rat, *Sigmodon hispidus*. *Ecology*, **58**, 31–45.

Rasa OAE (1973). Prey capture, feeding techniques, and their ontogeny in the African dwarf mongoose (*Helogale undulata refula*). *Zeitschrift für Tierspychologie*, **32**, 449–488.

Reading RP and Clark TW (1996). Carnivore reintroductions: an interdisciplinary examination. In JL Gittleman (ed.), *Carnivore behavior, ecology and evolution*, pp. 296–336. Cornell University Press, Ithaca, NY, USA.

Reid DG, Code TE, Reid ACH and Herrero SM (1994a). Food habits of the river otter in a boreal ecosystem. *Canadian Journal of Zoology*, **72**, 1306–1313.

Reid DG, Code TE, Reid ACH and Herrero SM (1994b). Spacing, movements and habitat selection of the river otter in boreal Alberta. *Canadian Journal of Zoology*, **72**, 1314–1324.

Reijnders PJH (1980). Organochlorine and heavy metal residues in harbour seals from the Wadden Sea and their possible effects on reproduction. *Netherlands Journal of Sea Research*, **14**, 30–65.

Reuther C (1980). Der Fischotter, *Lutra lutra* L. in Niedersachsen. *Naturschutz und Landschaftsforschung, Niedersachsen*, **11**, 1–182.

Reuther C (1991). Otters in captivity—a review with special reference to *Lutra lutra*. *Habitat, (Hankenbüttel)*, **6**, 269–308.

Reynolds J (2003). Silent killers in our rivers. *The Scotsman (Edinburgh)*, 1 December 2003, p. 5.

Richardson JS, Lissimore TJ, Healey MC and Northcote TG (2000). Fish communities of the lower Fraser River

(Canada) and a 21-year contrast. *Environmental Biology of Fishes*, **59**, 125–140.

Ricker WE (1975). Computation and interpretation of biological statistics of fish populations. *Bulletin of the Fisheries Research Board Canada*, **191**, 1–382.

Riedman ML and Estes JA (1988a). A review of the history, distribution and foraging ecology of sea otters. In GR VanBlaricom and JA Estes (eds), *The community ecology of sea otters*, pp. 4–21. Springer, Berlin, Germany.

Riedman ML and Estes JA (1988b). Predation on seabirds by sea otters. *Canadian Journal of Zoology*, **66**, 1396–1402.

Riedman ML and Estes JA (1990). The sea otter (*Enhydra lutris*): behavior, ecology and natural history. *United States Department of the Interior Fish and Wildlife Service (Washington) Biological Report*, **9**, 1–126.

Riedman ML, Estes JA, Staedler MM, Giles AA and Carlson DR (1994). Breeding patterns and reproductive success of California sea otters. *Journal of Wildlife Management*, **58**, 391–399.

Rieman B, Lee D, Burns D et al. (2003). Status of native fishes in the western United States and issues for fire and fuels management. *Forest Ecology and Management*, **178**, 197–211.

Robinson J (1969). Organochlorine insecticides and bird populations in Britain. In MW Miller and GC Berg (eds), *Chemical fall-out: current research on persistent pesticides*, pp. 113–173. Thomas, Springfield, IL, USA.

Rock KR, Rock ES, Bowyer RT and Faro JB (1994). Degree of association and use of a helper by coastal river otters, *Lutra canadensis*, in Prince William Sound, Alaska. *Canadian Field-Naturalist*, **108**, 367–369.

Romakkaniemi A, Perä I, Karlsson L, Jutila E, Carlsson U and Pakarinen T (2003). Development of wild Atlantic salmon stocks in the rivers of the northern Baltic Sea in response to management measures. *ICES Journal of Marine Science*, **60**, 329–342.

Rood JP (1986). Ecology and social evolution in the mongooses. In DI Rubenstein and RW Wrangham (eds), *Ecological aspects of social evolution*, pp. 131–152. Princeton University Press, Princeton, NJ, USA.

Roos A, Greyerz E, Olsson M and Sandegren F (2001). The otter (*Lutra lutra*) in Sweden—population trends in relation to DDT and total PCB concentrations during 1968–99. *Environmental Pollution*, **111**, 457–469.

Root TL, Price JT, Hall KR, Schneider SH, Rosenzweig C and Pounds JA (2003). Fingerprints of global warming on wild animals and plants. *Nature*, **421**, 57–60.

Rosas FCW, Zuanon JAS and Carter SK (1999). Feeding ecology of the giant otter, *Pteronura brasiliensis*. *Biotropica*, **31**, 502–506.

Rostain RR, Ben-David M, Groves P and Randall JA (2004). Why do river otters scent-mark? An experimental test of several hypotheses. *Animal Behaviour*, **68**, 703–711.

Route WT and Peterson RO (1991). An incident of wolf, *Canis lupus*, predation on a river otter, *Lutra canadensis*, in Minnesota. *Canadian Field-Naturalist*, **105**, 567–568.

Rowe-Rowe DT (1977). Prey capture and feeding behaviour of South African otters. *Lammergeyer*, **23**, 13–21.

Rowe-Rowe DT (1978). The small carnivores of Natal. *Lammergeyer*, **25**, 1–48.

Rowe-Rowe DT (1992). *The carnivores of Natal*. Natal Parks Board, Pietermaritzburg, South Africa.

Rowe-Rowe DT and Somers MJ (1998). Diet, foraging behaviour and coexistence of African otters and the water mongoose. *Symposia of the Zoological Society of London*, **71**, 215–227.

Ruiz-Olmo J (1995). Observations on the predation behavior of the otter *Lutra lutra* in NE Spain. *Acta Theriologica*, **40**, 175–180.

Ruiz-Olmo J (1998). Influence of altitude on the distribution, abundance and ecology of the otter (*Lutra lutra*). *Symposia of the Zoological Society of London*, **71**, 159–176.

Ruiz-Olmo J, Delibes M and Zapata SC (1998a). External morphometry, demography and mortality of the otter *Lutra lutra* (Linneo, 1758) in the Iberian peninsula. *Galemys*, **10**, 239–251.

Ruiz-Olmo J, Jiménez J and Margalida A (1998b). Capture and consumption of prey of the otter (*Lutra lutra*) in mediterranean freshwater habitats of the Iberian peninsula. *Galemys*, **10**, 209–226.

Ruiz-Olmo J, López-Martin JM and Delibes M (1998c). Otters and pollution in Spain. *Symposia of the Zoological Society of London*, **71**, 325–338.

Ruiz-Olmo J, Lopez-Martin JM and Palazon S (2001). The influence of fish abundance on the otter (*Lutra lutra*) populations in Iberian Mediterranean habitats. *Journal of Zoology, London*, **254**, 325–336.

Rumbold DG, Bruner MC and Mihalik MB (1996). Organochlorine pesticide in anhingas, white ibises and apple snails collected in Florida. *Archives of Environmental Contamination and Toxicology*, **30**, 379–383.

Saccheri I, Kuussaari M, Kankare M, Vikman P, Fortelius W and Hanski I (1998). Inbreeding and extinction in a butterfly metapopulation. *Nature*, **392**, 491–494.

Sample BE and Suter GW (1999). Ecological risk assessment in a large river-reservoir: 4. piscivorous wildlife. *Environmental Toxicology and Chemistry*, **18**, 610–620.

Samuel MD, Pierce DJ and Garton EO (1985). Identifying areas of concentrated use within the home range. *Journal of Animal Ecology*, **54**, 711–719.

Sandegren FE, Chu EW and Vandevere JE (1973). Maternal behavior in the Californian sea otter. *Journal of Mammalogy*, **54**, 668–679.

Sandell M (1989). The mating tactics and spacing patterns of solitary carnivores. In JL Gittleman (ed.), *Carnivore behavior, ecology and evolution*, pp. 164–182. Cornell University Press, Ithaca, NY, USA.

Sauer TM, Ben-David M and Bowyer RT (1999). A new application of the adaptive-kernel method: estimating linear home ranges of river otter, *Lutra canadensis*. *Canadian Field-Naturalist*, **113**, 419–424.

Schenck C (1997). *Vorkommen, Habitatsnutzung und Schutz des Riesenotters* (Pteronura brasiliensis) *in Peru*. Shaker, Aachen, Germany.

Schenck C, Staib E and Yasseri AM (1995). Unterwasserlaute bei Riesenottern (*Pteronura brasiliensis*). *Zeitschrift für Säugetierkunde*, **60**, 310–313.

Schmidt-Nielsen K (1983). *Animal physiology: adaptation and environment*. Cambridge University Press, Cambridge, UK.

Scholander PF, Walters V, Hock R and Irving L (1950). Body insulation in some Arctic and tropical mammals and birds. *Biological Bulletin*, **99**, 225–236.

Schomburgk R (1840). Information respecting botanical and zoological travellers. *American Natural History*, **5**, 282–288.

Sepulveda M, Bartheld JL, Reyes R and Medina G (2004). Space use of the southern river otter (*Lontra provocax*) on the upper part of the Quele River, Chile. *Abstracts of the Ninth International Otter Colloquium*, Frostburg State University, MD, USA.

Serfass TL (1995). Cooperative foraging by North American River Otters, *Lutra canadensis*. *Canadian Field-Naturalist*, **109**, 458–459.

Serfass TL, Rymon LM and Brooks RP (1990). Feeding relationships of river otters in northeastern Pennsylvania. *Transactions of the Northeast Section of the Wildlife Society*, **47**, 43–53.

Serfass TL, Brooks RP, Novak JM, Johns PE and Rhodes OE (1998). Genetic variation among populations of river otters in North America: considerations for reintroduction projects. *Journal of Mammalogy*, **79**, 736–746.

Serfass TL, Lovallo MJ, Brooks RP, Hayden AH and Mitcheltree DH (1999). Status and distribution of river otters in Pennsylvania following a reintroduction project. *Journal of the Pennsylvania Academy of Science*, **73**, 10–14.

Shannon JS (1989). Social organization and behavioral ontogeny of otters (*Lutra canadensis*) in a coastal habitat in northern California. *IUCN Otter Specialist Group Bulletin*, **4**, 8–13.

Shannon JS (1998). Behaviour of otters in a marine coastal habitat: summary of a work in progress. *IUCN Otter Specialist Group Bulletin*, **15**, 114–117.

Sheldon WG and Toll WG (1964). Feeding habits of river otter in a reservoir in central Massachusetts. *Journal of Mammalogy*, **45**, 449–455.

Sherrod SK, Estes JA and White CM (1975). Depredation of sea otter pups by bald eagles at Amchitka Island, Alaska. *Journal of Mammalogy*, **56**, 701–703.

Sidorovich VE (1997). *Mustelids in Belarus*. Zolotoy, Minsk, Belarus.

Sidorovich V, Kruuk H, Macdonald DW and Maran T (1998). Diets of semi-aquatic carnivores in northern Belarus, with implications for population changes. *Symposia of the Zoological Society of London*, **71**, 177–189.

Sidorovich V, Kruuk H and Macdonald DW (1999). Body size, and interactions between European and American mink (*Mustela lutreola* and *M. vison*) in Eatsern Europe. *Journal of Zoology, London*, **248**, 521–527.

Siegel S and Siegel NJC (1988). *Nonparametric statistics for the behavioral sciences*. McGraw-Hill, New York, USA.

Simmonds MP, Johnston PA, French MC, Reeve R and Hutchinson JD (1994). Organochlorines and mercury in pilot whale blubber consumed by Faroe islanders. *Science of the Total Environment*, **149**, 97–111.

Simpson VR (1997). Health status of otters (*Lutra lutra*) in south-west England based on postmortem findings. *Veterinary Record*, **141**, 191–197.

Simpson VR and Coxon KE (2000). Intraspecific aggression, cannibalism and suspected infanticide in otters. *British Wildlife*, **11**, 423–426.

Simpson VR, Bain MS, Brown R, Brown BF and Lacey RF (2000). Long-term study of vitamin A and polychlorinated hydrocarbon levels in otters (*Lutra lutra*) in south west England. *Environmental Pollution*, **110**, 267–275.

Sinclair ARE, Mduma S and Brashares JS (2003). Patterns of predation in a diverse predator–prey system. *Nature*, **425**, 288–290.

Sinha AA, Conaway CH and Kenyon KW (1966). Reproduction in the female sea otter. *Journal of Wildlife Management*, **30**, 121–130.

Siniff DB and Ralls K (1991). Reproduction, survival and tag loss in California sea otters. *Marine Mammal Science* **7**, 211–229.

Siniff DB, Williams TD, Johnson AM and Garshelis DL (1982). Experiments on the response of sea otters, *Enhydra lutris*, to oil contamination. *Biological Conservation*, **2**, 261–272.

Sivasothi N and Nor BHM (1994). A review of otters (Carnivora: Mustelidae: Lutrinae) in Malaysia and Singapore. *Hydrobiologia*, **285**, 151–170.

Sjöasen T (1996). Survivorship of captive-bred and wild-caught reintroduced European otters *Lutra lutra* in Sweden. *Biological Conservation*, **76**, 161–165.

Skaren C (1990). Fish farming and otters in Finland. *IUCN Otter Specialist Group Bulletin*, **5**, 28–33.

Skinner JD and Smithers RHN (1990). *The mammals of the Southern African sub-region*. University of Pretoria, Pretoria, South Africa.

Smit MD, Leonards PEG, Van Hattum B and de Jongh AWJJ (1994). *PCBs in European otter* (Lutra lutra) *populations*. Institute for Environmental Studies, Amsterdam, Netherlands.

Smith NJH (1981). Caimans, capybaras, otters, manatees, and man in Amazonia. *Biological Conservation*, **19**, 177–187.

Smith TG and Armstrong FAJ (1978). Mercury and selenium in ringed and bearded seal tissues from Arctic Canada. *Arctic*, **31**, 76–84.

Somers MJ (2000a). Foraging behaviour of Cape clawless otters (*Aonyx capensis*) in a marine habitat. *Journal of Zoology, London*, **252**, 473–480.

Somers MJ (2000b). Seasonal variation in the diet of Cape clawless otters (*Aonyx capensis*) in a marine habitat. *African Zoology*, **35**, 261–268.

Somers MJ (2001). *Habitat utilization of Cape clawless otters* Aonyx capensis. PhD Thesis, University of Stellenbosch, South Africa.

Somers MJ and Nel JAJ (1998). Dominance and population structure of freshwater crabs (*Potamonautes perlatus* Milne Edwards). *South African Journal of Zoology*, **33**, 31–36.

Somers MJ and Nel JAJ (2003). Diet in relation to prey of Cape clawless otters in two rivers in the Western Cape Province, South Africa. *African Zoology*, **38**, 317–326.

Somers MJ and Nel JAJ (2004). Movement patterns and home range of Cape clawless otters (*Aonyx capensis*), affected by high food density patches. *Journal of Zoology, London*, **262**, 91–98.

Somers MJ and Purves MG (1996). Trophic overlap between three syntopic semi-aquatic carnivores: Cape clawless otter, spotted-necked otter and water mongoose. *African Journal of Ecology*, **34**, 158–166.

Southwood TRE (1978). *Ecological methods*. Chapman and Hall, London, UK.

Spinola RM and Vaughan C (1995). Dieta de la nutria neotropical (*Lutra longicaudis*) en la estacion biologica La Selva, Costa Rica. *Vida Silvestre Neotropical*, **4**, 125–132.

Springer AM, Estes JA, Van Vliet GB *et al.* (2003). Sequential megafaunal collapse in the North Pacific Ocean: an ongoing legacy of industrial whaling? *Proceedings of the National Academy of Sciences of the USA*, **100**, 12223–12228.

Staib E (2002). *Oeko-Ethologie von Riesenottern* (Pteronura brasiliensis) *in Peru*. Shaker, Aachen, Germany.

Steneck RS, Graham MH, Bourque BJ *et al.* (2002). Kelp forest ecosystems: biodiversity, stability, resilience and future. *Environmnetal Conservation*, **29**, 436–459.

Stephens MN (1957). *The otter report*. Universities Federation for Animal Welfare, Potters Bar, London, UK.

Stephenson AB (1977). Age determination and morphological variation of Ontario otters. *Canadian Journal of Zoology*, **55**, 1577–1583.

Stone R (2003). Freshwater eels are slip-sliding away. *Science*, **301**, 221–222.

Storfer A (1999). Gene flow and endangered species translocations: a topic revisited. *Biological Conservation*, **87**, 173–180.

Stoskopf MK, Spelman LH, Summers PW, Redmond DP, Jochem WJ and Levine JF (1997). The impact of water temperature on core body temperature of North American river otters (*Lutra canadensis*) during simulated oil spill recovery washing protocols. *Journal of Zoo and Wildlife Medicine*, **28**, 407–412.

Strachan R, Birks JDS, Chanin PRF and Jefferies DJ (1990). *Otter survey of England 1984–1986*. Nature Conservancy Council, Peterborough, UK.

Stuart SN, Chanson JS, Cox NA *et al.* (2004). *Status and trends of amphibian declines and extinctions worldwide*. www.sciencemag.org/cgi/content/abstract/1103538

Stubbe M (1969). Zur Biologie und zum Schutz des Fischotters *Lutra lutra* (L.). *Archiv für Naturschutz und Landschaftsforschung*, **9**, 315–324.

Stubbe M (1980). Die Situation des Fischotters in der D.D.R. In C Reuther and A Festetics (eds), *Der Fischotter in Europa—Verbreitung, Bedrohung, Erhaltung*. Aktion Fischotterschutz, Oderhaus, Germany.

Sulkava R (1996). Diet of otters *Lutra lutra* in central Finland. *Acta Theriologica*, **41**, 395–408.

Taastrøm HM and Jacobsen L (1999). The diet of otters (*Lutra lutra* L.) in Danish freshwater habitats: comparisons of prey fish populations. *Journal of Zoology, London*, **248**, 1–13.

Tabor JE and Wight HH (1977). Population status of river otter in Western Oregon. *Journal of Wildlife Management*, **41**, 692–699.

Tarasoff FJ (1974). Anatomical adaptations in the river otter, sea otter and harp seal with reference to thermal regulation. In RJ Harrison (ed.), *Functional anatomy of marine mammals*, Vol. 2, pp. 111–142. Academic Press, London, UK.

Tarr RJQ, Williams PVG and Mackenzie AJ (1996). Abalone, sea urchins and rock lobster: a possible ecological shift that may affect traditional fisheries. *South African Journal of Marine Science*, **17**, 319–323.

Taylor PS and Kruuk H (1990). A record of an otter (*Lutra lutra*) natal den. *Journal of Zoology, London*, **222**, 689–692.

Tembrock G (1957). Zur Ethologie des Rotfuchses (*V. vulpes* L.) unter besonderer Berücksichtigung der Fortpflanzung. *Zoologische Garten NF*, **23**, 289–532.

Testa JW, Holleman DF, Bowyer RT and Faro JB (1994). Estimating populations of marine river otters in Prince William Sound, Alaska, using radiotracer implants. *Journal of Mammalogy*, **75**, 1021–1032.

Tiedeman R, Suchentrunk F, Cassens I and Hartle G (2000). Mitochondrial DNA variation in the European otter (*Lutra lutra*) and the use of spatial autocorrelation analysis in conservation. *Journal of Heredity*, **91**, 31–35.

Timmis WH (1971). Observations on breeding the Oriental short-clawed otter *Amblonyx cinerea* at Chester Zoo. *International Zoo Yearbook*, **11**, 109–111.

Tinbergen N (1951). *The study of instinct*. Clarendon Press, Oxford, UK.

Tinbergen N (1960). The evolution of behaviour in gulls. *Scientific American*, **12**, 118–130.

Tinbergen N (1963). On aims and methods of ethology. *Zeitschrift für Tierpsychologie*, **20**, 410–433.

Toweill DE and Tabor JE (1982). River otter. In JA Chapman and GA Feldhamer (eds), *Wild mammals of North America*, pp. 688–703. John Hopkins University Press, Baltimore, MD, USA.

Traas TP, Luttik R, Klepper O et al. (2001). Congener-specific model for polychlorinated biphenyl effects on otter (*Lutra lutra*) and associated sediment quality criteria. *Environmental Toxicology and Chemistry*, **20**, 205–212.

Turnbull-Kemp PSJ (1960). Quantitative estimations of populations of river crab, *Potamon (Potamonautes) perlatus* (Edw.), in Rhodesian trout streams. *Nature*, **185**, 481.

Twelves J (1983). Otter (*Lutra lutra*) mortalities in lobster creels. *Journal of Zoology, London*, **201**, 585–588.

Udevitz MS and Ballachey BE (1998). Estimating survival rates with age-structure data. *Journal of Wildlife Management*, **62**, 779–792.

Udevitz MS, Bodkin JL and Costa DP (1995). Detection of sea otters in boat-based surveys of Prince William Sound, Alaska. *Marine Mammal Science*, **11**, 59–71.

Uryu Y, Malm O, Thornton I, Payne I and Cleary D (2001). Mercury contamination of fish and its implications for other wildlife of the Tapajos Basin, Brazilian Amazon. *Conservation Biology*, **15**, 438–446.

Valdimarsson SK, Metcalfe NB, Thorpe JE and Huntingford FA (1997). Seasonal changes in sheltering: effect of light and temperature on diel activity in juvenile salmon. *Animal Behaviour*, **54**, 1405–1412.

Van Der Zee D (1981). Prey of the Cape clawless otter (*Aonyx capensis*) in the Tsitsikama Coastal National Park, South Africa. *Journal of Zoology, London*, **194**, 467–483.

Van Niekerk CH, Somers MJ and Nel JAJ (1998). Freshwater availability and distribution of Cape clawless otter spraints along the south-west coast of South Africa. *South African Journal of Wildlife Research*, **28**, 68–72.

van Wijngaarden A and van de Peppel J (1970). De otter, *Lutra lutra* (L.) in Nederland. *Lutra*, **12**, 1–72.

Van Zyl RF, Mayfield S, Pulfrich A and Griffiths CL (1998). Predation by west coast rock lobsters (*Jasus lalandii*) on two species of winkle (*Oxystele sinensis* and *Turbo cidaris*). *South African Journal of Zoology*, **33**, 203–209.

Van Zyll de Jong CG (1972). A systematic review of the Nearctic and Neotropical river otters (Genus *Lutra*, Mustelidae, Carnivora). *Research in Ontario Museum of Life Sciences*, **80**, 1–104.

Van Zyll de Jong CG (1987). A phylogenetic study of the Lutrinae (Carnivora; Mustelidae) using morphological data. *Canadian Journal of Zoology*, **65**, 2536–2544.

VanBlaricom GR (1984). Relationships of sea otters to living marine resources in California: a new perspective. In V Lyle (ed.), *Ocean Studies Symposium*, Asilomar, California, p. 361. California Department of Fish and Game, Sacramento, CA, USA.

VanBlaricom GR and Estes JA (eds) (1988). *The community ecology of sea otters*. Springer, Berlin, Germany.

VanWagenen RF, Foster MS and Burns F (1981). Sea otter predation on birds near Monterey, California. *Journal of Mammalogy*, **62**, 433–434.

Verberne G and de Boer JN (1976). Chemo-communication among domestic cats. *Zeitschrift für Tierpsychologie*, **42**, 86–109.

Verwoerd DJ (1987). Observations on the food and status of the Cape clawless otter (*Aonyx capensis*) at Betty's Bay, South Africa. *South African Journal of Zoology*, **22**, 33–39.

Videler JJ (1981). Swimming movements, body posture and propulsion in cod *Gadus morhua*. *Symposia of the Zoological Society of London*, **48**, 1–27.

Wang-Andersen G, Skaare JU, Prestrud P and Steinnes E (1993). Levels and congener pattern of PCBs in Arctic fox, *Alopex lagopus*, in Svalbard. *Environmental Pollution*, **82**, 269–275.

Watson HC (1978). *Coastal otters in Shetland*. Vincent Wildlife Trust, London, UK.

Watson LH and Lang AJ (2003). Diet of Cape clawless otters in Groenvlei Lake, South Africa. *South African Journal of Wildlife Research*, **33**, 135–137.

Watt JP (1991). *Prey selection by coastal otters* (*Lutra lutra* L.). PhD Thesis, University of Aberdeen, UK.

Watt JP (1993). Ontogeny of hunting behaviour of otters (*Lutra lutra* L.) in a marine environment. *Symposia of the Zoological Society of London*, **65**, 87–104.

Watt J (1995). Seasonal and area-related variations in the diet of otters *Lutra lutra* on Mull. *Journal of Zoology, London*, **237**, 179–194.

Watt J, Krause B and Tinker TM (1995). Bald eagles kleptoparasitizing sea otters at Amchitka Island, Alaska. *Condor*, **97**, 588–590.

Watt J, Siniff DB and Estes JA (2000). Inter-decadal patterns of population and dietary change in sea otters at Amchitka Island, Alaska. *Oecologia*, **124**, 289–298.

Wayne RK, Modi WS and O'Brien SJ (1986). Morphologic variability and asymmetry in the cheetah (*Acinonyx jubatus*), a genetically uniform species. *Evolution*, **40**, 78–85.

Wayre P (1974). Otters in western Malaysia, *Otter Trust Annual Report*, **1974**, 16–38.

Wayre P (1979). *The private life of the otter*. Batsford, London, UK.

Webb JB (1975). Food of the otter (*Lutra lutra*) on the Somerset levels. *Journal of Zoology*, **177**, 486–491.

Webb J and Hawkins AD (1989). The movements and spawning behaviour of adult salmon in the Gurnock Burn, a tributary of the Aberdeenshire Dee, 1986. *Scottish Fisheries Research Report*, **40**, 1–42.

Weber J-M (1990). Seasonal exploitation of amphibians by otters (*Lutra lutra*) in north-east Scotland. *Journal of Zoology, London*, **220**, 641–651.

Weber J-M and Roberts L (1990). A bacterial infection as a cause of abortion in the European otter, *Lutra lutra*. *Journal of Zoology, London*, **219**, 688–690.

Wendell F (1994). Relationship between sea otter range expansion and red abalone abundance and size distribution in central California. *Californian Fish and Game*, **80**, 45–56.

Wendell FE, Hardy RA, Ames JA and Burge RT (1986). Temporal and spatial patterns in sea otter, *Enhydra lutris*, range expansion and in the loss of pismo clam fisheries. *California Fish and Game*, **72**, 197–212.

Westemeier RL, Brawn JD, Simpson SA *et al.* (1998). Tracking the long-term decline and recovery of an isolated population. *Science*, **282**, 1695–1698.

Westin L and Aneer G (1987). Locomotor activity patterns of nineteen fish and five crustacean species from the Baltic Sea. *Environmental Biology of Fishes*, **20**, 49–65.

Wheeler A (1978). *Key to the fishes of northern Europe*. Frederick Warne, London, UK.

White PCL, Gregory KW, Lindley PJ and Richards G (1997). Economic values of threatened mammals in Britain: a case study of the otter *Lutra lutra* and the water vole *Arvicola terrestris*. *Biological Conservation*, **82**, 345–354.

Widdowson EM (1981). The role of nutrition in mammalian reproduction. In D Gilmore and B Cook (eds), *Environmental factors in mammal reproduction*, pp. 145–159. Macmillan, London, UK.

Willemsen GF (1992). A revision of the Pliocene and Quarternary Lutrinae from Europe. *Scripta Geologica*, **101**, 1–115.

Williams DL, Simpson VR and Flindall A (2004b). Retinal dysplasia in wild otters (*Lutra lutra*). *Veterinary Record* **155**, 52–56.

Williams T (1986). Thermo-regulation of the North American mink during rest and activity in the aquatic environment. *Physiological Zoology*, **59**, 293–305.

Williams TD, Allen DD, Groff JM and Glass RL (1992). An analysis of California sea otter (*Enhydra lutra*) pelage and integument. *Marine Mammal Science*, **8**, 1–18.

Williams TM (1989). Swimming by sea otters: adaptations for low energetic cost locomotion. *Journal of Comparative Physiology A*, **164**, 815–824.

Williams TM (1998). Physiological challenges in semi-aquatic mammals: swimming against the energetic tide. *Symposia of the Zoological Society of London*, **71**, 17–29.

Williams TM, Kastelein RA, Davis RW and Thomas JA (1988). The effects of oil contamination and cleaning on sea otters (*Enhydra lutris*). I. Thermoregulatory implications based on pelt studies. *Canadian Journal of Zoology*, **66**, 2776–2781.

Williams TM, Estes JA, Doak DF and Springer AM (2004a). Killer appetites: assessing the role of predators in ecological communities. *Ecology*, **85**, 3373–3384.

Williamson H (1927). *Tarka the otter*. Putnam, London, UK.

Wirth T and Bernatchez L (2003). Decline of North Atlantic eels: a fatal synergy? *Proceedings of the Royal Society of London, B*, **270**, 681–688.

Wise MH, Linn IJ and Kennedy CR (1981). A comparison of the feeding biology of mink *Mustela vison* and otter *Lutra lutra*. *Journal of Zoology, London*, **195**, 181–213.

Wlodek K (1980). Der Fischotter in der Provinz Pomorze Zachodnie (West-Pommern) in Polen. In C Reuther and A Festetics (eds), *Der Fischotter in Europa—Verbreitung, Bedrohung, Erhaltung*, pp. 187–194. Aktion Fischotterschutz, Oderhaus, Germany.

Woodroffe G (2001). *The otter*. Mammal Society, London, UK.

Woolington JD (1984). *Habitat use and movements of river otters at Kelp Bay, Baranoff Island, Alaska*. MSc Thesis, University of Alaska, Fairbanks, AL, USA.

Wren CD (1984). Distribution of metals in tissues of beaver, raccoon and otter from Ontario, Canada. *Science of the Total Environment*, **34**, 112–114.

Wren CD (1986). A review of metal accumulation and toxicity in wild mammals. 1. Mercury. *Environmental Research*, **40**, 210–244.

Wren CD (1991). Cause–effect linkages between chemicals and populations of mink (*Mustela vison*) and otter

(*Lutra canadensis*) in the Great Lakes area. *Journal of Toxicology and Environmental Health*, **33**, 549–585.

Wren CD, Fischer KL and Stokes PM (1988). Levels of lead, cadmium and other elements in mink and otter from Ontario, Canada. *Environmental Pollution*, **52**, 193–202.

Yoxon P (1999). *Geology and otter distribution on Skye*. PhD Thesis, Open University, Milton Keynes, UK.

Zijlstra M and Van Eerden MR (1995). Pellet production and the use of otoliths in determining the diet of cormorants *Phalacrocorax carbo sinensis*: trials with captive birds. *Ardea*, **83**, 123–131.

Zippin C (1958). The removal method of population estimation. *Journal of Wildlife Management*, **22**, 82–90.

Index

Bold numbers refer to illustrations
For general topics also check sub-heading of individual species of otter

abalone 113, 118, 135, 192
activity timing 217
age structure 193
aggression 86–8
Anguilla anguilla, see eel
Aonyx capensis, see Cape clawless otter
Aonyx cinereus, see small-clawed otter
Aonyx congicus, see Congo clawless otter
Aptocyclus ventricosus, see lumpsucker
Arvicola terrestris, see water vole
Astacus fluviatilis, see crayfish
Atilax paludosis, see water mongoose
Austromobius pallipes, see crayfish

badger, Eurasian 221
bald eagle 2, 13, 97, 145, 191, 206
beaver lodges 9, 48, **49**, 75–6
beaver, as food 112
biomass (otters) 177
birds as food 103, 114, 115, 118, 142
body temperature 162–4
Bufo bufo, see toad
bullrout 103–6, **123**, 124–7
butterfish 103–7, **122**, 124–7, 128

cadmium 201
caiman 68, 96, 191, 207
Cancer magister, see crab, Dungeness
cannibalism 73, 207
Cape clawless otter **23**, **147**
  activity times 150
  competition 190–1
  conservation 234, 235, 237
  couch 16
  cub dispersal 96
  damage to fisheries 227
  dens 76–7
  diet 24, 118, 138
  dives 147
  feeding patches 153
  foraging 23–4, 147, 152
  freshwater need 45
  geographic distribution **24**
  gestation 25
  groups 69–70
  habitat 41, 45, 52–3
  home ranges 69–70
  pollution 53
  prey availability 136
  reproduction 189
  scats, *see* spraints
  size 24
  social 24–5, 69–70, **72**
  spraints 82
  tracks **174**
  trapping 231
  washing in freshwater 171
Carcinus maenas, see crab
catfish 115, 116, 117–18, 134, 150
Catostomus macrocheilus, see sucker
cichlids 26, 52, 115–16, 117, 134, 136, 138
Ciliata mustela, see rockling
clam 113, 135, 144–**5**
climate change 189, 211, 224
cod 102–3, 108
commensals of otters 192
communication 78–86, 98, 220
competition, between otters 25, 51, 53, 55, 82, 85–6, 107, 117, 145, 189–90, 192, 209, 213, 218, 220
  with other carnivores 191–2, 209
  with people 192, 226
condition index 197
Conger conger, see conger eel
conger eel 102–3, 124
Congo clawless otter **25**
  conservation 237
  diet 118–9
  foraging 25, 147–8
  geographic distribution **26**
  habitat 53
  size 26
  social 70
conservation 233–8
consumption 111, 119
co-operative behaviour 55, 71, 143, 152
core areas 7, 58–64, 67–8, 70–1, **72**, 83, 115, 214, 217, 219
coyote 207
crab, Dungeness 113
  freshwater 109, 116, 117, 118, 119, 136, **137**
  shore **102**, 104, 107, 111, 123, 128–9, 137

crab-eating mongoose 191
crayfish **108**, 109–10, 111, 112, 116, 120, 132, 138
crocodile 25, 26,28, 69, 70, 71, 207, 219, 220, 235
Cyclopterus lunpus, see lumpsucker
cyprinids 109, 111, 112, 117, 131, 136, 143, 218

DDT 3, 8, 200–1
delayed implantation 8, 13, 16, 25, 182, 186, 188
dens, *see* holts
density 54, 177, 221, 230
dentine layers for ageing **193**
depth of dives 41, 74, 125–7, 152–4, 165–6
development 90–6, 159–61
dieldrin 3, 8, 200–1
diet 119
diet methodology 100–2,
  individual differences 106–7, 119
  sex differences 106–7, 119
disease 13, 32, 198–9, 208, 221–3, 225, 237
dive times 141
DNA analyses 35–6, 62, 180–1, 212
dogfish 102–4, 123, 124, **129**
dolphin, associate with otters 192
domestic dog 206

earthworms as food 118–9
ecosystems 137, 221, 225, 233–7
eel 95, 109–10, 111, 120, 124, **130**, 138, 151, 189
eelpout 103–6, **121**, 124–7, 128, 141
effective population size 55
effects on prey 217
elephant seal 181
energetics 155–60, 216
Enhydra lutra, see sea otter
Enhydra macrodonta 30
Enhydriodon 30
Enhydriterium 30
Eurasian otter **1**, **6**, **57**,**184**
  abandoning cubs 92, 98, **182**, 185
  activity times 148–9
  age structure 193–5
  aggression 58–60, 86–8, 198
  agriculture 44–5
  altitude 39

Eurasian otter (cont.)
  biomass 176
  blastocyst 182
  body condition 197
  body temperature 162–3
  cannibalism 73, 90, 207
  census 173–6, 178
  competition 190
  conservation management 234–5, 237
  copulation 89–90
  consumption 111, 132–3
  core areas 58–9
  couches **48**, 75, 214
  cubs 181–**2**
  damage to fisheries 226–7
  damage to poultry 227
  decline in numbers 137–8
  density 8, 44, 45–9, 176–8, 201, 204–6
  depths 41, 153–5
  development foraging 160
  diet 99–100, 102–10, 117, 129, 138
  diet, analysis 101, **102–3**
  diet, diversity 110
  diet, individual differences 106–7
  diet, seasonal 104, 110, 128
  diet, sex difference 106–7
  digestion speed 143
  diving 141, 153–5, 156
  diving success cubs 94–5
  effects on prey 132–4
  feeding patches 42, 152–3
  fishing behaviour 141–3
  fishing duration 164
  food intake 111
  foraging energetics 156–9
  foraging success 150–1
  fur trade 229
  genetics 180–1
  geographic distribution **6**
  geology 45
  gestation 8, 182, 186
  grooming 141, 165–**6**, 167, 170, 204
  group ranges 58
  habitat selection 41, 46–7
  habitat, sex difference 46, 48
  holt (den) **42**, 48, 72–6, 90–1
  holt density **44**
  home range 56–62
  infanticide 90
  injuries 87
  litter size 181–2
  longevity 193

  marine habitat 41–2, 43–5
  mortality 185–6, 193–5
  oil spills 203–4
  parental behaviour 7–8, **59**, 90–6, **91**
  play with prey 2, 95
  pollution 178, 180, 200–3
  population limitation 204–6
  population sizes 177–8, 204
  predation 206
  prey capture rates 156–9
  prey depth 126
  prey seasonality 126–8
  prey selection 128–30
  prey sizes 94, 109–10
  provisioning cubs 93–5
  public interest 234
  recruitment 181–6, 189
  reintroduction 233
  resting sites, see couches
  seasonal mortality 128, 198, 205
  seasonal breeding 128, 182–4
  sexual behaviour 88–9
  size 7, 59
  social 7, 56–62, **72**, 219–20
  speed 156
  spraints 7, **36**, 39–40, 79, **80**, 83–6, 100, 128, **174**
  starvation 204–5
  stream width 47
  survival 194–5
  teaching cubs 95, 217
  temperature effects 156–60
  thermo-insulation 165–8
  throat patch **31**, 56
  tracks **35**
  trapping 230–1
  urban 225–6
  vocalizations 88, 93
  washing in freshwater 170–1
evolution, of otters 29–30, **31**, 212, 213
Exxon Valdez 179, 196, 203–4, 209, 236

faeces, see spraints
field methods 33–8, **38**, 56
fish farms 8, 16, 23, 25, 28, 53, 73, 192, 218, 226–7, 234
fish trapping 124
flatfish 124
foraging 215–8
foraging patches 214
foraging success 150–2, 161
freshwater along sea coasts 40, 45, 53, 170–2

frog 108, 112, 117, **132**, 138, 143
fulmar 103
fur maintenance 165–72
fur trade 135, 226, 229–30, 233

Gadus morhua, see cod
Gasterosteus aculeatus, see stickleback
genetics 5, 31, 180–1, 221, 232
geographic variability 30
giant otter **13**, **36**, **67**, **146**, **187**, **222**, **238**
  activity times 150
  biomass 176
  campsites 13, 14, 80–82, **81**
  cannibalism 90, 207
  competition 15, 190
  conservation 237
  consumption 115
  cubs **96**, **188**
  den **76**
  density 176–7
  development of foraging 161
  diet 14–15, 114–15,
  dolphins 192
  foraging behaviour 145–**6**
  fur trade 229–30
  geographic distribution **14**
  gestation 188
  grooming 67, 97
  habitat 50–1
  home range 67
  infanticide 90
  litter size 188
  longevity 196
  parental 14, 96–7
  poaching 236
  pollution 204
  predation 96, 207
  reproduction 188
  scent, see campsites
  sexual behaviour 89
  size 14, 66
  social 14, 66–8, **72**
  spraints, see campsites
  territory 67
  tourism 236
  vocalizations 88, 98
grooming 165–70
group living 219–20
guillemot 103

habitat 39–40, 53–4, 212–5
habitat change 224–5, 237
habitat selection, sex difference 46, 48, 54

Index

hairy-nosed otter
  conservation 237
  diet 20, 117
  geographic distribution **18**
  habitat 52
  rhinarium **18**
  social 68
  sprainting **18**
Haliotis, see abalone
Herpestes urva, see crab-eating mongoose
holts 7, 16, 42–8, 60, 72–7, 80, 85, 90–2, 148, 152, 170–1, 177–8, 182, 185–6, 213, 231
home range 32, 38, 42, 57–62, 67–8, 70, 73, 74, 76, 82, 153, 175, 177, 204, 214, 219
Hoplias, see wolf fish
huillin, see southern river otter
human population 224
hunting 228

immobilization 34
infanticide 90,
  see also cannibalism
Inia geoffrensis, see dolphin
intra-specific fighting 198
intra-specific variation 219

Jasus lalandii, see lobster

killer whale 136, 137, 208–9, 225, 235
kokanee 112

Lake Victoria 52, 117, 136, 138
Lates niloticus, see Nile perch
lead (Pb) 201
life expectancy 8, 10, 181, 193–6, 209, 220
limiting factors 32, 137, 218
ling 102–4
lobster 118, 136
Lontra canadensis, see river otter
Lontra felina, see marine otter
Lontra genus 29–30
Lontra longicaudis, see neotropical otter
Lontra provocax, see southern river otter
lumpsucker 102–4, 113, 123, 124–7, 145
Lutra genus 29–30
Lutra licenti 29
Lutra lutra, see Eurasian otter
Lutra maculicollis,
  see spotted-necked otter

Lutra sumatrana, see hairy-nosed otter
Lutrinae 29, 31
Lutrugale perspicillata, see smooth otter

mackerel 107
marine otter **18, 116**
  activity times 150
  competition 190
  conservation 237
  dens 76
  density 177
  diet 18, 116
  dives 147
  foraging behaviour 146–7, 152
  fur maintenance 172
  geographic distribution **18**
  habitat 41, 45
  size 18
  social 18, 68
megafauna collapse 179, 209, 222
mercury 8, 15, 189, 201–4, 221–2, 225
metabolism 18, 156–60, 201
mink, American 34, 38, 89, 133, 165, 172, 191–2, 201–3, 227, 233
Mionictis 29
molluscs, as food 112, 113, 116, 117, 118, 135, 144, 147
Molva molva, see ling
mortality 8, 13, 21, 28, 31–2, 87, 174–86, 189, 196–210, 220–2, 226
  seasonal 128–9
Myxocephalus bubalis, see bullrout

natal holt 73, 90–2, 182, 185–6, 213
neotropical otter **15, 217**
  acitivity times 150
  competition 190, 191
  conservation 237
  den 16, 76
  diet 16, 115–6
  foraging behaviour 146
  freshwater need 45
  fur trade 229
  geographic distribution **16**
  habitat 49–50
  predation 207
  size 16
  social 16, 68
  spraints 80
  tourism 236

Nile perch 138
North American otter, see river otter

octopus 102
oil spill 204, **235**, see also Exxon Valdez
Oncorhunchus nerka, see kokanee
Orca, see killer whale
organochlorines 3, 8, 200–3, 225, 236
Orkney 44
oxygen consumption 35, 156–9, 204

Pantanal 13, 50, 134, 190, 191
paralytic shellfish poisoning 144
parental behaviour 90–8, 103, 220
patch fishing 152
PCB 8, 200–3, 225, 236
Pholis gunnellus, see butterfish
phylogeny, of otters 29–30, **31**
piranha 3, 14–16, 17, 34, 50, 51, 68, 134, 190, 115, 116, 220
platypus 162
Pollachius pollachius, see pollack
Pollachius virens, see saithe
pollack 102–6, 124–7
pollution and populations 3, 180, 189, 200–4, 211–12, 221, 225, 236–7
population limitation 204–10
Potamon, see crab, freshwater
Potamonautes, see crab, freshwater
predation 4, 12, 32, 55, 64, 66, 68, 92, 96, 113–14, 133–8, 142–4, 191–2, 196, 199, 212, 206–9, 218–222, 225–7, 237
prey:
  activity 123, 125
  availability 12, 32, 46, 66, 77, 85, 92, 100, 114, 119, 126–7, 128, 131, 134, 136, 153, 158–9, 182–4, 189, 204–9, 216–22, 236
  biomass 132, 134, 205
  calorific values 94, 107, 110, 111, 158
  density 127
  depth 126
  effects by otters 132–4
  seasonality 126–7, 131–2
  selection 128–30, 217
Prince William Sound 2–3, 8, 42–5, 64–6, 98, 112–13, 143, 176, 179, 186, 196, 203–4, 206, 207, 208, 236

Procambarius clarkii, see crayfish
Pteronura brasiliensis, see giant otter

rabbit 109, 142–3
radio-tracking 33, 35, 40, 46, 48, 53, 60–2, 63, 68, 69, 71, 76, 111, 144, 155, 158, 163, 174–8, 205, 232–4
Rana, see frogs
recruitment 182–9, and food availability 183–5
reintroductions 9, 199, 223, 227, 232–3
reproduction, seasonality 128
resource dispersion 32, 56, 214, 137
river otter (North American) **2**, 8, **9**, **63**, **169**, **216**
  abandoning cubs 96, 160
  activity times 149
  age structure 193
  anti-predator 64
  body temperature 163
  competition, 189–91
  conservation 237
  cooperation 143
  core area 64
  couch 16
  cub dispersal 91, 96
  damage to fish ponds 227
  delayed implantation 10, 186
  den 2, 76
  density 10, 176
  depth 41
  development of foraging 160
  diet 9, 112
  dive times 143
  fishing behaviour 8, 143–4
  foraging success 152
  freshwater need 45
  fur trade 229
  geographic distribution **10**
  gestation 10
  grooming 64, 170
  groups 64, 219–20
  habitat use 41, 42, **43**, 48–9
  home range 64
  life expectancy 10
  litter size 186
  locomotion 9
  longevity 195
  metabolism 159
  mortality 195, 198–9
  oil spills 203
  parental 90, 96
  pedal glands 86
  people proximity 226

pollution 202–3
populations 178
predation 206
prey availability 134
reintroductions 232–3
reproduction 186
scent marking, *see* spraints
size 8
social 9–10, 62–4, **72**
spraints 80, 86
thermo-insulation 165
trapping 198–9, 231
washing in freshwater 171
rockling, five-bearded 103–6, **122**, 124–7, 128
Rubondo 26, **52**, 70, 117, 148

saithe 102–3, 121, 124–7, 128
Salmo salar, see salmon
Salmo trutta, see brown trout
salmon **109**, 110, 130, 138, 142–3
salmonids 120, 130–1
Satherium 30
Saxidomus giganteus, see clam
scent marking, *see* sprainting
Scomber scombrus, see mackerel
Scottish study area 38–9
Scyliorhinus canicula, see dogfish
sea cat, see marine otter
sea otter **207**, **219**, 4, 11, 65, **113**, **114**, **145**, **169**
  capture 231
  carrying capacity 206
  competition 189–91
  conservation 235, 237
  cooperation 145
  copulation 66, 89
  cub dispersal 98, 160
  cub mortality 187
  damage to fisheries 192, 228
  delayed implantation 186
  density 20, 54, 176, 206–9
  depth of dives 155
  development of foraging 160–1
  diet 112–4
  disease 199
  dive times 144
  diving 12
  drinking 40
  effects on communities 134–6
  equilibrium density 206
  exploitation 11–12
  food consumption 114
  foraging behaviour 12, 144, 151–2
  fur trade 135–6, 181, 229, 233
  genetics 181, 232

  geographic range 11, **12**
  gestation 13, 186
  grooming 12, 40, 88, 98, 169
  habitat use 41–3
  home range 65–6
  individual diets 114, 145–6
  kelp use 42–3
  key-stone species 42–3, 114, 134–6, 137
  kleptoparasitism 145
  metabolism 159
  mortality 195–6, 199
  oil spills 203–4
  parental **97**–8
  people proximity 226
  pollution 169
  populations 178–80
  predation 206–9
  prey availability 134–6
  rafts 64–5
  reproduction 186–8
  sex segregation 65
  sexual behaviour 89–90
  size 11
  social 64–6, **72**, 98
  subspecies 181
  survey methods 173
  survival 196
  territory 66, 88
  thermo-insulation 165, 168–9
  time budget 98
  tool-use 12, 144
  translocation 233
  vocalizations 98
  weaning 187
sea scorpion 103–7, **122**, 124–7
sea urchin 113–**4**, 134–6, 144, 206, 209
selenium 201, 202–3
sex difference in habitat 46, 48, 218
sexual behaviour 83–4, 86, 87, 88–90, 181, 186
shark predation 70, 199, 207–8, 220, 235
Shetland 1–3, 6, 36–7, 137–8, 177, 230
small-clawed otter **22**
  competition 190
  conservation 237
  diet 23, 117
  foraging behaviour 147
  geographic distribution **22**
  groups 69
  habitat 52
  litter size 188
  metabolism 159

parental **97**
scats 82
size 21
social 22–3, 69
zoos 228
smooth otter **20**
  activity times 150
  competition 190
  conservation 237
  co-operation 20, 69, 147
  cub dispersal 96
  density 21
  diet 117
  foraging behaviour 147
  fur trade 229–**30**
  geographic distribution **21**
  gestation 188
  groups **68**, 69
  habitat 51–2
  holts 52, 77
  home range 68
  parental 90
  prey availability 136
  reproduction 188
  size 21
  social 21, 68–9, **72**
  spraints 82
  sub-species *maxwelli* 3
  territory 68
  use in fisheries 228
smooth-coated otter, see smooth otter
social life 218–20
social systems 24, 55–6, 61–73, **72**, 98, 219
southern river otter **17**
  competition 190, 191
  conservation 237
  diet 17, 116
  geographic distribution **17**
  habitat 45, 51
  holts 51
  social 68
  spraints 80

spatial organization 32, 56–7, 61, 62, **72**, 78–9, 105, 153, 211
spotted-necked otter **27**
  anti-predator 70
  competition 190–1
  conservation management 234, 235, 237
  damage to fisheries 28, 192, 227
  dens 77
  density 177
  diet 117–8, 138
  dives 148
  fishing 26–8
  foraging behaviour 148
  foraging success 152
  geographic distribution **28**
  groups 70–1
  habitat 52
  holts 52
  home range 70–1
  predation 207
  prey availability 136
  size 26
  social 28, 70–1, **72**
  spraints 82, 101
  territory 71
sprainting 217
  behaviour 14, **19**, 79–86, 217
  communication 79–86
  diet analysis, use in 35, 100–2, 174
  habitat use indicators 39–40, 83, 174
  seasonality 40, 83–4, 128
  surveys 16, 40, 174–5
starvation 13, 196–9, 204–6
stickleback 107, 109, 123, 124–7, 128
Strongylocentrotus, see sea urchin
sucker 112, 134, 138
surveys 43–4, 83, 173–6, 178
swamp otter, *see* Congo clawless otter
swimming speed 156

tagging 34
Taurulus bubalis,
  see sea scorpion
temperature, water, effects of 85, 156–61, 162–4
territory 7, 14, 24, 32, 42, 55, 56–8, 66, 83, 84, 86, 88, 89, 184, 188, 213–14
thermo-insulation 40, 165–70, 172, 213, 216
threats to populations **236**
tide, effects 125
Tinbergen, Niko 78
tilapia 27, 117–18, 136
toad 108, 132, 138
tool use 12, 144, 160
tourism 236
toxoplasmosis 199, 208, 210, 221, 225
tracks **35**, 38, **49**, 51
trapping 5, 9–10, 34, 178, 198–9, 229, 230–1, 233
traps,
  for fish **124**–30, 184
  for otters 10, 34–5, 60, 142, **231**–2, 233
  Shetland 'otter house' **230**
trout 38, 53, 109, 111, 118, 130, 132, 138, 149, 218, 227

use of otters:
  in fisheries 228
  as food 228
  as medicin 228

water mongoose 191
water vole 109
wolf 206
wolf fish 115, 116, 134, 150
wrasse 108

zinc-65: 35, 132, 175–6
Zoarces viviparus, see eelpout
zoos 228